Advances in Bearing Lubrication and Thermal Sciences

Advances in Bearing Lubrication and Thermal Sciences

Editor

Ke Yan

MDPI • Basel • Beijing • Wuhan • Barcelona • Belgrade • Manchester • Tokyo • Cluj • Tianjin

Editor
Ke Yan
School of Mechanical
Engineering
Xi'an Jiaotong University
Xi'an
China

Editorial Office
MDPI
St. Alban-Anlage 66
4052 Basel, Switzerland

This is a reprint of articles from the Special Issue published online in the open access journal *Lubricants* (ISSN 2075-4442) (available at: www.mdpi.com/journal/lubricants/special_issues/bearing_lubrication_thermal_sciences).

For citation purposes, cite each article independently as indicated on the article page online and as indicated below:

LastName, A.A.; LastName, B.B.; LastName, C.C. Article Title. *Journal Name* **Year**, *Volume Number*, Page Range.

ISBN 978-3-0365-5342-9 (Hbk)
ISBN 978-3-0365-5341-2 (PDF)

Contents

About the Editor

Ke Yan

Dr. Yan is currently a professor at Xi'an Jiaotong University. He has won the National Science Fund for Outstanding Young Scholars and the honor of Outstanding Young Talents of Universities in Shaanxi Province, China. Prof. Yan is mainly engaged in the research of digital design and intelligent sensing of high-performance rolling bearings. He presided over more than 20 national key R&D projects, National Natural Science Foundation of China, and enterprise cooperation projects. He has published more than 80 academic papers in well-known journals, and applied for/authorized more than 30 national invention patents. He won the first prize for technological invention in universities of Shaanxi Province, China. At present, Prof. Yan serves as the editorial board member of {Journal of Tribology}, {Bearing,} and other journals. Served as a member of the Technical Committee of China Bearing Industry Association. Prof. Yan has established in-depth cooperative research with world-renowned SKF, NSK, and the top ten enterprises in China's bearing industry, and the research work has been applied to support bearings for major equipment in wind power, high-speed rail, and machine tools.

Preface to "Advances in Bearing Lubrication and Thermal Sciences"

Bearings are regarded as the "joints"of machines and are the most widely used mechanical components, which directly affect the safe and stable operation of major equipment. During high-speed and heavy-load operation, friction and heat generation increase between rolling elements, cages, and rings. Therefore, in the analysis and evaluation of many bearing properties, "thermal science"has always been regarded as one of the core issues of rolling bearings. Statistics show that about 80% of the failures of rolling bearings are caused by local heating problems. At this time, lubrication technology is becoming the key to reducing bearing friction and wear, enhancing bearing heat dissipation, and extending bearing life.

Over the last century, significant research has been carried out on various aspects of bearing thermal analysis and corresponding lubrication techniques. However, as bearing speeds continue to increase, complex operating conditions present more challenges for bearing thermal analysis. At the same time, the continuous cross-integration of materials, sensors, big data, and emerging technologies have led to the continuous expansion of bearing lubrication technology.

This reprint focuses on the hot issue of bearing lubrication and thermal analysis, and brings together many cutting-edge studies, such as bearing multi-body dynamics, bearing tribology, new lubrication and heat dissipation structures, bearing self-lubricating materials, thermal analysis of bearing assembly process, bearing service state prediction, etc.

Ke Yan
Editor

lubricants

MDPI

Article

A Study on Bearing Dynamic Features under the Condition of Multiball–Cage Collision

Shuaijun Ma, Xiaohong Zhang, Ke Yan *, Yongsheng Zhu and Jun Hong

Key Laboratory of Education Ministry for Modern Design and Rotor-Bearing System, Xi'an Jiaotong Univercity, Xi'an 710049, China; msj821@stu.xjtu.edu.cn (S.M.); xh_zhang1206@stu.xjtu.edu.cn (X.Z.); yszhu@mail.xjtu.edu.cn (Y.Z.); jhong@mail.xjtu.edu.cn (J.H.)

* Correspondence: yanke@mail.xjtu.edu.cn

Abstract: Cage stability directly affects the dynamic performance of rolling bearing, which, in turn, affects the operating state of rotating equipment. The random collision between the rolling elements and the cage pocket is the main reason for cage instability. In this paper, from the perspective of the relative sliding velocity between the rolling elements and the bearing raceway, the interactions of the rolling elements and the cage pockets were analyzed, and the four zones with different collision features were defined. On this basis, and on the basis of the bearing dynamics model, the interaction of two adjacent rolling elements and the cage pockets in the a'–b' area is discussed, and the peak impact force of the adjacent two balls and the cage pockets was investigated in terms of the rotation speed, radial load, acceleration/deceleration, and materials. When the ball runs close to the loaded zone, the probability of multiball random collision increases, which leads to an increase in the cage instability. At the entrance of the loaded zone, the peak impact force has the greatest impact on the cage stability during the acceleration process. Compared to the radial load applied to the bearing, the peak impact force is more sensitive to the bearing speed changes. The multiball collision analysis method provides a new idea for the research of cage stability.

Keywords: cage instability; multiball random collision; bearing dynamic model; peak impact force

Citation: Ma, S.; Zhang, X.; Yan, K.; Zhu, Y.; Hong, J. A Study on Bearing Dynamic Features under the Condition of Multiball–Cage Collision. *Lubricants* **2022**, *10*, 9. https://doi.org/10.3390/lubricants10010009

Received: 23 November 2021
Accepted: 23 December 2021
Published: 11 January 2022

1. Introduction

Rolling bearing has been widely applied to high-speed rotating systems, such as aeroengines, high-speed precision spindles, etc. The main reasons for bearing failure are cage instability and the frictional heating induced by the sliding between the rolling elements and the raceway [1]. The cage instability may cause large fluctuations in the bearing torque and can increase the impact force between the cage pocket and the rolling elements, which finally results in excessive wear, and even in the direct breakage of the cage and the bearings [2,3].

According to relevant statistics, up to 25% of the bearing failures of aeroengines are caused by sliding and cage instability [4]. At present, more in-depth research is being conducted on sliding, while there are few investigations on the dynamic performance of the cage in terms of the intermittent collision between the rolling elements and the cage pocket. In the early stages, because of a lack of pertinent theories and immature simulation technology, cage stability was mainly studied through experiments. First, Kingsbury [5,6] found that there is a relationship between the cage motion and torque fluctuation in the instrument bearing, and that cage whirling is mainly caused by the friction between the balls and the cage. Stevens [7] experimentally analyzed the effects of different lubricants and cage structural parameters on the cage stability, and found that the cage instability mainly presented as a radial high-frequency vibration. On the basis of previous experiments, Genfle [8] found that there are many factors affecting the bearing cage stability, such as the viscous resistance of the lubricants, the amounts of the lubricants, the cage geometric size, and the bearing operating temperature. In the above experimental investigations, it was observed that some of the structure parameters and operating conditions have a certain relationship with the bearing cage stability, but the rules of the influence of these factors on the stability need to be studied further.

With the development of quasidynamics and the dynamics analysis of rolling bearing, the analytical model has been constantly improved, which provides the possibility for an analysis of the bearing cage stability. Walters [9] first established the bearing dynamic model and obtained cage instantaneous speed. Although there was no profound study on the motion of the cage, the research provided a basis for the development of bearing dynamics analysis. On the basis of this, many scholars have been studying the factors that affect cage stability, from different viewpoints. Kannel [10] established a dynamic model with a low-DOF cage by refining the contact model between the ball and the cage pocket, and qualitatively analyzed the influence of the cage–land clearance, the ball–cage pocket clearance, and the lubricant viscosity on the cage stability. Gupta [11–14] established a 6-DOF complete dynamic model, and the influence of the cage–land clearance, the ball–cage pocket clearance, and the lubricating oil parameters on the cage stability were systematically studied. By simplifying the dynamic model, Meeks et al. [15,16] improved the model solving speed. On the basis of the peak impact between the rolling element and the cage pocket, the structural parameters of the cage were optimized. On the basis of a numerical simulation model of bearing, the contact state and the deformation between the rolling element and the cage is discussed in-depth by Houpert [17–20]. In order to analyze the instability of the cage, Rivera [21] established a simplified mechanical model and compared the influencing factors on the cage stability, including the sliding velocity between the rolling elements and the cage, and the sliding velocity between the guide ring and the cage. In addition, on the basis of a simplified lubricant drug model, Boesiger et al. [22] established a dynamic model with 2-DOF cage, and a new integral algorithm was employed to solve it. Through the combination of simulation and experimental analyses, it was found that the cage instability was highly sensitive to the friction between the rolling element and the cage pocket. By setting up a simple cage instability model, based on the previous work, Takashi Nogi et al. [23] further found that the increase in the cage friction coefficient may cause unstable motion. In addition to the purely rigid bearing dynamic model, a few scholars have used the commercial dynamic software, ADAMS, to carry out cage dynamic analysis, and some establish dynamics to analyze the stability of the cage by making the cage flexible. Sakaguchi et al. [24,25] analyzed the force between the cage pocket and the rolling elements by using the commercial software, ADAMS, and found that the interaction force has a large influence on the trajectory of the cage in the export of the loaded zone. Weinzapfel [26] and Ashtekar [27] established a dynamic model of ball bearing with a flexible cage by dispersing the cage into a flexible body composed of finite elements, which showed that the flexible cage significantly reduced the ball–cage pocket force and the time needed for the bearing to reach a steady state of operation.

From the perspective of experiments and theory, the abovementioned scholars have carried out relevant investigations into the cage instability under different structural parameters and operating conditions, and the rules of the influence of some of the parameters on the instability were obtained, which is of significance to understanding the cage instability. However, the current work mainly analyzed the impact factors on the cage stability, while a further discussion on the cage instability mechanism is lacking. The reason for this is that the turbulent collision between the rolling elements and the cage pocket presents a certain degree of complexity and randomness, due to the change in the relative velocity between the rolling elements and the raceway. To this end, from the perspective of the relative sliding velocity between the rolling elements and the raceway, the mechanism of the collision between the rolling elements and the cage pocket was analyzed. Taking the deep-groove ball bearing as the research object, a dynamic model with a low-DOF bearing was established. For the collision of the two rolling elements, the cage rotation was divided into four stages, in which the peak impact force has an important influence on the cage stability when the ball enters the loaded zone. Taking the peak impact force near the entrance of the loaded zone as the object, the influence of the operating parameters on the cage instability, such as the rotational speed, radial load, acceleration and deceleration, and the material of the rolling elements, were all analyzed.

2. Bearing Dynamic Model

In order to establish the cage dynamics model for multiball collision simulation, this paper simplifies the complex contact and motion relationship of the deep-groove ball bearing during actual operation, and some assumptions are made, as follows:

1. The motions of the rolling elements, cage, and inner ring are limited to the radial plane of the bearing, without considering the axial movement;
2. The contact between the rolling elements and the other parts satisfies the Hertz contact theory without considering plastic deformation;
3. All parts are treated as rigid bodies, and the center of mass coincides with the centroid during service;
4. The center of the outer ring coincides with the bearing center, which is the coordinate origin;
5. A spring model is set between the rolling element and the cage pocket [19], and the inner ring rotational acceleration is always constant (100 rad/s^2);
6. Only one rotational degree of freedom of the cage is considered. The contact force between the ball and the pocket is considered, and the friction is ignored;
7. The gap between the ball and the pocket is ignored, and the contact deformation coefficient between them is treated as a constant value.

2.1. Interaction Force between Bearing Components

The relationship among the bearing's internal components is described by Figure 1. The outer ring is fixed, the inner ring rotates at a constant speed, w_i, and w_c denotes the angle velocity of the cage. The revolution and rotational speed of the j_{th} rolling element are w_{mj} and w_{bj}. Moreover, d_i is the diameter of the inner groove bottom, while the diameter of the outer groove bottom is d_o. The pitch diameter of the bearing is d_m, which is the mean of d_i and d_o. D_w represents the diameter of the ball. ψ_j is the orbital angle of the j_{th} rolling element, and ψ_{cj} is the orbital angle of the cage pocket corresponding to the j_{th} rolling element.

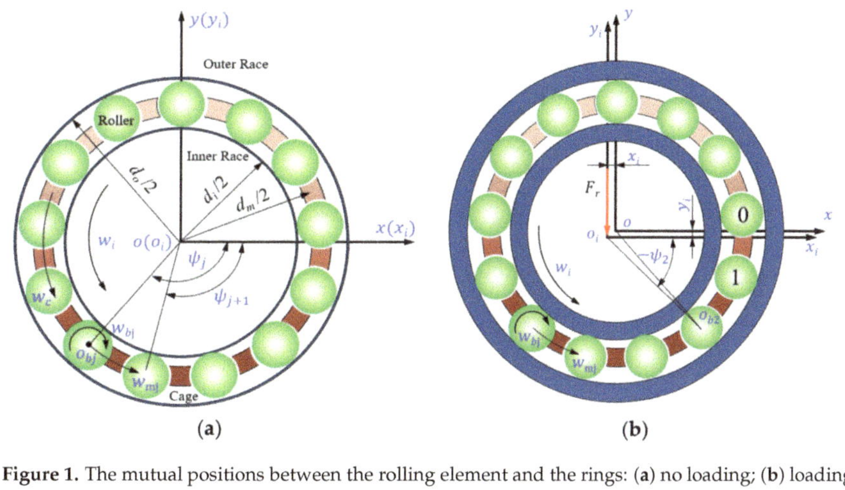

Figure 1. The mutual positions between the rolling element and the rings: (**a**) no loading; (**b**) loading.

2.1.1. The Force between the Rolling Elements and the Rings

Because of the radial load at operation, the rolling element deforms, which can be determined by the transient coordinates of the inner ring and the rolling element. Under the assumption, the mass center, O_{bj}, of the j_{th} rolling elements is (x_{bj}, y_{bj}) at a certain moment, while the mass center, O_i, of the inner ring is (x_i, y_i), relative to the bearing center, O.

As shown in Figure 1, the contact deformations between the rolling element and the inner or outer rings can be expressed as Equation (1) or Equation (2), respectively.

$$\delta_{ij} = \left[-\left| \overline{OO_{bj}} - \overline{OO_i} \right| + \frac{1}{2}(d_i + D_w) \right]_+ \tag{1}$$

$$\delta_{oj} = \left[\left| \overline{OO_{bj}} \right| - \frac{1}{2}(d_o - D_w) \right]_+ \tag{2}$$

where the subscript, '+', indicates that, when the value in the bracket is greater than 0, deformation occurs because of the contact, and when it is less than or equal to 0, it means that no contact or collision occurs, and the deformation is set to 0.

According to the assumption, the contact between the rolling element and the rings can be treated by the Hertz contact theory; therefore, the contact force between the j_{th} rolling element and the inner or outer rings are computed as Equation (3) or Equation (4), respectively:

$$Q_{ij} = K_i \delta_{ij}^{1.5} \tag{3}$$

$$Q_{ij} = K_i \delta_{ij}^{1.5} \tag{4}$$

where K_i represents the deformation coefficient between the rolling element and the inner ring, while K_o represents the coefficient between the rolling element and the outer ring [28].

At this time, the oil film traction force between the rolling element and the inner or outer rings are calculated by the following equations:

$$T_{ij} = \mu_{ij} Q_{ij} \tag{5}$$

$$T_{oj} = \mu_{oj} Q_{oj} \tag{6}$$

where the coefficient μ of the oil film traction can be calculated as follows:

$$\mu = (A + Bs)e^{Cs} + D \tag{7}$$

The parameters of the lubricating oil are obtained from [29], which include A, B, C, and D; s represents the sliding–rolling ratio (the ratio of the sliding velocity to the rolling velocity) at the contact point, $s = v/u$, as shown in Figure 2, which is calculated as follows:

$$s_{ij} = 2 \cdot \frac{(w_i - w_{mj})d_i - w_{bj}D_w}{(w_i - w_{mj})d_i + w_{bj}D_w} \tag{8}$$

$$s_{oj} = 2 \cdot \frac{-w_{mj} \cdot d_o + w_{bj} \cdot D_w}{-w_{mj} \cdot d_i - w_{bj} \cdot D_w} \tag{9}$$

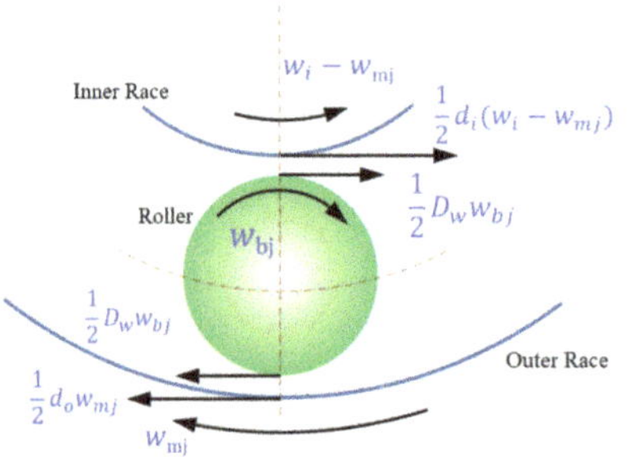

Figure 2. Velocity analysis between the rolling element and the raceway.

2.1.2. The Force between the Rolling Elements and the Cage

During bearings operation, the revolution speed of the rolling element in the non-loaded zone is slower than in the loaded zone because of the lack of an inner ring traction force to maintain the speed. Therefore, the rolling elements need to be driven by the cage; thus, the impact force between the rolling element and the cage pocket is negative. Since the inner ring can provide sufficient traction in the loaded zone, the revolution speed of the rolling element increases. At this time, the cage is driven by the rolling element, so the impact force is positive, as shown in Figure 3.

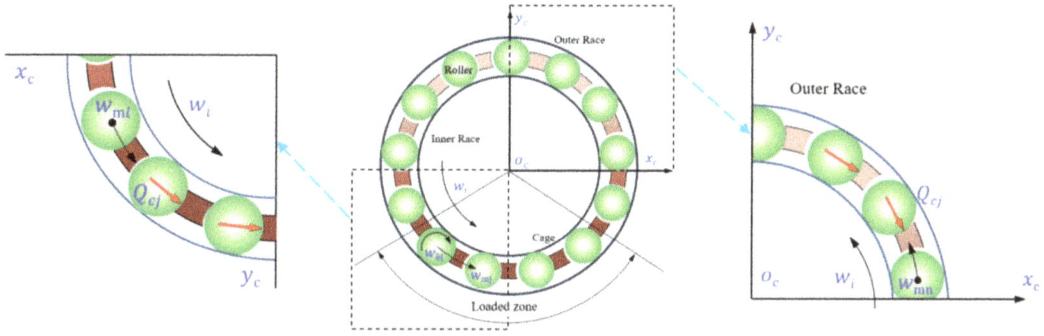

Figure 3. Interaction between the rolling element and the cage pocket.

In this paper, the spring model is used to simulate the impact force between the rolling element and the cage pocket [19]. The impact force between the j_{th} rolling element and the corresponding pocket can be written as Equation (10):

$$Q_{cj} = \begin{cases} 0.5K_c(\psi_j - \psi_{cj})d_m \text{ Coliding with the front side of pocket} \\ 0.5K_c(\psi_{cj} - \psi_j)d_m \text{ Coliding with the rear side of pocket} \end{cases} \tag{10}$$

where K_c is the deformation coefficient between the rolling element and the cage pocket, which can be obtained according to [30,31].

2.2. Mechanical Equilibrium Equations

As is shown in Figure 4, the inner ring mainly suffers the contact force and traction force of the rolling elements, while the cage is mainly subjected to the impact force of the rolling elements; thus, the quasistatic model of the bearing can be initially established.

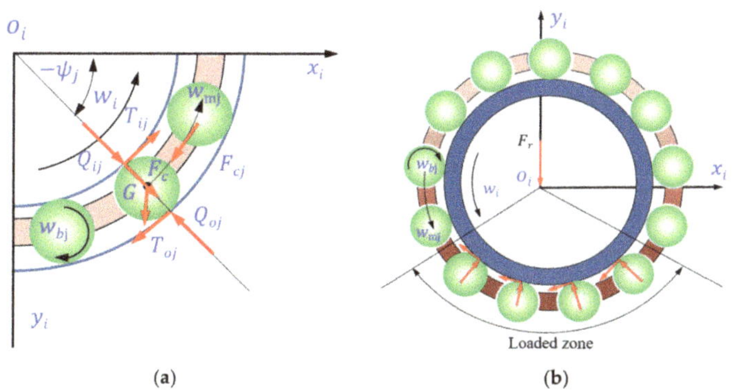

Figure 4. Force diagram acting on the bearing element and the inner ring: (**a**) the j_{th} rolling element; (**b**) the inner ring.

The nonlinear equilibrium equations of the force on the rolling element can be written as:

$$0 = Q_{ij} - Q_{oj} - F_{cj} + G\cos\psi_j \quad (11)$$

$$0 = T_{ij} - T_{oj} - Q_{cj} + G\sin\psi_j \quad (12)$$

where F_{cj} is the centrifugal force on the rolling element from [28], and G names the gravity of the rolling element.

The nonlinear equilibrium equations of the inner ring are as follows:

$$0 = \sum_{j=1}^{Z}(-Q_{ij}\sin\psi_j + T_{ij}\cos\psi_j) \quad (13)$$

$$F_r = \sum_{j=1}^{Z}(\,Q_{ij}\cos\psi_j + T_{ij}\sin\psi_j) \quad (14)$$

where F_r is the radial load acting on the bearing, and Z is the number of balls.

On the basis of quasistatic model, the dynamic model of the bearing is established considering the bearing dynamic information of each component.

The differential equations of the j_{th} rolling element are given as:

$$m_b\ddot{r}_j = Q_{ij} - Q_{oj} - F_{cj} + G\cos\psi_j \quad (15)$$

$$\frac{1}{4}m_b d_m^2\dot{w}_{mj} = \frac{1}{2}(T_{ij} - T_{oj} - Q_{cj} + G\sin\psi_j)d_m \quad (16)$$

$$\frac{1}{5}m_b D_w^2\dot{w}_{bj} = (T_{ij} + T_{oj})D_w \quad (17)$$

where m_b is the mass of the rolling element; r_j represents the vector from the center of the bearing to the j_{th} rolling element; and $|r_j| = \sqrt{x_{bj}^2 + y_{bj}^2}$.

The differential equation of the cage is as follow:

$$\frac{1}{4}m_b d_m^2\dot{w}_c = \frac{1}{2}d_m(Q_{c1} + Q_{c2} + Q_{c3} + \cdots\cdots + Q_{cZ}) \quad (18)$$

where m_c is the mass of the cage, and w_c represents the rotational speed of the cage.

The differential equations of the inner ring can be written as follows:

$$m_i\ddot{x}_i = \sum_{j=1}^{Z}(-Q_{ij}\sin\psi_j + T_{ij}\cos\psi_j) \quad (19)$$

$$m_i\ddot{y}_i = \sum_{j=1}^{Z}(\,Q_{ij}\cos\psi_j + T_{ij}\sin\psi_j) \quad (20)$$

where m_i is the mass of the inner ring, and x_i and y_i represent the displacements of the inner ring in the x_i direction and the y_i direction, respectively.

2.3. Model Solution and Verification

Since the above bearing dynamic model is composed of nonlinear differential equations with strong nonlinearity, with consideration to the convergence and computational efficiencies, the fourth-order Runge–Kutta method was employed to solve the equations. In order to ensure the accuracy of the solution, the Newton–Raphson iteration method was used to solve the quasistatics model first, the results of which were input as the initial values of the differential equations. The specific solution process is shown in Figure 5.

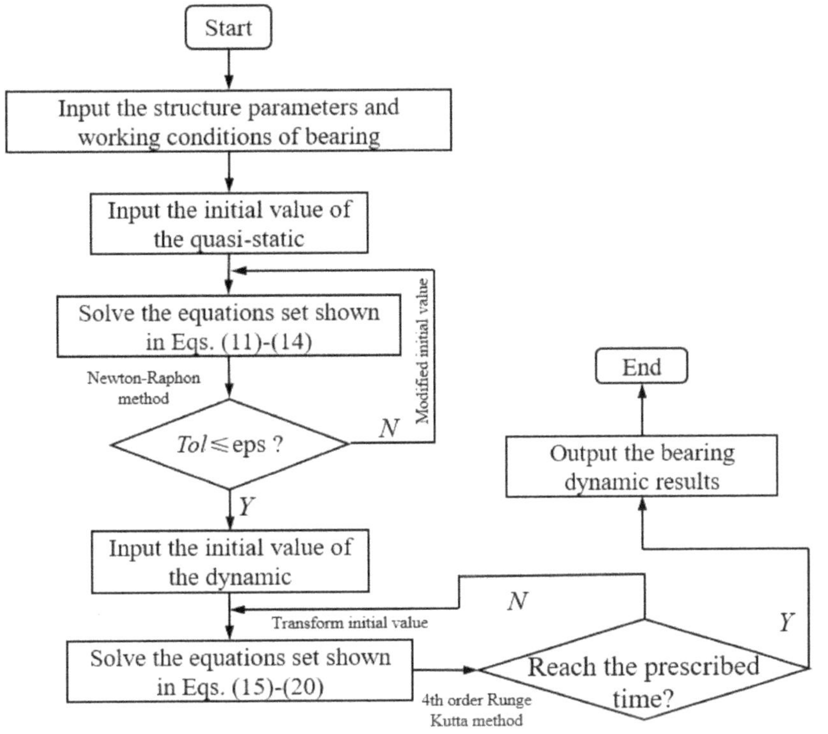

Figure 5. Flow chart for the solving process.

From Figure 5, the bearing structural parameters and the working condition that needed to be solved were input first. Then, the Newton–Raphson iteration method was used to solve the quasistatic equations of the bearing, namely, Equations (11)–(14). If the error is greater than the set target value, the initial value is modified so that the equation can continue to be solved. The result of the solution is taken as the initial value of the dynamics equations, as long as the error is small enough. The fourth-order Runge–Kutta method was used to solve Equations (15)–(20), and the motion state of the bearing of each component at this moment can be obtained. Each calculated time is used to determine whether the prescribed time was reached. If it was not reached, the solution state of each part at this moment is taken as the initial value of the next moment. If it was reached, the simulation analysis results are output.

In order to verify the accuracy, the mathematical results of this model were compared with the experimental data [32]. According to the experiment conditions [32], the radial forces were 1500 N and 2100 N, respectively, while the rotational speed of the inner ring was gradually increased, from 4000 r/min to 10,000 r/min. Then, the correctness of the model was verified by the sliding rate of the cage, which can be expressed as:

$$\Delta w_c = \frac{\left| w_c - w^l{}_c \right|}{w^l{}_c} \times 100\% \tag{21}$$

where $w^l{}_c$ represents the theoretical cage speed, $w^l{}_c = n_i \cdot (1 - D_w/d_m)/2$, and w_c is the actual cage speed.

As shown in Figure 6, the sliding rate of the cage increases with the increase in the bearing rotational speed: the greater the radial force, the smaller the sliding rate.

The mathematical and the experimental results are basically the same, which proves the correctness of the dynamic model proposed in this paper.

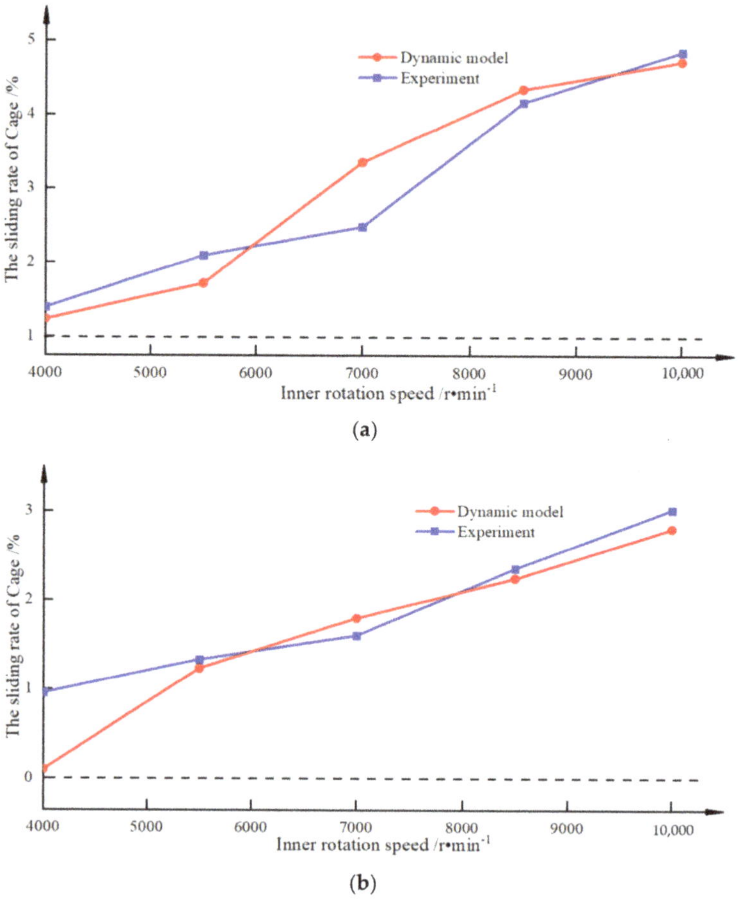

Figure 6. Comparison of the mathematical and experimental results: (**a**) Fr = 1500 N; (**b**) Fr = 2100 N.

3. Collision Characteristics between the Ball and the Cage

In order to analyze the collision characteristics between the rolling element and the cage pocket, the deep-groove ball bearing was taken as the research object, of which the structural parameters are shown in Table 1. The change process of the impact force between the cage pocket and the rolling element was analyzed, and the influence of the impact force of the adjacent two rolling elements on the cage stability was further discussed.

3.1. Interaction of a Single Ball and the Cage Pocket

In order to analyze the interaction between the balls and the bearing cage, the bearing radial load was set as 1500 N, the inner ring initial speed was 3000 r/min, and the angular acceleration was 100 rad/s^2. By solving the dynamics model in the above Section 2, the contact forces between the ball and the inner/outer rings, and the impact force between the ball and the cage pocket can be obtained, respectively.

Table 1. Structural parameters of the 6016.

Name	Value
Inner diameter/mm	80
Outer diameter/mm	125
Bearing width/mm	22
Pitch diameter/mm	102.198
Cage outer diameter/mm	108.3
Cage inner diameter/mm	102.448
Cage width/mm	18.3
Number of balls	14
Ball diameter/mm	13.474

Figure 7 shows the changes in the contact forces of the rolling elements. The blue curve represents the contact force between the ball and the outer ring, the green curve represents the contact force between the ball and the inner ring, and the red curve represents the impact force between the ball and the cage pocket. There are some key points on the collision between the ball and the cage pocket, as seen in Figure 7. The a point is the stable starting point of the collision between the ball and the rear side of the cage pocket; the b point is located at the starting point of the loaded zone; the c point is the ending point of the collision between the ball and the front side of the cage pocket; and the d point is the starting point of the collision between the ball and the front side of the cage pocket. When the ball is rotating in the a–b region, the impact force between the ball and the rear side of the pocket is stable. When the ball is running in the b–c region, the impact force between the ball and the rear side of the pocket is gradually reduced, and is eventually converted from the rear to the front. When the ball is located in the c–d region, the impact force between the ball and the front side of the pocket initially increases, and then decreases. When the ball is in the d–a region, the magnitude of the impact force between the ball and the rear side of the pocket changes significantly.

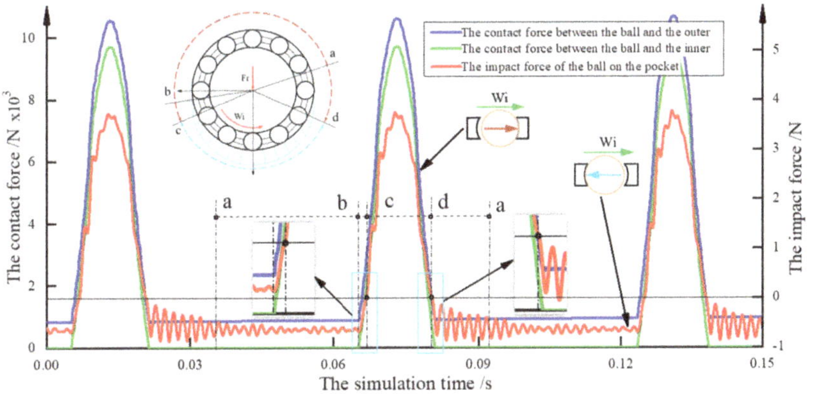

Figure 7. The trend diagram of the contact force and impact.

The above phenomenon is mainly caused by the time-varying relative sliding velocity between the ball and the raceway, as is shown in Figure 8. The impact force between the ball and the rear side of the cage pocket is relatively stable when the ball runs in the a–b zone. This is because, in the nonloaded zone, the contact force between the ball and the inner raceway is slight. In addition, the interaction of the ball and the outer raceway is enhanced because of the centrifugal force. The static outer raceway has a braking effect on the motion of the ball, resulting in an increase in the sliding velocity between the ball and the raceway. At this stage, the orbital speed of the ball is mainly maintained by the cage, which leads to a stable impact force between the rear side of the pocket and the ball. When the ball is situated in the b–c zone, the contact position of the ball and the cage is gradually transferred from the rear side of the pocket to the front, and the impact force decreases as well. This is because the contact force between the ball and the inner ring recovers and gradually increases, which inhibits the sliding between the ball and the raceway to a certain extent, resulting in the increase in the ball orbital speed. Eventually, the impact force between the ball and the rear side of the pocket gradually decreases. When the ball is in the c–d zone, the impact force between the ball and the front of the pocket initially increases, and then decreases. This is because the ball–raceway traction force produced by the contact behavior between the ball and the inner ring inhibits the ball from sliding into the loaded zone, but the traction force is too small to stop the sliding completely. With the increase in the contact force, the ball–raceway traction force increases gradually on the left side of the loaded zone, and the ball orbital speed greatly increases. On the right side of the loaded zone, the orbital speed of the ball increases slightly because the ball–raceway traction force decreases gradually as the contact force decreases, which leads to the change in the impact force between the ball and the front side of the pocket in the loaded zone. When the ball is located in the d–a zone, the impact force between the ball and the rear side of the cage pocket changes greatly. This is because the ball–raceway traction force produced by the contact behavior decreases gradually. Meanwhile, the ball orbital speed has been increasing in the load region, resulting in the increased magnitude of the ball centrifugal force, which further aggravates the sliding of the ball. Finally, the above reasons lead to the collision with the rear side of the pocket, and the magnitude of the impact force is large.

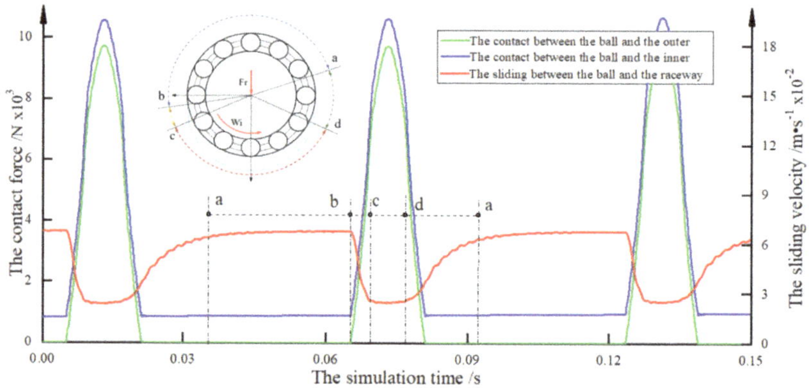

Figure 8. The changes in the ball–raceway contact force and ball sliding.

3.2. *Impact Force between Two Balls and the Cage*

The impact force between two balls and the cage pockets is discussed in this section. As is shown in Figure 9, the a′ point occurs where the collision between the front ball and the cage pocket is converted from the front side to the rear, and the b′ point denotes that the collision between the back ball and the cage pocket has been converted from the front side to the rear. The c′ point appears where the collision between the front ball and the cage pocket is converted from the front side to the rear, and the d′ point means that the interaction has been converted from the front side to the rear. When both of the two balls are running in the a′–b′ zone, the impact force between the front ball and the pocket changes from the rear side to the front, while the back ball maintains the contact with the rear side. The cage will suffer an impact load because of the opposite collision directions. When the two balls are located in the b′–c′ zone, the impact force between the back ball and the pocket changes from the rear side to the front, while the front ball maintains the contact with the front of the pocket. Meanwhile, the impact direction of the two balls is the same, and they push the cage forward together. When the two balls are in the c′–d′ zone, the impact force between the front ball and the pocket changes from the front side to the rear, while the back ball maintains the contact with the rear side of the pocket. Although the impact direction is opposite, the impact force of the two balls acts on the same beam of the pocket, reducing the cage impact load. When the two balls are situated in the d′–a′ section, the impact force between the back ball and the pocket changes from the front side to the rear, while the front ball maintains the contact with the rear side of the pocket. At this time, the direction of the two forces is the same, which is opposite to the cage speed. In summary, when the two balls are located in the a′–b′ zone, a relatively larger collision occurs on the cage because of the opposite impact force acting on the different beams, which may have a significant influence on the cage stability. When the two balls are in the b′–c′ zone, the cage is pushed by the two balls, which can help to maintain the cage stability. When the two balls are running in the c′–d′ zone, the impact force decreases, and the stability of the cage is improved. When the two balls are situated in the d′–a′ zone, the cage is braked by the two balls because of the opposite impact direction. On the basis of the above analysis, this paper focuses on the impact force when the two balls are located in the a′–b′ zone.

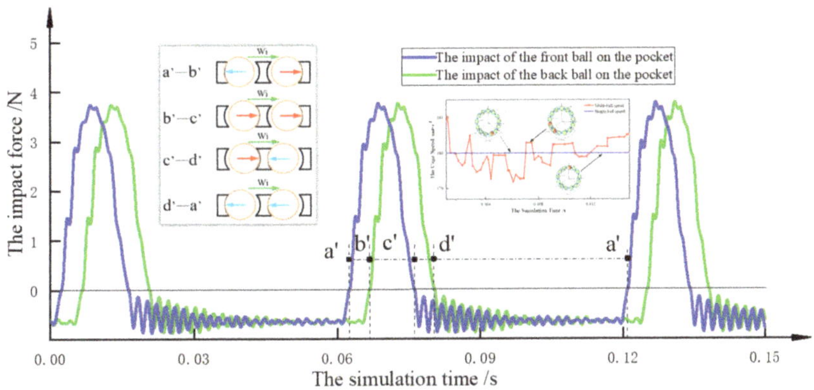

Figure 9. An analysis of the impact between the two balls and the cage.

4. Influencing Factors Analysis

In order to further investigate the influence of the different factors on the cage stability, on the basis of the analysis in Section 3.2, the deep-groove ball bearing, 6016, in Section 3, was taken as the research object. The peak impact force of the rolling elements in the a′–b′ stage was employed as the evaluation index to evaluate the cage stability. After consulting the relevant design data of this bearing, it was discovered that the basic dynamic capacity of a rolling element–raceway contact is 47.5 kN. The amounts of 5% and 25% of the bearing's basic dynamic capacity were taken as the light-load and heavy-load operating conditions, respectively, and 1000 r/min, 5000 r/min, and 8000 r/min were taken as the low-speed, medium-speed, and high-speed, respectively. Factors including the rotational speed and radial load, the acceleration and deceleration of the inner ring, and the materials of the rolling element were all analyzed.

4.1. Radial Load and Rotational Speed

In order to analyze the influence of the rotational speed applied to the inner ring and radial load on the impact force between the ball and the cage pocket in the a′–b′ zone, a dynamic simulation of the rolling bearing under different speeds and loads was carried out. Moreover, the change in the peak impact force under the same rotational speed and different radial loads (or under different speeds and the same radial load) was studied while the inner ring rotational acceleration was constant (100 rad/s^2).

The change in the peak impact force under the same radial load and different rotating speeds is shown in Figure 10.

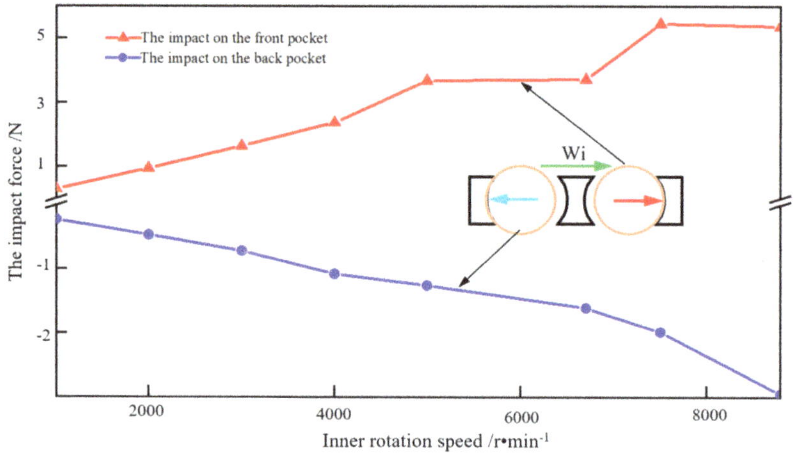

(a)

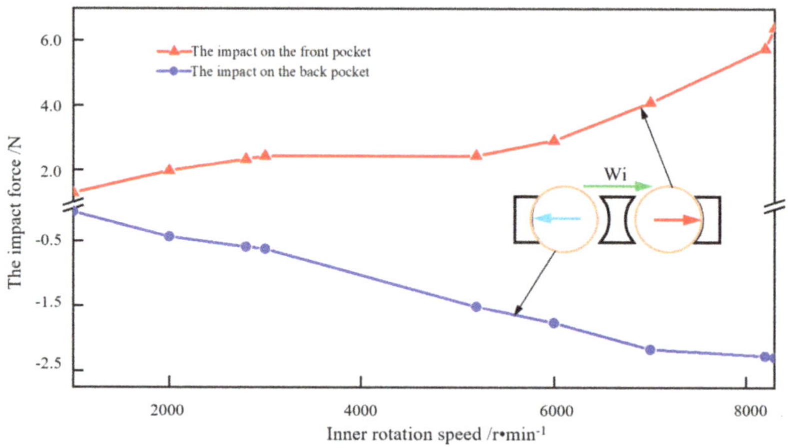

(b)

Figure 10. The effect of the rotational speeds on the impact force: (**a**) Fr = 2375 N; (**b**) Fr = 11,875 N.

Under the same radial load, with the increase in the rotating speed, the peak impact force (the impact forces between the balls and the front side of the pocket, and that of the rear) gradually increases. This is because, with the gradual increase in the rotating speed, the centrifugal force on the ball gradually increases, and the difference in the contact force between the inner ring and the outer ring increases during the ball running in the a′–b′ zone, which results in the increase in the ball sliding. The fluctuation in the ball orbital speed will finally cause the impact force between the ball and the front side of the pocket to increase. At the same time, the contact force between the ball and the outer ring in the unloaded zone increases gradually and is converted into the traction force. Thus, the orbital speed of the ball decreases in the unloaded zone, resulting in the increase in the impact force between the ball and the rear side of the pocket.

The rule of the peak impact force under the same rotating speed and different radial loads is shown in Figure 11.

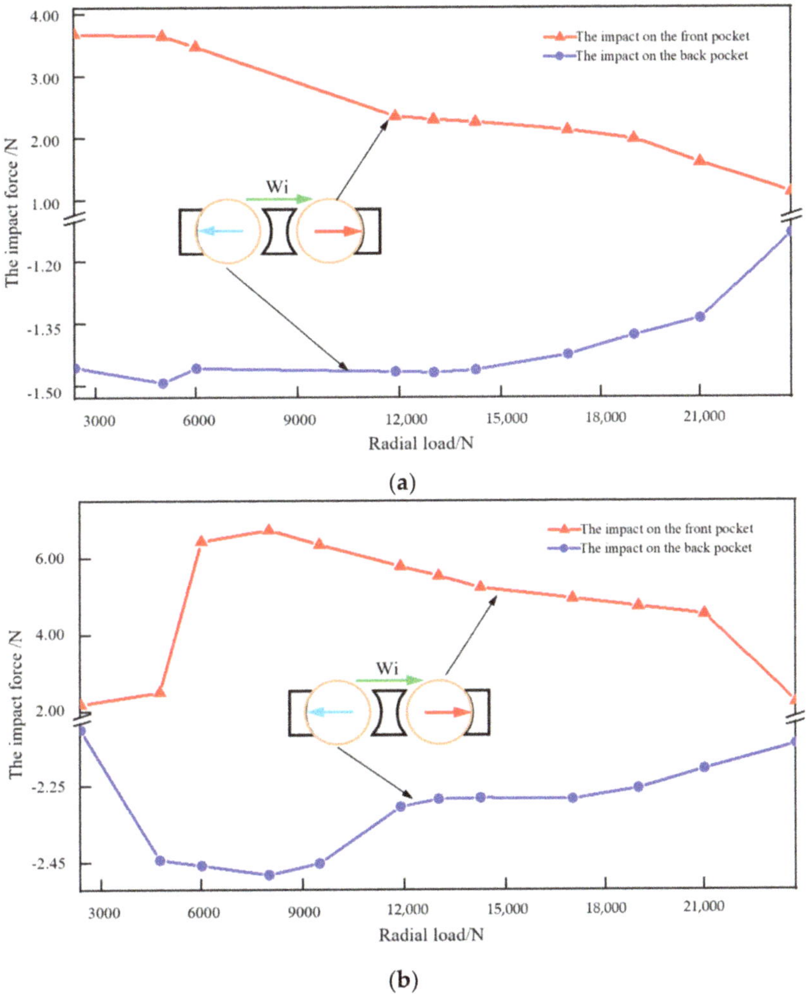

Figure 11. The effect of the radial force on the impact force: (**a**) $n = 5000$ r/min; (**b**) $n = 8000$ r/min.

Under the condition of the same rotating speed, with the increase in the radial load applied to the inner ring, the impact force increases gradually in the early stage, and then decreases. This is because, when the load increases in the early stage, it is not enough to overcome the sliding of the ball in the loaded zone. As the radial load increases, the traction force increases. Meanwhile, the increase in the radial load means an increase in the loaded zone, and, finally, an increase in the range of the traction force, which leads to the increase in the fluctuations of the ball orbital speed and the increase in the impact force. When the radial load increases to a certain extent, it is enough to overcome the ball sliding. At this time, with the gradual increase in the radial load, the fluctuation of the ball orbital speed decreases, resulting in the decrease in the impact force.

4.2. Inner Ring Speed Acceleration

In order to analyze the influence of the inner ring acceleration on the peak impact force between the ball and the pocket, a dynamic simulation of the rolling bearing under the different acceleration conditions was carried out. Moreover, the influence of the different accelerations on the impact force under a light load, heavy load, and a low rotational speed (1000 r/min) were analyzed. When the rotational speeds of the inner ring and the radial load are constant, the speed acceleration applied to the inner ring gradually increases from 50 rad/s^2 to 150 rad/s^2. The variation curve of the impact force is shown in Figure 12.

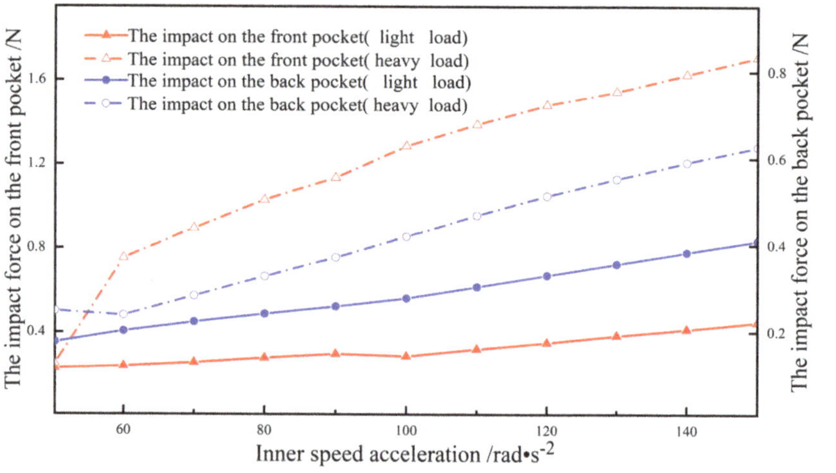

Figure 12. The effect of the speed acceleration on the impact force.

When the rotational speed of the bearing is 1000 r/min, the peak impact force between the ball and the pocket increases with the increase in the inner speed acceleration, regardless of the light load applied to the inner ring or the heavy load. This is because the increases in the inner ring acceleration of the inner ring increases while the contact force between the ball and the inner ring remains unchanged, which aggravates the sliding between the ball and the raceway. In addition, the coefficient of the traction eventually increases, and it increases the fluctuation range of the ball orbital speed, resulting in an increase in the peak impact force. Under a low-speed condition, the peak impact force under the heavy-load condition is larger than under the light-load condition. This is because, with a heavy radial load, the contact force between the ball and the raceway in the loaded zone is enough to overcome the sliding of the ball, while it is not with a light radial load. Moreover, with the inner speed acceleration increasing, the traction force between the ball and the raceway in the heavy-load condition increases more than in the light-load condition, which causes a larger fluctuation range of the ball orbital speed with heavy-load bearing. Therefore, the peak impact force between the ball and the pocket under the heavy load is greater.

4.3. Inner Ring Speed Deceleration

The influence of the inner ring deceleration on the peak impact force under different load and inner ring speed (8000 r/min) was studied. The speed deceleration applied to the inner ring increased gradually from -150 rad/s^2 to -50 rad/s^2, while the rotational speed and the radial load was constant. The variational curve of the peak impact force is shown in Figure 13.

Contrary to the acceleration state, in the process of inner ring deceleration, the greatest influence on the cage instability appears at the exit of the loaded zone. Because of the effect of deceleration, the impact force between the ball and the rear side of the cage pocket mainly occurs in the loaded zone, while the impact force with the front side of the pocket mainly occurs in the unloaded zone. Therefore, the peak impact force of the adjacent balls on the cage pocket is just opposite, at the loaded zone exit, which may reduce the stability of the cage. The peak impact force gradually decreases with the increase in the deceleration. This is because the deceleration gradually increases, and its value gradually decreases. The reason for the change is the same as that for the speed acceleration change in Section 4.2.

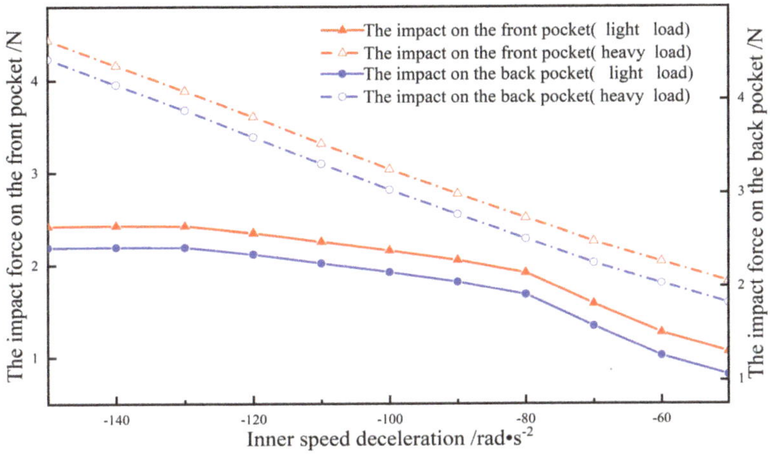

Figure 13. The effects of the speed deceleration on the impact force.

4.4. The Different Ball Material

In order to analyze the influence of the ball material on the peak impact force between the ball and the pocket, a dynamic simulation of the rolling bearing with different roller materials was carried out. Ceramic material and bearing steel were the focuses of the investigation, which are widely used in bearings. The ceramic density was about 3.2 g·cm^{-3}, while the density of the bearing steel was 7.8 g·cm^{-3}.

The effects of different roller materials on the peak impact force were studied under the condition of the same radial load and different rotational speeds (or the same rotational speed and different radial loads). The variation curves of the peak impact forces of the bearings with different ball materials is shown in Figure 14, with the changes in the rotation speeds and radial loads.

It can be seen that the peak impact force of the ceramic ball bearings is smaller than that of the steel ball. This is because the density of ceramic is lower than that of bearing steel. At the same rotational speed, the centrifugal force of the ceramic ball is smaller than that of the steel balls. On the basis of the above analysis, small centrifugal force leads to less contact force between the balls and the bearing raceway, resulting in a lighter sliding of the ceramic ball. Furthermore, the smaller the fluctuations in the ball's orbital speed, the smaller the peak impact force. Under the condition of the same rotational speed and different loads, the peak impact force of the ceramic ball bearing is almost unchanged. This is because the load has little effect on the peak impact force at the same rotational speed, while the ceramic has little effect on the peak impact force because of its low density.

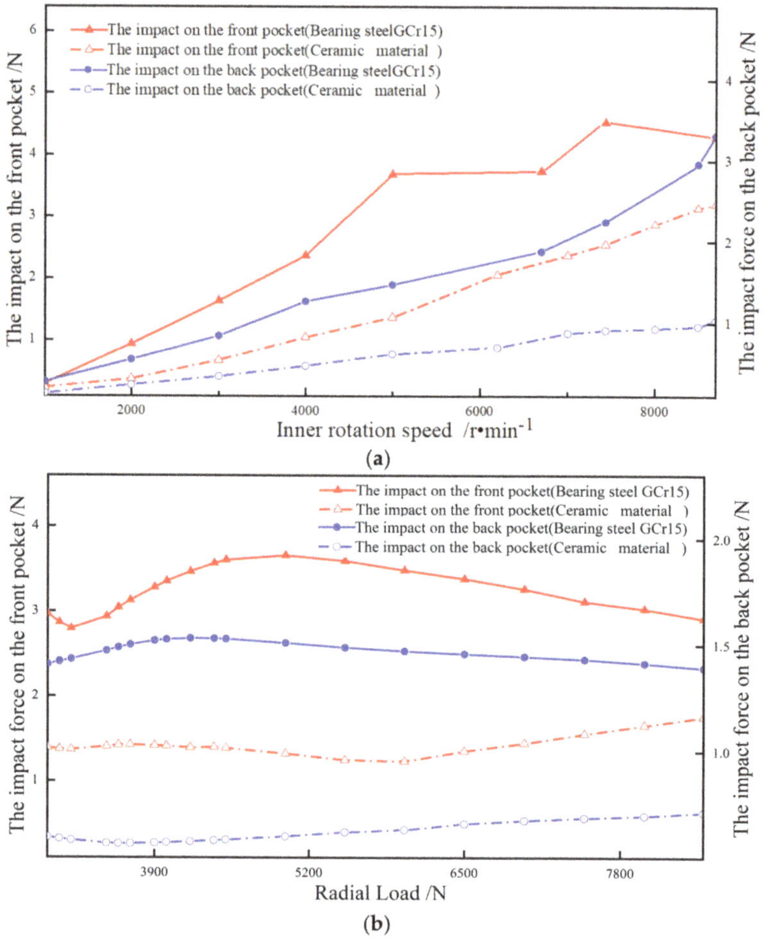

Figure 14. The effect of the material on the impact force: (**a**) Fr = 2375 N; (**b**) n = 5000 N.

5. Conclusions

This work established a dynamic model of deep-groove ball bearing, and the factors affecting the cage stability were analyzed from the perspective of the collision between the balls and the pockets. The phenomenon of multiball collision tends to occur when balls near the loaded zone, which may increase the cage instability. Moreover, taking the peak impact force near the loaded zone as the evaluation index, the influences of the radial load, rotational speed, acceleration, deceleration, and material were investigated. The main conclusions are presented as follows:

1. From the perspective of the relative sliding velocity between the ball and the raceway, the operation of the ball is divided into four zones. When the balls run in the a′–b′ zone and the c′–d′ zone, the multiball collision effect is obvious. The multiball collision can easily increase the stage instability because of the opposite impact directions of the adjacent two balls when the balls are in the a′–b′ zone;
2. Compared to the radial force, the peak impact force is more sensitive to the rotational speed. With the increase in the rotational speed, the sliding between the balls and the raceway becomes more serious, resulting in an increase in the peak impact force;

3. The influence of bearing inner ring acceleration and deceleration on the cage instability is the opposite. The peak impact force on the a′–b′ zone can easily cause the cage instability under acceleration, while the deceleration is in the c′–d′ zone because of the direction of the relative sliding velocity.

Author Contributions: Conceptualization, S.M., K.Y. and J.H.; Data curation, S.M. and J.H.; Formal analysis, S.M. and Y.Z.; Funding acquisition, J.H.; Investigation, S.M., X.Z. and K.Y.; Methodology, S.M. and Y.Z.; Project administration, K.Y. and J.H.; Resources, J.H.; Software, S.M.; Supervision, X.Z. and K.Y.; Validation, X.Z. and Y.Z.; Visualization, Y.Z., Y.Z. and J.H.; Writing—original draft, S.M. and K.Y.; Writing—review & editing, K.Y., Y.Z. and J.H. All authors have read and agreed to the published version of the manuscript.

Funding: This research received was founded by the National Outstanding Youth Science Fund Project of the National Science Foundation of China (52022077).

Acknowledgments: The authors would like to thank the National Outstanding Youth Science Fund Project of the National Science Foundation of China (52022077) for the financial support.

Conflicts of Interest: The authors declare no conflict of interest.

References

1. Zhang, T.; Chen, X.; Gu, J.; Li, Q. Progress of research on cage stability of high-speed angular contact ball bearings, Hangkong Xuebao/Acta Aeronaut. *Astronaut. Sin.* **2018**, *39*, 22026. [CrossRef]
2. Sathyan, K.; Gopinath, K.; Lee, S.-H.; Hsu, H.-Y. Bearing Retainer Designs and Retainer Instability Failures in Spacecraft Moving Mechanical Systems. *Tribol. Trans.* **2012**, *55*, 503–511. [CrossRef]
3. Liu, D.L.; Jiang, T.; He, Y.H. Discussion on failure problems of aero-bearing. *J. Fail. Anal. Prev.* **2015**, *10*, 324–330.
4. Liu, X.H. Dynamics Analysis Model of High-Speed Rolling Bearings and Dynamic Performance of Cages. Ph.D. Thesis, Dalian University Technology, Dalian, China, 2011.
5. Kingsbury, E.P. Torque variations in instrument ball bearings. *ASLE Trans.* **1965**, *8*, 435–441. [CrossRef]
6. Kingsbury, E.; Walker, R. Motions of an Unstable Retainer in an Instrument Ball Bearing. *J. Tribol.* **1994**, *116*, 202–208. [CrossRef]
7. Stevens, K.T. Experimental observations on torque variation caused by ball bearing cage instabilities (spin tests). In Proceedings of the Second Space Tribology Workshop, Risley, UK, 15–17 October 1980; Volume 21, pp. 101–111.
8. Gentle, C.R.; Pasdari, M. Measurement of Cage and Pocket Friction in a Ball Bearing for Use in a Simulation Program. *ASLE Trans.* **1985**, *28*, 536–541. [CrossRef]
9. Walters, C.T. Dynamics of ball bearings. *J. Lubr. Technol.* **1971**, *93*, 1–10. [CrossRef]
10. Kannel, J.W.; Bupara, S.S. A Simplified Model of Cage Motion in Angular Contact Bearings Operating in the EHD Lubrication Regime. *J. Lubr. Technol.* **1978**, *100*, 395–403. [CrossRef]
11. Gupta, P.K. Vibration Lubrication Technology of ball bearings. *J. Lubr. Technol. Trans. ASME* **1977**, *99*, 284–289. [CrossRef]
12. Gupta, P.K. Frictional Instabilities in Ball Bearings. *Tribol. Trans.* **1988**, *31*, 258–268. [CrossRef]
13. Gupta, P.K. On the Frictional Instabilities in a Cylindrical Roller Bearing. *Tribol. Trans.* **1990**, *33*, 395–401. [CrossRef]
14. Gupta, P.K. Modeling of Instabilities Induced by Cage Clearances in Ball Bearings. *Tribol. Trans.* **1991**, *34*, 93–99. [CrossRef]
15. Meeks, C.R.; Ng, K.O. The Dynamics of Ball Separators in Ball-Bearings I: Analysis. *ASLE Trans.* **1985**, *28*, 277–287. [CrossRef]
16. Meeks, C.R. The Dynamics of Ball Separators in Ball Bearings—Part II: Results of Optimization Study. *ASLE Trans.* **1985**, *28*, 288–295. [CrossRef]
17. Houpert, L. Piezoviscous-Rigid Rolling and Sliding Traction Forces, Application: The Rolling Element–Cage Pocket Contact. *J. Tribol.* **1987**, *109*, 363–370. [CrossRef]
18. Houpert, L. CAGEDYN: A Contribution to Roller Bearing Dynamic Calculations Part I: Basic Tribology Concepts. *Tribol. Trans.* **2009**, *53*, 1–9. [CrossRef]
19. Houpert, L. CAGEDYN: A Contribution to Roller Bearing Dynamic Calculations Part II: Description of the Numerical Tool and Its Outputs. *Tribol. Trans.* **2009**, *53*, 10–21. [CrossRef]
20. Houpert, L. CAGEDYN: A Contribution to Roller Bearing Dynamic Calculations. Part III: Experimental Validation. *Tribol. Trans.* **2010**, *53*, 848–859. [CrossRef]
21. Rivera, M.P. Bearing-Cage Frictional Instability—A Mechanical Model. *Tribol. Trans.* **1991**, *34*, 117–121. [CrossRef]
22. Boesiger, E.A.; Donley, A.D.; Loewenthal, S. An Analytical and Experimental Investigation of Ball Bearing Retainer Instabilities. *J. Tribol.* **1992**, *114*, 530–538. [CrossRef]
23. Nogi, T.; Maniwa, K.; Matsuoka, N. A Dynamic Analysis of Cage Instability in Ball Bearings. *J. Tribol.* **2018**, *140*, 011101. [CrossRef]
24. Sakaguchi, T.; Kaoru, U. Dynamic Analysis of Cage Behavior in a Cylindrical Roller Bearing. *Spec. Suppl. Ind. Mach.* **2004**, *71*, 8–17.
25. Sakaguchi, T.; Harada, K. Dynamic Analysis of Cage Behavior in a Tapered Roller Bearing. In Proceedings of the World Tribology Congress III, Washington, DC, USA, 12–16 September 2005; pp. 165–166. [CrossRef]

26. Weinzapfel, N.; Sadeghi, F. A Discrete Element Approach for Modeling Cage Flexibility in Ball Bearing Dynamics Simulations. *J. Tribol.* **2009**, *131*, 021102. [CrossRef]
27. Ashtekar, A.; Sadeghi, F. A New Approach for Including Cage Flexibility in Dynamic Bearing Models by Using Combined Explicit Finite and Discrete Element Methods. *J. Tribol.* **2012**, *134*, 041502. [CrossRef]
28. Harris, T.A. *Rolling Bearing Analysis*, 5th ed.; John Wiley and Sons, Inc.: New York, NY, USA, 2006.
29. Wang, Y.S. Study of Rheological Behavior of Aviation Lubricating Oil and Its Effect on Lubrication Performance. Ph.D. Thesis, Harbin Institute of Technology, Harbin, China, 2006.
30. Tu, W.; Shao, Y.; Mechefske, C.K. An analytical model to investigate skidding in rolling element bearings during acceleration. *J. Mech. Sci. Technol.* **2012**, *26*, 2451–2458. [CrossRef]
31. Wang, Y. Dynamic Analysis of Angular Contact Ball Bearing-Rotor System during Start up and Shut down. *Jixie Gongcheng Xuebao/Chin. J. Mech. Eng.* **2018**, *54*, 9–16. [CrossRef]
32. Zhang, W.H. Study on the Dynamics Simulation and Semi-Physical Experiment of Cylindrical Roller Bearing. Ph.D. Thesis, Northwestern Polytechnical University, Xi'an, China, 2006.

lubricants MDPI

Article

Analysis of Grooves Used for Bearing Lubrication Efficiency Enhancement under Multiple Parameter Coupling

Chao Wang [1], Mingkai Wang [1] and Linbo Zhu [2,*]

[1] Key Laboratory of Education Ministry for Modern Design and Rotor-Bearing System, Xi'an Jiaotong University, Xi'an 710049, China; jannik@stu.xjtu.edu.cn (C.W.); mkwang@stu.xjtu.edu.cn (M.W.)

[2] School of Chemical Engineering and Technology, Xi'an Jiaotong University, Xi'an 710049, China

* Correspondence: linbozhu@mail.xjtu.edu.cn

Abstract: Adding axial groove structures to the surface of bearing inner ring is an effective way to enhance the bearing lubrication efficiency. In this paper, the angular contact ball bearing H7006C was taken as the research object, and through visual simulations and quantitative experiments, systematic analyses and discussions were carried out to find out the relationship between groove-enhancing performance and working conditions. Firstly, simulation models of standard bearing and groove-added bearing were established. By combining the Volume of Fluid (VOF) model, the enhancing mechanism of grooves was found. Secondly, the trend of groove-enhancing performance with the change of rotation speed was studied through simulations and quantitative experiments. On this basis, through multiple parameter coupling, the effects of oil supply amount and nozzle diameter on the groove performance were discussed. The results show that the bottom oil layer is the key for grooves to achieve the lubrication efficiency enhancement, and its distribution can reflect the groove-enhancing performance. The groove width that best adapts to the change of working conditions increases with the increase of oil supply amount and nozzle diameter. To maintain the stability of bottom oil layer, the nozzle diameter should be larger than the groove width. This research is of great significance to the application of grooves in the lubrication efficiency enhancement of high-speed bearing.

Keywords: bearing lubrication enhancement; groove structure; multiple parameter coupling

Citation: Wang, C.; Wang, M.; Zhu, L. Analysis of Grooves Used for Bearing Lubrication Efficiency Enhancement under Multiple Parameter Coupling. *Lubricants* **2022**, *10*, 39. https://doi.org/10.3390/lubricants10030039

Received: 19 February 2022
Accepted: 5 March 2022
Published: 7 March 2022

1. Introduction

The complex motions and collisions of bearing assemblies can seriously affect bearing performance [1]. Collisions can generate much heat, leading to a large temperature rise. Keeping bearing well-lubricated can reduce bearing friction loss, control the temperature rise, and improve bearing performance [2]. Among the common lubrication methods, jet lubrication is widely used in high-speed working conditions for its high lubrication efficiency [3]. In jet lubrication, oil or oil-air mixture is normally injected from the nozzle mounted on the side of the bearing into the bearing cavity, then adheres to the bearing assemblies or flows into the contact area. However, high-speed and high-pressure vortices form when the bearing rotates at high speed, preventing oil from flowing into the contact area, resulting in the decrease of bearing lubrication efficiency [4,5]. In order to improve the bearing lubrication, several studies from the perspective of the optimization of oil supply parameters have been carried out. Zhang [6] has studied the influence of different oil supply amount, number of nozzles, injection speed, and rotation speed on the two-phase flow field in the bearing cavity and compared the friction torque of the bearing under the nozzles of different diameters. Taking the bearing temperature as the index, Li [7] and Zheng [8] have studied the effects of the number of nozzles, rotational speed, and nozzle diameter on the bearing lubrication efficiency under jet lubrication through experiments.

Although changing the oil supply parameters is a feasible way to improve the bearing lubrication efficiency, the oil transfer into the contact area is still restricted by the strong air flow. Therefore, the way of the optimization of the oil supply structure to improve lubrication efficiency is gaining much attention. A novel guide-type nozzle designed by Guo [9,10] can improve the utilization rate of oil for using the capillary force to reduce the influence of the air core inside the nozzle, so oil at the outlets is of higher velocity and easy to cross the air curtain inside the bearing cavity. NSK Ltd. has manufactured Spinshot II nozzle, this type of nozzle injects oil to the surface of the inner ring, which reduces the influence of intense air flow on the oil transfer into the bearing contact area [11]. A similar type of nozzle has also been designed by NTN Corporation [12], with this nozzle oil injected into the counterbore of the inner ring and then transported into the bearing cavity by centrifugal force. Wang [13] compared the oil-air flow inside the bearing cavity between the traditional nozzle and Spinshot II nozzle, the results show that with Spinshot II more oil accumulates near both contact areas of the bearing. However, the inner-ring-oriented nozzles are still restricted, while oil has high flow resistance when flowing on the inner ring surface for its liquid-solid contact with the surface. Several scholars have proposed a method based on groove structure to further improve lubrication efficiency [14–18]. Yan [15] established a whole bearing simulation model with some axial groove structures added to the non-contact area of the bearing inner ring, then compared the lubrication state of the new bearing and the standard one. The results show that the one with grooves has more oil adhering to the ball surface. The flow process of oil in the axial groove has also been studied, it can be divided into three stages dominated by the injection velocity, capillary force, and viscous force, respectively. Through bearing lubrication performance experiments, the validity of using groove to improve the lubrication efficiency is also verified [16]. Based on this, studies on the influences of injection angle, injection speed, and groove width on the groove-enhancing capacity have been carried out [17,18].

However, previous studies are all single-factor analyses. In fact, the factors interact with each other and influence groove performance together. In this paper, through the oil flow simulations and the quantitative experiments, the basic relationship between rotational speed and groove enhancement performance is found, the influences of bearing oil supply amount and nozzle size on grooves are discussed, and the changing law of groove width is found.

2. Analysis of Lubrication Efficiency Enhancing Mechanism of Groove

2.1. Bearing Simulation Model

The H7006C angular contact ball bearing was treated as the research object in this paper. The axial grooves were evenly added to the non-contact area on the inner ring surface, as seen in Figure 1a. Due to its periodic geometry, only 1/17 of the model was taken to accelerate the calculation speed. Meanwhile, in order to monitor the flow of lubricating oil on the inner ring surface more intuitively and quantify the amount of oil entering the raceway, the bearing cage and ball were ignored. Through the extraction of fluidic domain, the simulation models of the H7006C bearing with groove and the standard H7006C bearing were obtained, as shown in Figure 1b.

The geometrical parameters and of the nozzle and groove and their positional relationship are shown in Figure 1c. The main parameters of the grooves include the groove width w and the groove depth h. In this paper, the research mainly focuses on the groove width w, the groove depth h is set as 0.2 mm. The nozzle whose diameter d is 0.5 mm is located on the side of the bearing, and the axial distance between nozzle and bearing side D is 0.5 mm and height H is 0.3 mm, and the injection angle is 30° to the horizontal. The simulation models were divided into unstructured cells by ICEM CFD, and the meshes near the non-contact area were refined to increase the calculation accuracy.

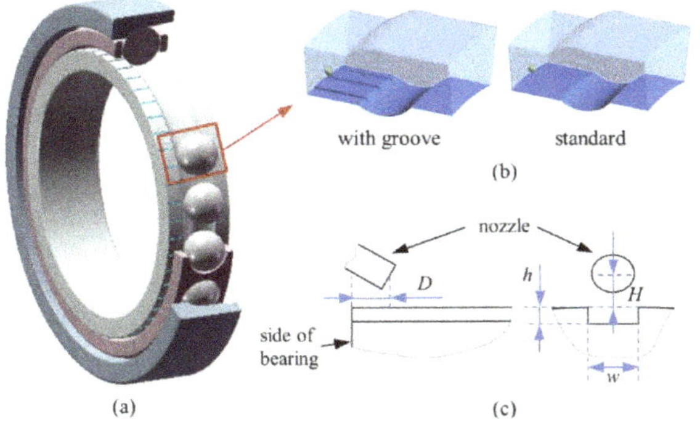

Figure 1. The bearing lubrication simulation model. (**a**) geometry model, (**b**) simulation model, (**c**) parameters and position relationship.

2.2. Boundary Conditions

The calculations were carried out with Fluent. The inner ring of the bearing was set as a rotating wall, and the outer ring and nozzle wall were set as a stationary wall. The nozzle was set as mass-flow-inlet with oil amount m. The other boundaries were all set as pressure-outlet. The set-up of boundary conditions is shown in Figure 2.

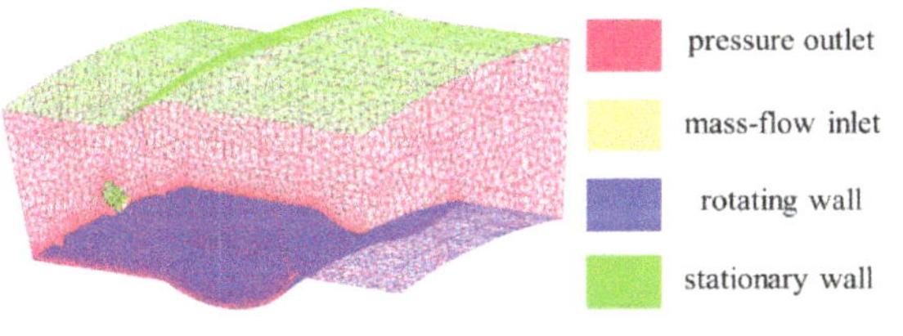

Figure 2. CFD mesh and boundary conditions.

The Volume of Fluid (VOF) model was used to track the oil-air interface, air was considered as the main phase, while the secondary phase is oil. The physical properties of oil are listed in Table 1. The standard k-ε turbulent model was used due to the intense air flow motion inside the bearing cavity.

Table 1. Physical properties of oil.

Physical Properties	Value	Physical Properties	Value
Density ρ (kg/m^3)	876	Surface tension γ (n/m)	0.04
Viscosity μ (kg/m·s)	0.058	Contact angle θ_{CA} (°)	52.8

2.3. Oil-Flow-Enhancing Mechanism of Grooves

Figure 3a shows the oil flow distribution on the inner ring surface of standard bearing and grooved bearing with m = 4 mL/s and bearing rotating speed n = 10,000 rpm. It can be found that more oil flows into the grooved bearing raceway, which shows that grooves can certainly improve the axially flowing capacity of oil. Figure 3b shows the oil distribution on the axial section of the groove of different bearings. Compared with the standard bearing, oil flows longer, and its velocity decreases more slowly. The oil distribution in the groove can be divided into two oil layers. The oil in the groove with a lower axial velocity is named as "bottom oil layer". The oil with a higher axial velocity, which flows on the top of "bottom oil layer", is named as "top oil layer". Although the oil in the bottom oil layer cannot flow into the bearing raceway, it changes the solid-liquid contact when oil flows on the standard bearing into fluid contact and decreases oil flowing resistance.

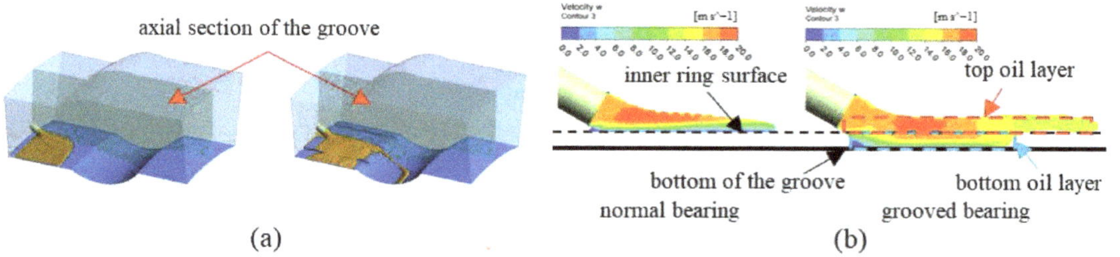

Figure 3. Oil distribution of bearing with m = 4 mL/s and n = 10,000 rpm. (**a**) Oil flow distribution on both bearings, (**b**) oil distribution on the groove axial section.

Since the bottom oil layer in the groove is the key to achieving the enhancement of bearing lubrication efficiency, its distribution area may affect its oil-flow-enhancing capacity. The oil distribution on the bottom wall of the groove was taken as the index D_{oil} to evaluate the enhancing capacity of the groove on bearing lubrication efficiency, which is defined as follows:

$$D_{oil} = \frac{A_{oil}}{A_{groove}} \tag{1}$$

where: A_{oil} is the area of the oil distribution on the bottom wall.

A_{groove} is the area of the bottom wall.

Meanwhile, the amount of oil that flows into the bearing raceway was taken as the other index M_{oil}. Both indexes were acquired through CFD-post, and the methods are shown in Figure 4.

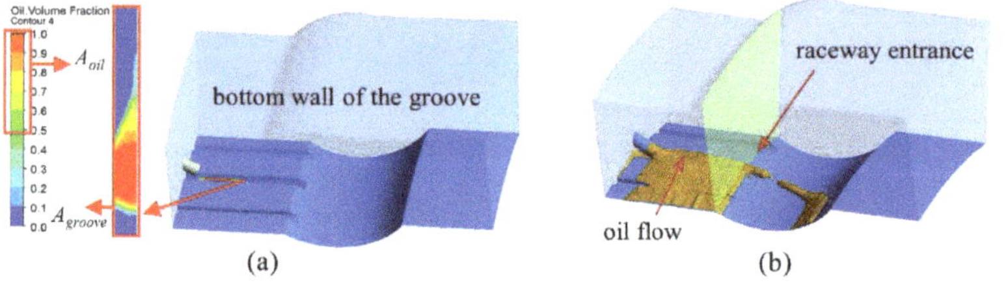

Figure 4. Index acquisition methods. (**a**) D_{oil}, (**b**) M_{oil}.

A_{oil} is the sum of the areas where the oil volume fraction is no less than 0.5. It should be stressed that if no bottom oil layer exists in the groove, the index D_{oil} was taken as 0. For instance, as shown in Figure 5, the oil distribution on the axial section of the groove is similar to that of the standard bearing, which indicates that only a low amount of oil can be enhanced by the groove. In this case, oil is more likely to spread and deflect circumferentially, which is hard for the bottom oil layer to form.

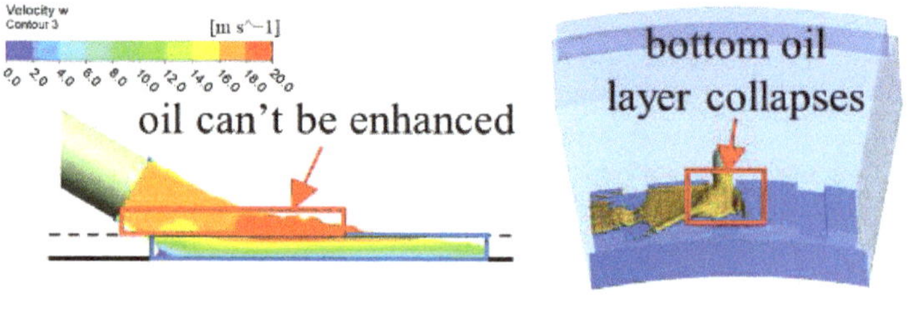

Figure 5. The oil flow collapses inside the groove.

3. Effect of Rotation Speed on the Enhancing Capacity of Grooves

3.1. Simulation Results and Discussion

Figure 6 shows both indexes of different groove widths with the change of bearing rotation speed, where the bars stand for index M_{oil} and curves stand for D_{oil}. When the bearing rotation speed is 1000 rpm, the wider the groove is, the larger M_{oil} is. This is because more oil can be filled in the groove with the increase of the groove width, thereby improving the efficiency on the oil flow. However, M_{oil} decreases with the increase of the rotation speed due to the increased centrifugal force and the deflection effect. Besides, the wider grooves show a more obvious downward trend. The reason is that as the groove width increases, the restriction of the groove side walls on the oil flow is reduced, oil is more susceptible to the deflection of the bearing, resulting in the decrease of the stability of the bottom oil layer, and the drag reduction effect on the top oil layer is greatly reduced, which can be seen in Figure 4.

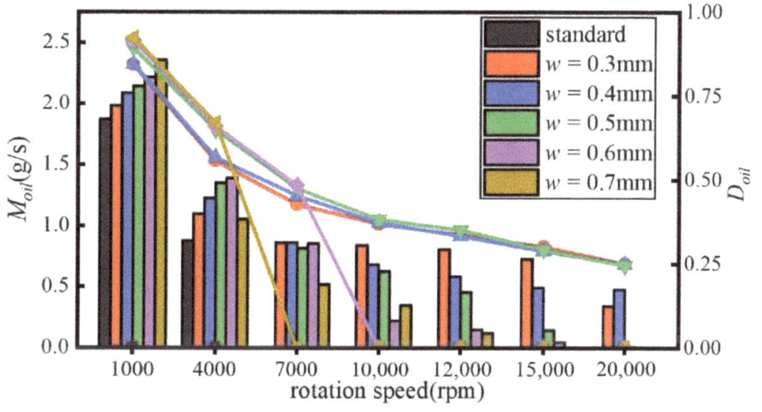

Figure 6. Change in the enhancing capacity of different grooves.

From the curves that show the change of D_{oil} with rotation speed, it can be seen that the at lower rotation speed, D_{oil} is larger with wider grooves, but when rotation speed increases to 10,000 rpm, no more bottom oil layer exists in the grooves with width w 0.6 or 0.7 mm. The downward trend of D_{oil} is gentler when the width w is 0.3 or 0.4 mm. It can be seen that the downward trend of M_{oil} is similar to that of D_{oil}, which indicates that the distribution of the bottom oil layer can reflect the enhancing capacity of the grooves. In the working conditions described above, the groove with width w = 0.3 mm behaves well and can adapt to the changes of rotation speed best. This kind of width is called "the most adaptive width" in the following part of the article. Besides, for each certain rotation speed, there is an optimal width that has the best enhancing capacity, this is called "the optimal width" in the following part of the article.

3.2. Experimental Verification

3.2.1. Experimental System

The experimental system is shown in Figure 7a. It was made up of a motorized spindle, mechanical spindle, electronic balance, oil supply structure, and a simplified bearing inner ring. The simplified bearing inner ring, called "simplified ring" in the following part of the article, is divided into the upstair area and the downstair area, as shown in Figure 7b. The upstair area seen as the non-contact area has grooves and the downstair area seen as the bearing raceway can collect oil by using oil-absorbing tissue. The ring was installed on the end of the mechanical spindle. The nozzle of oil supply structure was set at the top of the ring. The inclination angle between the nozzle and the horizontal plane was 30°. During the experiments, first, the spindle speed was set to the experimental speed and then the oil supply structure was turned on. After 30 min, the increased weights of tissues were treated as the results, which reflects the flowing capacity of oil. To improve the accuracy of the results, seven experiments were carried out for each rotation speed of each ring, five from which with the smallest variance were selected, and the average value of them was taken as the final result.

Figure 7. Experiment system and the simplified ring. (**a**) experiment system (**b**) simplified ring.

3.2.2. Results and Discussion

Several simplified rings were manufactured, including one ring without groove as the standard bearing, and three rings with different grooves widths, respectively 0.3, 0.5, and 0.7 mm. The increased weights of oil on these four rings were achieved when the spindles rotated at the speed of 1000, 3000, and 5000 rpm., as shown in Figure 8. The results show that at each rotation speed, the weight increase of rings with groove widths of 0.3 and 0.5 mm is larger than the weight increase of the ring without groove, which shows that proper grooves can enhance the oil flow capacity and let more oil flow into bearing raceway. Secondly, at the same speed, as the groove width increases, the weight increase shows a trend of first increasing and then decreasing, indicating for each rotation speed, there is an optimal width for the grooves. The decreasing trend of weight increase of the ring with a groove width of 0.5 mm is sharper than that of the ring with a groove width of 0.3 mm with the change of speed. This result shows a similar phenomenon to the simulation one that the decline of M_{oil} of wider grooves is larger than that of narrower grooves.

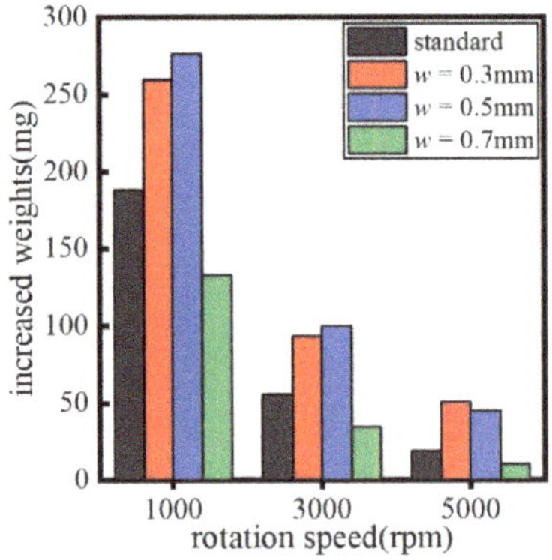

Figure 8. The increased weights of tissue on different rings.

4. Influence of Oil Supply Situation on the Enhancing Capacity of the Grooves

4.1. Influence of Oil Supply Amount on the Groove Width under Different Rotation Speed

Since oil supply is related to the bearing lubrication performance, it may affect the enhancing capacity of the grooves. The bars and the curves in Figure 9 represent the indexes M_{oil} and D_{oil} at oil supply amounts of 4 mL/s and 8 mL/s, respectively. As can be seen, at each rotation speed, the amount of oil flowing into the bearing raceway increases with the increase of oil supply amount. The reason is when the oil supply amount increases, the oil sprayed from the nozzle has higher kinetic energy, which can help resist friction loss, so that the oil flowing on the non-contact area surface may have a larger axially flowing speed, thus improving its axial flowing capacity. Besides, more oil flowing into the groove is beneficial to the form of the bottom oil layer and provides a better drag reduction effect for the top oil layer. In addition, it can be seen from the downward trend that the most suitable groove width value increases with the increase of the oil supply amount, from 0.3 mm at 4 mL/s to 0.5 mm at 8 mL/s.

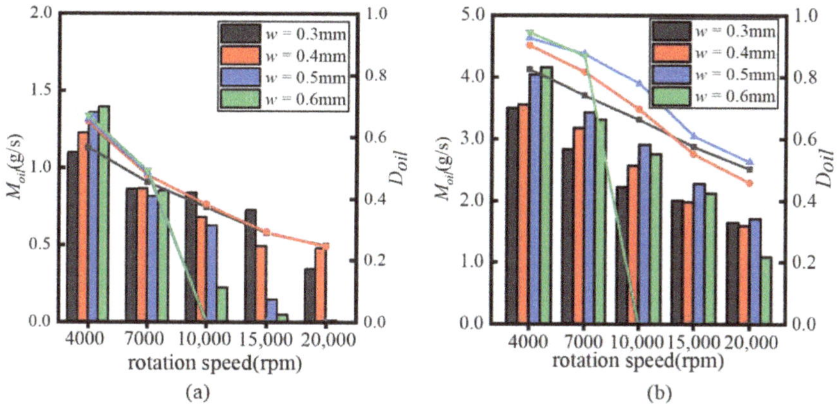

Figure 9. M_{oil} and D_{oil} under different oil supply amounts. (**a**) Oil supply amount m = 4 mL/s, (**b**) Oil supply amount m = 8 mL/s.

For D_{oil}, the results show that as the oil supply amount increases, D_{oil} at each groove width increases. This is because the increase in the oil supply amount can increase the oil amount of the bottom oil layer in the groove and can supplement the oil that leaves the groove structure due to the deflection of the bearing, which effectively improves the dynamic stability of the bottom oil layer. Besides, the bottom oil layer can form in the groove with w = 0.6 mm at high rotation speed with oil supply amount increasing and its upward trend is more obvious than that of groove with w = 0.3 mm, which is owing to the larger capacity of wider groove for oil to feed in. Therefore, the improvement of oil-flow-enhancing capacity of wider grooves seems more obvious than the narrower ones.

4.2. Influence of Nozzle Diameter d on the Groove Width under Different Rotation Speed

The change of the nozzle diameter d will affect the oil flow area, so the enhancing capacity of the groove may also change. The indexes M_{oil} and D_{oil} at nozzle diameter of 0.4 and 0.6 mm are shown by bars and curves of Figure 10, respectively. The wide groove is better at low speed, but the downward trend is greater. Besides, M_{oil} of each groove width at each rotation speed decreases with the increase of nozzle diameter. The reason is that there is a negative relationship between the nozzle diameter and the sprayed velocity when the oil supply amount is a fixed value, oil has lower kinetic energy if it is sprayed from larger nozzles, after it enters the grooves, it will get easier to flow out of the grooves due to the deflection caused by the bearing rotation, resulting in a shorter bottom oil layer and a weaker drag reduction efficiency.

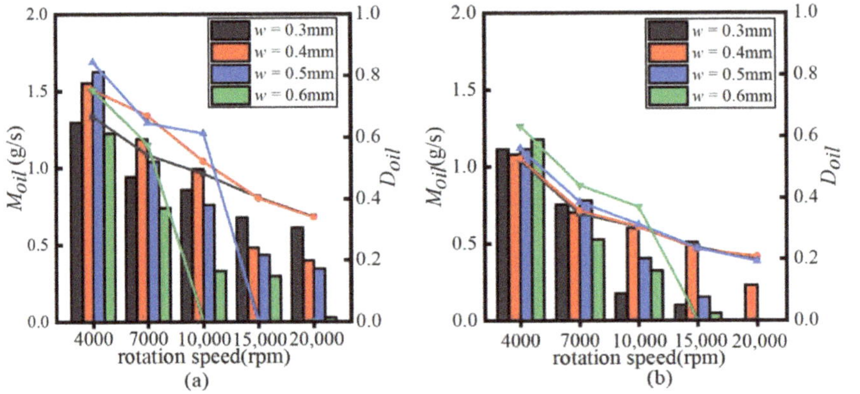

Figure 10. M_{oil} and D_{oil} under different oil supply amount. (**a**) Nozzle diameter d = 0.4 mm, (**b**) nozzle diameter d = 0.6 mm.

For D_{oil}, at a low rotation speed, the smaller nozzle has a larger D_{oil} because it sprays faster oil, which helps the expansion of the bottom oil layer. However, with the increase of rotation speed, the bottom oil layer of the smaller nozzle is less stable. When the nozzle diameter is larger than 0.5 mm, the bottom oil layer can form and remain stable in the groove of width 0.5 mm, but it becomes unstable when the nozzle diameter is 0.4 mm. Besides, the increase of nozzle diameter results in the increase of stability of the bottom oil layer in wider grooves, the oil layer under the 0.6 mm nozzle does not collapse until 15,000 rpm, while under other diameters, it collapses at 10,000 rpm. The downward trend of D_{oil} at each nozzle diameter is also similar to that of M_{oil}. When nozzle diameter d is 0.4 and 0.5 mm, the most adaptive width of groove is 0.3 mm, but when nozzle diameter d is 0.6 mm, the most adaptive width of groove increases to 0.4 mm. From the curves, can we learn that at high rotation speed D_{oil} of groove of width 0.3 mm almost equals that of groove of width 0.4 mm, ideally the oil which can be enhanced can be seen in Figure 11, however, for a fixed oil supply amount, the amount of enhanced oil decreases with the increase of nozzle diameter, this amount becomes the lowest when groove width is 0.3 mm, resulting in a significant shrink of M_{oil} at high speed. To maintain the stability of the bottom oil layer, it is suggested that the groove width should be no more than the nozzle diameter.

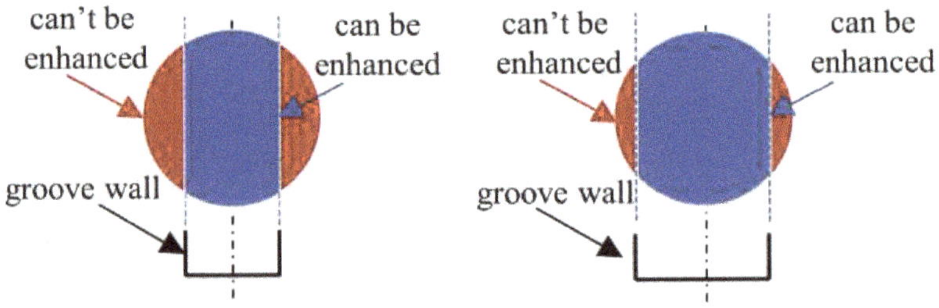

Figure 11. Simplified schematic diagram of enhanced oil.

5. Conclusions

By analyzing the flow characteristics of oil on the surface of different bearings, the enhancing mechanism of groove was analyzed. The enhancing capacity of different grooves was compared through simulations and experiments and the coupling effects of rotation speed, oil supply amount, nozzle diameter on the enhancing capacity were discussed. The conclusions are as follows:

1. The oil distribution in the groove can be divided into the top oil layer and the bottom oil layer. The bottom oil layer is the key for groove to achieve oil flow enhancement.
2. The trend of the experimental results is in good agreement with the trend of the simulation results. Compared with that of the groove of width 0.5 mm, the enhancing capacity of groove of width 0.3 mm seems more adaptive to the increase of rotation speed from 1000 to 5000 rpm.
3. The most adaptive width increases with the increase of oil supply amount and nozzle diameter. To achieve better enhancing capacity, the groove width should be no bigger than the nozzle diameter.

Author Contributions: Conceptualization, C.W., M.W. and L.Z.; Methodology, C.W.; Software, L.Z.; Validation, C.W. and L.Z.; Formal analysis, C.W. and M.W.; Investigation, C.W. and M.W.; Resources, L.Z.; Data curation, C.W.; Writing—original draft preparation, C.W.; Writing—review and editing, L.Z.; Visualization, C.W. and M.W.; Supervision, L.Z.; Project administration, L.Z.; Funding acquisition, L.Z. All authors have read and agreed to the published version of the manuscript.

Funding: This research was funded the National Key R&D Program of China (No. 2020YFB2007900).

Institutional Review Board Statement: Not applicable.

Informed Consent Statement: Not applicable.

Conflicts of Interest: The authors declare no conflict of interest.

References

1. Ma, S.; Zhang, X.; Yan, K.; Zhu, Y.; Hong, J. A Study on Bearing Dynamic Features under the Condition of Multiball–Cage Collision. *Lubricants* **2022**, *10*, 9. [CrossRef]
2. Li, H.; Shin, Y.C. Analysis of bearing configuration effects on high speed spindles using an integrated dynamic thermo-mechanical spindle model. *Int. J. Mach. Tools Manuf.* **2004**, *44*, 347–364. [CrossRef]
3. Pinel, S.I.; Signer, H.R.; Zaretsky, E.V. Comparison between oil-mist and oil-jet lubrication of high-speed, small-bore, angular-contact ball bearings. *Tribol. Trans.* **2001**, *44*, 327–338. [CrossRef]
4. Yan, K.; Wang, Y.; Zhu, Y.; Hong, J.; Zhai, Q. Investigation on heat dissipation characteristic of ball bearing cage and inside cavity at ultra high rotation speed. *Tribol. Int.* **2016**, *93*, 470–481. [CrossRef]
5. Yan, K.; Dong, L.; Zheng, J.; Li, B.; Wang, D.; Sun, Y. Flow performance analysis of different air supply methods for high speed and low friction ball bearing. *Tribol. Int.* **2018**, *121*, 94–107. [CrossRef]
6. Zhang, R. Investigation on the Two-Phase Flow and Heat Transfer Inside High-speed Ball Bearings for in Vehicular Transmission. Master's Thesis, Beijing Institute of Technology, Beijing, China, 2016; pp. 36–54.
7. Li, Z.H. Effect of Operation Parameters on the Performance of Rolling Bear under Oil-air Lubrication. Master's Thesis, Qingdao Technological University, Qingdao, China, 2016; pp. 24–28.
8. Zheng, D.Z.; Gu, L.; Wang, L.Q.; Wei, Y.Q. Effect of lubrication oil supply parameters on the performance of high-speed rolling bearing. *J. Harbin Inst. Technol.* **2006**, *38*, 11–14.
9. Liu, M.Y.; Guo, F.; Jiao, Y.H.; Peng, S.L.; Wang, X. A novel guide-type nozzle for oil-air lubrications. *China Mech. Eng.* **2018**, *29*, 1284–1288.
10. Jiao, Y.H.; Peng, S.L.; Liu, M.Y.; Peng, S.L.; Wang, X. FLUENT Simulation of a Novel Guide-Type Nozzle for Oil-Air Lubrication. *Mach. Des. Manuf.* **2019**, *29*, 238–241.
11. NSK. Super Precision Bearings. [EB/OL]. Available online: https://www.nsk-literature.com/en/super-precision-bearings (accessed on 14 January 2022).
12. NTN. Precision Rolling Bearings. [EB/OL]. Available online: https://www.ntn.com.cn/feedl/pdf/CAT.NO.2260C.pdf (accessed on 14 January 2022).
13. Wang, B.M.; Bai, C.; Nan, Y.; Wu, Y. Influences of Nozzle Structure on Air Curtain Effect in Oil-air Lubrication of Angular Contact Ball Bearings. *China Mech. Eng.* **2021**, *32*, 2197–2202.
14. Liu, H.B.; Liu, G.P.; Hao, J.H.; Yang, M.K. Oil penetration mechanism induced by different bearing wall grooves. *J. Aerosp. Power* **2019**, *34*, 1127–1136.
15. Ge, L.F.; Yan, K.; Wang, C.; Zhu, Y.; Hong, J. A novel method for bearing lubrication enhancement via the inner ring groove structure. *J. Phys. Conf. Ser.* **2021**, *1820*, 012092. [CrossRef]
16. Ge, L.; Wang, C.; Yan, K.; Zhu, Y.; Hong, J. Design of groove structures for bearing lubrication enhancement based on the flow mechanism analysis. *Tribol. Int.* **2021**, *158*, 106950. [CrossRef]

17. Ge, L.F.; Chen, F.; Yan, K.; Hong, J.; Zhu, Y.S. Research on Lubrication Efficiency Enhancement of High-speed Bearing under Oil-starvation Conditions. *J. Mech. Eng.* **2021**, 1–8. Available online: http://kns.cnki.net/kcms/detail/11.2187.TH.20211116.1650.030.html (accessed on 18 February 2022).
18. Wang, C.; Yan, K.; Zhu, Y.S.; Hong, J. Design of Groove Structure on Bearing Ring Surface for Lubrication Enhancement. *J. Xi'an Jiaotong Univ.* **2022**, 1–8. Available online: http://kns.cnki.net/kcms/detail/61.1069.T.20210830.1027.002.html (accessed on 18 February 2022).

lubricants

MDPI

Article

A Novel Model for Evaluating the Operation Performance Status of Rolling Bearings Based on Hierarchical Maximum Entropy Bayesian Method

Liang Ye [1], Yusheng Hu [2,3,*], Sier Deng [1], Wenhu Zhang [1], Yongcun Cui [1] and Jia Xu [2,3]

1 School of Mechatronics Engineering, Henan University of Science and Technology, Luoyang 471003, China; 9945070@haust.edu.cn (L.Y.); dse@haust.edu.cn (S.D.); zwh@haust.edu.cn (W.Z.); 9945020@haust.edu.cn (Y.C.)
2 State Key Laboratory of Air-Conditioning Equipment and System Energy Conservation, Gree Electric Appliances Co., Ltd., Zhuhai 519070, China; jidianyihao@163.com
3 Guangdong Key Laboratory of Refrigeration Equipment and Energy Conservation Technology, Gree Electric Appliances Co., Ltd., Zhuhai 519070, China
* Correspondence: cmi@cn.gree.com

Abstract: Information such as probability distribution, performance degradation trajectory, and performance reliability function varies with the service status of rolling bearings, which is difficult to analyze and evaluate using traditional reliability theory. Adding equipment operation status to evaluate the bearing operation performance status has become the focus of current research to ensure the effective maintenance of the system, reduce faults, and improve quality under the condition of traditional probability statistics. So, a mathematical model is established by proposing the hierarchical maximum entropy Bayesian method (HMEBM), which is used to evaluate the operation performance status of rolling bearings. When calculating the posterior probability density function (PPDF), the similarities between time series regarded as a weighting coefficient are calculated using overlapping area method, membership degree method, Hamming approach degree method, Euclidean approach degree method, and cardinal approach degree method. The experiment investigation shows that the variation degree of the optimal vibration performance status can be calculated more accurately for each time series relative to the intrinsic series.

Keywords: rolling bearing; performance degradation; variation degree; probability density function; similarities between time series

Citation: Ye, L.; Hu, Y.; Deng, S.; Zhang, W.; Cui, Y.; Xu, J. A Novel Model for Evaluating the Operation Performance Status of Rolling Bearings Based on Hierarchical Maximum Entropy Bayesian Method. *Lubricants* **2022**, *10*, 97. https://doi.org/10.3390/lubricants10050097

Received: 10 April 2022
Accepted: 5 May 2022
Published: 13 May 2022

1. Introduction

As the key components of rotating machinery, whether rolling bearings are in normal working condition directly affects the running state of the host. As the running environment of rolling bearings becomes more and more complex, changeable, and harsh, the performance analysis and evaluation of rolling bearings are facing serious challenges. Therefore, performance degradation analysis, evaluation, and fault diagnosis are very urgent and necessary for rolling bearings, which will directly affect the safety and stability of the whole host system [1–6]. The nonlinear contact and collision between the components (inner race, outer race, rolling element, and cage) of rolling bearings lead to nonlinear and complex dynamic characteristics in the performance degradation progress. Most attention focused on rolling bearing concerns its performance degradation index, so it is generally necessary to evaluate the degradation state of rolling bearing vibration performance effectively.

The performance maintaining relative reliability (PMRR) is used to characterize the performance degradation degree of rolling bearings. Performance maintaining reliability (PMR) is the probability of a rolling bearing running at the optimum performance status, which can be expressed as a function [7]. During the period of optimal vibration performance, there is almost no possibility of performance failure. Not only is the vibration data value small, but also the data fluctuation is not violent. Owing to the interaction and coupling among internal and external factors, it is difficult to solve PMR using the existing dynamic equations and statistical theory.

Adding equipment operation status to evaluate the performance reliability has become the focus of current research to ensure the effective maintenance of the system, reduce faults, and improve quality under the condition of traditional probability statistics [8]. The similarity between time series plays an important role in the analysis and evaluation of performance degradation, which can be calculated using the overlapping area method, membership degree method, Hamming approach degree method, Euclidean approach degree method, and cardinal approach degree method [9–11]. Therefore, this factor should be considered in the dynamic evaluation process of PMR bearing vibration performance. When calculating the PPDF, the similarity between time series can be regarded as a weighting coefficient, so the variation degree of the optimal vibration performance can be calculated more accurately.

Traditional performance degradation evaluation methods often need complete performance data and assume that the data are deterministic. Weibull distribution plays an important role in the research of bearing reliability and bearing life in the traditional theoretical system. Many scholars build models using Weibull distribution or compare the test results with the calculation results using Weibull distribution, which greatly promotes the development of bearing performance reliability [12–15]. However, owing to the high reliability and long performance degradation period of rolling bearings, it is difficult to obtain complete vibration performance data in the experiment, and the collected test data have the typical characteristics of poor information. Poor information occurs when the characteristic information of objects researched is incomplete or inadequate [16–19]. Moreover, the vibration data of rolling bearings are uncertain in the testing process because of the uncertainty factors such as the system error of the testing device, the changeability of the service environment, and so on, which has great limitations for the analysis and evaluation of bearing performance reliability [20,21].

In the existing data-analysis methods, the analysis result of using a single method often has one-sidedness in some aspects, so the evaluation results of PMR can be obtained more comprehensively by fusing several different methods. The bootstrap method can simulate the probability distribution of data samples through re-sampling [22–25], which can separate the systematic errors in dynamic evaluation process by using the nuclear concept, but it needs to take advantage of the prior information of some rules of data sampling. The maximum entropy principle is used to calculate the probability distribution of data samples while making the subjective estimation error minimum [17,24,26,27]. Bayesian theory fully combines the prior information with the current sample information to obtain the posterior sample information [28–30]. The current real-time updating method is based on Bayesian theory, which only uses the observed data sample to update the prior probability distribution, but cannot apply other available information such as the moments of parameters or functions of moments in the probability distribution. Therefore, a variety of methods can be selected for fusion according to the research needs to make up for the limitations of a single method. Ye et al. [7] initially put forward a new concept—accuracy maintaining reliability (AMR) of super-precision rolling bearings, effectively fusing gray bootstrap method and maximum entropy method to predict the failure degree of a bearing successfully maintaining its optimum service accuracy status in the future. However, in the process of calculation, the accuracy threshold was given in advance, which leads to artificial subjective error. So, it is necessary to study a more general real-time performance reliability evaluation method.

In view of this, this paper proposes a HMEBM to establish the reliability evaluation model for evaluating the operation performance status of rolling bearings under the condition that the prior sample information of vibration performance time-series is unknown. The ideas are as follows: based on the vibration acceleration data collected during the service period of rolling bearings, the maximum entropy method was used to calculate the

PDF of different time series. Then, the Bayesian method was applied to obtain the PPDF of different time series. Similarities between time series were calculated using the overlapping area method, membership degree method, Hamming approach degree method, Euclidean approach degree method, and cardinal approach degree method, which were regarded as weighting coefficients to calculate the PMR of different time series more accurately. Bootstrap method and maximum entropy method were used to calculate the estimated true value and estimated interval of PMRR. Finally, the failure probability of the rolling bearings maintaining the optimal vibration performance status was analyzed. The flow diagram of the proposed method is shown in Figure 1.

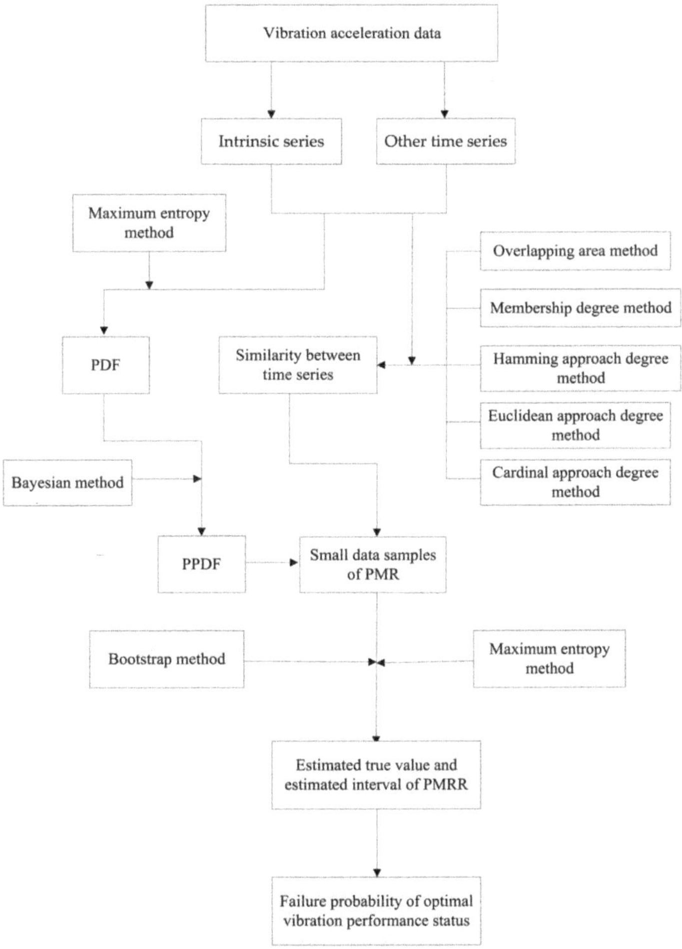

Figure 1. Flow diagram of proposed method.

2. Mathematical Models

During the service period of rolling bearings, the vibration acceleration data are periodically recorded. The time variable is defined as t, and w time series are obtained as $\boldsymbol{X}_w$.

$$\mathbf{X}_w = (x_w(1), x_w(2), \cdots, x_w(k), \cdots, x_w(N)); k = 1, 2, \cdots, N; w = 1, 2, \ldots, r \quad (1)$$

where w stands for the order number of time series $\boldsymbol{X}_w$; r is the number of time series; $x_w(k)$ is the kth performance data in time series $\boldsymbol{X}_w$; N is the number of original data.

The intrinsic series is the time series where the performance data are recorded during the optimum vibration performance status for rolling bearings, which is recorded as the first time series and expressed by $\boldsymbol{X}_1$.

2.1. Solving PDF

Transmuting the performance data into continuous information, the PDF $f_w(x)$ of the time series $\boldsymbol{X}_w$ with maximum entropy is defined as

$$f_w(x) = \exp(c_{0w} + \sum_{i=1}^{j} c_{iw} x^i) \tag{2}$$

where c_{0w} is the first Lagrange multiplier and c_{iw} is the $(i + 1)$th Lagrange multiplier for the time series $\boldsymbol{X}_w$; i is the order number of origin moment, $i = 1, 2, \ldots, j$; j stands for the highest origin moment order; generally, $j = 5$.

According to the maximum entropy principle, the optimal estimation of the density function based on sample information can be obtained, and the main idea of maximum entropy is that the solution is the most "unbiased" among all feasible solutions, as follows:

$$H_w(x) = -\int_{\Omega_w} f_w(x) \ln f_w(x) \mathrm{d}x \rightarrow \max \tag{3}$$

where $H_w(x)$ is the information entropy and Ω_w represents the feasible domain for the data sample of the time series $\boldsymbol{X}_w$, and $\Omega_w = [x_{\mathrm{min}w}, x_{\mathrm{max}w}]$; $x_{\mathrm{min}w}$ and $x_{\mathrm{max}w}$ are the lower-bound value and upper-bound value in the time series $\boldsymbol{X}_w$; $\ln f_w(x)$ is the logarithmic value of $f_w(x)$.

Equation (3) satisfies the constraint conditions

$$\int_{\Omega w} f_w(x) \mathrm{d}x = 1 \tag{4}$$

$$\int_{\Omega_w} x^i f_w(x) \mathrm{d}x = m_{iw} \tag{5}$$

where m_{iw} stands for the ith order origin moment for the data sample of the time series $\boldsymbol{X}_w$, and $m_{0w} = 1$.

The entropy can reach its maximum by adjusting $f_w(x)$, and the PDF $f_w(x)$ can be obtained by using the Lagrange multiplier method.

The Lagrange function $L_w(x)$ can be expressed as

$$L_w(x) = H_w(x) + (c_{0w} + 1)[\int_{\Omega_w} f_w(x) \mathrm{d}x - m_{0w}] + \sum_{i=1}^{j} c_{iw}[\int_{\Omega_w} x^i f_w(x) \mathrm{d}x - m_{iw}] \tag{6}$$

The first Lagrange multiplier can be given by

$$c_{0w} = -\ln[\int_{\Omega_w} \exp(\sum_{i=1}^{j} c_{iw} x^i) \mathrm{d}x] \tag{7}$$

Other j Lagrange multipliers should satisfy the constraint condition

$$1-\frac{\int\limits_{\Omega_w} x^i\exp(\sum\limits_{i=1}^{j} c_{iw}x^i)\mathrm{d}x}{m_{iw}\int\limits_{\Omega_w}\exp(\sum\limits_{i=1}^{j} c_{iw}x^i)\mathrm{d}x}=0 \tag{8}$$

To ensure solution convergence, the original data interval is mapped to interval [−e, e] by the substitution of the variable. Let

$$x = a_w t + b_w \tag{9}$$

where a_w and b_w are mapping parameters for the data sample of the time series $\boldsymbol{X}_w$; $t\in[-e, e]$, e has a value of 2.71828.

Based on the mapping of the original data in the interval [−e, e], the PDF in Equation (2) can be obtained as

$$f_w(t) = \exp[c_{0w} + \sum_{i=1}^{j} c_{iw}(a_w t + b_w)^i] \tag{10}$$

In order to be more adaptable to researchers' habits, the variable t is replaced by the variable x.

2.2. Parameter Estimation

Set a significant level and let $\alpha\in(0,1)$. The confidence level P is given by

$$P = (1-\alpha)\times 100\% \tag{11}$$

Assume that the maximum entropy estimated interval is $[x_{\mathrm{L}w}, x_{\mathrm{U}w}]$ under confidence level P. The lower boundary value $x_{\mathrm{L}w}$ should satisfy

$$\frac{1}{2}\alpha = \int_{x_{\mathrm{min}w}}^{x_{\mathrm{L}w}} f_w(x)\mathrm{d}x \tag{12}$$

The upper boundary value $x_{\mathrm{U}w}$ should satisfy

$$1-\frac{1}{2}\alpha = \int_{x_{\mathrm{min}w}}^{x_{\mathrm{U}w}} f_w(x)\mathrm{d}x \tag{13}$$

2.3. Calculating PPDF

Consider the data sample in the intrinsic series $\boldsymbol{X}_1$ as the prior sample, which is obtained during the period of the optimal vibration performance status for rolling bearings. In order to study the variation process of the vibration performance reliability for rolling bearings, the PPDF for each time series is constructed according to Bayesian statistics as

$$hyf_w(x) = \frac{f_1(x)f_w(x)}{\int\limits_{\Omega_{1w}} f_1(x)f_w(x)\mathrm{d}x} \tag{14}$$

where $hyf_w(x)$ is the PPDF of the wth time series $\boldsymbol{X}_w$; Ω_{1w} is the intersection of the feasible regions of the data samples for the intrinsic series $\boldsymbol{X}_1$ and the wth time series $\boldsymbol{X}_w$.

The similarity between time series plays an important role in the analysis and evaluation of bearing vibration performance variation. Therefore, this factor should be considered in the dynamic evaluation process of PMR of bearing vibration performance. When calculating the PPDF, the similarity between time series can be regarded as a weighting coefficient so that the variation degree of the optimal vibration performance can be calculated more accurately for each time series relative to the intrinsic series. Moreover, the result of analysis using a single method often has one-sidedness in some aspects, so the evaluation results of PMR can be obtained more comprehensively by fusing several different methods.

2.4. Overlapping Area Method

The PMR for vibration performance of rolling bearings is used to express the possibility that rolling bearings run at the optimal vibration performance status. The PMR for the intrinsic series is defined as $R_1 = 1$. The PMR R_w (1) for the wth time series is expressed using the overlapping area method by

$$R_w(1) = S_{1w} = \int_{x=x_{1\mathrm{L}}}^{x_{1w}} hyf_1(x)\mathrm{d}x + \eta_w \int_{x=x_{1w}}^{x_{2w}} hyf_w(x)\mathrm{d}x + \int_{x=x_{2w}}^{x_{1\mathrm{U}}} hyf_1(x)\mathrm{d}x \tag{15}$$

where S_{1w} is the overlapping area for the PPDF of the wth time series and the intrinsic series; $x_{1\mathrm{L}}$ and $x_{1\mathrm{U}}$ are the lower and upper bound values of confidence intervals for the PPDF of intrinsic series, respectively; x_{1w} and x_{2w} are the abscissa values of the intersections for the PPDF of the wth time series and the intrinsic series, and $x_{1w} < x_{2w}$; η_w is the overlapping area for the PDF of the wth time series and the intrinsic series.

$$\eta_w = \int_{x={x^*}_{1\mathrm{L}}}^{{x^*}_{1w}} f_1(x)\mathrm{d}x + \int_{x={x^*}_{1w}}^{{x^*}_{2w}} f_w(x)\mathrm{d}x + \int_{x={x^*}_{2w}}^{{x^*}_{\mathrm{U}1}} f_1(x)\mathrm{d}x \tag{16}$$

where ${x^*}_{1\mathrm{L}}$ and ${x^*}_{1\mathrm{U}}$ are the lower- and upper-bound values of confidence intervals for the PDF of intrinsic series, respectively; ${x^*}_{1w}$ and ${x^*}_{2w}$ are the abscissa values of the intersections for the PDF of the wth time series and the intrinsic series, and ${x^*}_{1w} < {x^*}_{2w}$.

2.5. Membership Degree Method

$\mathbf{X}_w$ is mapped into [0,1] by mapping formula to ensure that all data are fuzzy numbers.

$$z_w(k) = \frac{x_w(k) - x_{w\min}}{x_{w\max} - x_{w\min}} \tag{17}$$

Let $\mathbf{Z}_w$ be a fuzzy subset on the finite field $\boldsymbol{Q}$. The elements between columns have different attributes, and the elements in the same column have the same attributes. $\mathbf{Z}_w$ is described as

$$\mathbf{Z}_w = (z_w(1), z_w(2), \ldots, z_w(k), \ldots, z_w(N)) \tag{18}$$

When studying the conformity degree of $\mathbf{Z}_w$ ($w = 1,2, \ldots ,r$) relative to $\mathbf{Z}_1$, define the absolute difference as

$$\Delta_w(k) = |z_w(k) - z_1(k)| \tag{19}$$

$$\Delta k_{\max} = \max_w \Delta_w(k) \tag{20}$$

The membership functions of elements with the same attributes are established as

$$\mu_{wk} = \mu_{wk}(z_w(k), z_1(k)) = 1 - \frac{\Delta_w(k)}{\Delta k_{\max}} \tag{21}$$

The average membership degree is given as

$$\mu_w = \frac{1}{N}\sum_{k=1}^{N} \mu_{wk}, w = 1, 2, \ldots, r \tag{22}$$

The membership degree can reflect the similarity degree between data series. The larger the average membership degree μ_w is, the more significant the relationship between

$\mathbf{Z}_w$ and $\mathbf{Z}_1$ is; and on the contrary, the less significant the relationship between $\mathbf{Z}_w$ and $\mathbf{Z}_1$ is.

The PMR $R_w(2)$ for the wth time series is expressed using the membership degree method by

$$R_w(2) = \int_{x=x_{1L}}^{x_{1w}} hyf_1(x)\mathrm{d}x + \mu_w \int_{x=x_{1w}}^{x_{2w}} hyf_w(x)\mathrm{d}x + \int_{x=x_{2w}}^{x_{1U}} hyf_1(x)\mathrm{d}x \tag{23}$$

2.6. Approach Degree Method

Let $\mathbf{X}_1$ and $\mathbf{X}_w$ be fuzzy subsets on the finite field $\mathbf{Q}$. $\mu_{1k}\in[0,1]$ and $\mu_{wk}\in[0,1]$ are the membership degrees of $\mathbf{X}_1$ and $\mathbf{X}_w$, respectively. Minkowski distance d_{pk} is defined as

$$d_{pk} = d_p(\mathbf{X}_1, \mathbf{X}_w) = \left(\frac{1}{N}\sum_{k=1}^{N}|\mu_{1k} - \mu_{wk}|^p\right)^{\frac{1}{p}} \tag{24}$$

where N is the number of elements in sets $\mathbf{X}_1$ and $\mathbf{X}_w$; p is a constant, and generally is considered to be a positive integer.

When p = 1, Minkowski distance becomes Hamming distance; when p = 2, Minkowski distance becomes Euclidean distance.

The approach degree between two fuzzy subsets $\mathbf{X}_0$ and $\mathbf{X}_i$ is defined as

$$\beta = \beta_p(\mathbf{X}_1, \mathbf{X}_w) = \frac{1}{2}\left[\mathbf{X}_1 \circ \mathbf{X}_w + \left(1 - \mathbf{X}_1 \,\widehat{\circ}\, \mathbf{X}_w\right)\right] \tag{25}$$

where

$$\mathbf{X}_1 \circ \mathbf{X}_w = \bigvee_{k=1}^{N}(\mu_{1k} \wedge \mu_{wk}) \tag{26}$$

$$\mathbf{X}_1 \,\widehat{\circ}\, \mathbf{X}_w = \bigwedge_{k=1}^{N}(\mu_{1k} \vee \mu_{wk}) \tag{27}$$

The Hamming approach degree β_{1w} is given as

$$\beta_{1w} = \beta_1(\mathbf{X}_1, \mathbf{X}_w) = 1 - \frac{1}{N}\sum_{k=1}^{N}|\mu_{1k} - \mu_{wk}| \tag{28}$$

The PMR $R_w(3)$ for the wth time series is expressed using the Hamming approach degree method by

$$R_w(3) = \int_{x=x_{1L}}^{x_{1w}} hyf_1(x)\mathrm{d}x + \beta_{1w} \int_{x=x_{1w}}^{x_{2w}} hyf_w(x)\mathrm{d}x + \int_{x=x_{2w}}^{x_{1U}} hyf_1(x)\mathrm{d}x \tag{29}$$

Euclidean approach degree β_{2w} is defined as

$$\beta_{2w} = \beta_2(\mathbf{X}_1, \mathbf{X}_w) = 1 - \left(\frac{1}{N}\sum_{k=1}^{N}|\mu_{1k} - \mu_{wk}|^2\right)^{1/2} \tag{30}$$

The PMR $R_w(4)$ for the wth time series is expressed using the Euclidean approach degree method by

$$R_w(4) = \int_{x=x_{1L}}^{x_{1w}} hyf_1(x)\mathrm{d}x + \beta_{2w} \int_{x=x_{1w}}^{x_{2w}} hyf_w(x)\mathrm{d}x + \int_{x=x_{2w}}^{x_{1U}} hyf_1(x)\mathrm{d}x \tag{31}$$

Cardinal approach degree β_{3w} is given as

$$\beta_{3w} = \beta_3(\mathbf{X}_1, \mathbf{X}_w) = \frac{2\sum_{k=1}^{N}(\mu_{1k} \wedge \mu_{wk})}{\sum_{k=1}^{N}\mu_{1k} + \sum_{k=1}^{N}\mu_{wk}} \tag{32}$$

The PMR $R_w(5)$ for the wth time series is expressed using the Cardinal approach degree method by

$$R_w(5) = \int_{x=x_{1L}}^{x_{1w}} hyf_1(x)\mathrm{d}x + \beta_{3w}\int_{x=x_{1w}}^{x_{2w}} hyf_w(x)\mathrm{d}x + \int_{x=x_{2w}}^{x_{1U}} hyf_1(x)\mathrm{d}x \tag{33}$$

The approach degree describes the similarity degree between $\boldsymbol{X}_1$ and $\boldsymbol{X}_w$. The larger the approach degree is, the more significant the relationship between $\boldsymbol{X}_1$ and $\boldsymbol{X}_w$ is; on the contrary, the less significant the relationship between $\boldsymbol{X}_1$ and $\boldsymbol{X}_w$ is.

2.7. Dynamic Evaluation of PMR

According to the above five values of R_w, small data samples of PMR of rolling bearing vibration performance are obtained for each time series.

$$R_w = (R_w(1), R_w(2), \ldots, R_w(5)) = (R_w(\gamma)); \gamma = 1,\ 2, \ldots, 5; w = 1, 2, \ldots, r; \tag{34}$$

where R_w is the data sample of PMR for the wth time series; $R_w(\gamma)$ is the γth data in the PMR data sample for the wth time series.

Using the bootstrap method, B bootstrap re-sampling samples of size z, namely the bootstrap re-sampling samples $R_{w\text{bootstrap}}$, can be obtained by an equiprobable sampling as

$$R_{w\text{Bootstrap}} = (R_{w1}, R_{w2}, \ldots, R_{w\theta}, \ldots, R_{wB}) \tag{35}$$

where $R_{w\theta}$ is the θth bootstrap re-sampling sample, θ = 1, 2, … , B; B is the times of the bootstrap re-sampling, and also the number of bootstrap samples, with

$$R_{w\theta} = [R_{w\theta}(\mathbf{\Theta})]; \mathbf{\Theta} = 1, 2, \ldots, z \tag{36}$$

where $R_{w\theta}(\Theta)$ is the Θth data in the θth bootstrap re-sampling sample of PMR for the wth time series.

The maximum entropy method is used to calculate the PDF of the generated sample $R_{w\text{bootstrap}}$. According to the PDF, the true value and upper and lower bound values are estimated for the data sample of PMR of each time series.

The PDF of the PMR data sample can be calculated as

$$f(R_w) = \exp[c^*{}_{0w} + \sum_{i=1}^{j} c^*{}_{iw}(a^*{}_w R_w + b^*{}_w)^i)] \tag{37}$$

where $c^*{}_{0w}$ is the first Lagrange multiplier and $c^*{}_{iw}$ is the (i + 1)th Lagrange multiplier for the time series R_w; i is the order number of origin moment, i = 1,2, … ,j; j stands for the highest origin moment order; generally, j =5.where $a^*{}_w$ and $b^*{}_w$ are mapping parameters for the data sample of the time series R_w.

The estimated true value of the PMR is obtained as

$$R_{w0} = \int_S R_w f(R_w)\mathrm{d}R_w \tag{38}$$

Set a significant level and let $\alpha \in (0,1)$. The maximum entropy estimated interval is given as

$$[R_{w\mathrm{L}}, R_{w\mathrm{U}}] = [R_{w\frac{\alpha}{2}}, R_{w_{1-\frac{\alpha}{2}}}] \tag{39}$$

with

$$\frac{\alpha}{2} = \int_{R_{w\min}}^{R_{w\frac{\alpha}{2}}} f(R_w)\mathrm{d}R_w \tag{40}$$

$$1 - \frac{\alpha}{2} = \int_{R_{w\min}}^{R_{w_{1-\frac{\alpha}{2}}}} f(R_w)\mathrm{d}R_w \tag{41}$$

where $R_{w\mathrm{L}}$ is the lower bound value and $R_{w\mathrm{U}}$ is the upper bound value of the PMR for the wth time series.

The variation probability of the PMR for different time series relative to the intrinsic series is defined by

$$P_w(\gamma) = 1 - R_w(\gamma) \tag{42}$$

According to the concept of relative error in measurement theory, the failure degree of rolling bearings running at the optimal vibration performance status—that is, the PMRR for the vibration performance of rolling bearings—is expressed by

$$d_w(\gamma) = \frac{R_w(\gamma) - R_{10}}{R_{10}} \times 100\% \tag{43}$$

where R_1 is the PMR for the intrinsic series of rolling bearings.

$d_w(\gamma) < 0$ indicates that the PMR for the time interval corresponding with the wth time series is less than the PMR for the time interval corresponding with the intrinsic series. The smaller $d_w(\gamma)$ is, the greater the failure probability of the rolling bearings maintaining the optimal vibration performance status to work for the time interval corresponding with the wth time series.

$$d_{w0} = \frac{R_{w0} - R_{10}}{R_{10}} \times 100\% \tag{44}$$

$$d_{w\mathrm{L}} = \frac{R_{w\mathrm{L}} - R_{10}}{R_{10}} \times 100\% \tag{45}$$

$$d_{w\mathrm{U}} = \frac{R_{w\mathrm{U}} - R_{10}}{R_{10}} \times 100\% \tag{46}$$

where d_{w0} is the estimated true value, $d_{w\mathrm{L}}$ is the lower-bound value and $d_{w\mathrm{U}}$ is the upper-bound value of the PMRR for the wth time series.

The flow diagram of the proposed method is shown in Figure 1.

3. Experimental Verification

3.1. Case 1

This is a strength lifetime test on the vibration performance of rolling bearings. Experimental data are collected in the whole life-cycle of rolling bearings in Hangzhou Bearing Test & Research Center. The test machine model is an ABLT-1A. This machine mainly consists of a test head seat, test head, transmission system, loading system, lubrication system, and a computer control system. The physical drawings of the testing machine and test head are shown in Figures 2 and 3a–c. The test bearings and support bearings are angular contact ball bearings with grade P2 and type 7008AC provided by Luoyang Bearing Science &Technology Co., Ltd. (Luoyang, China). The inner ring, outer ring, and rolling elements are made of high carbon chromium bearing steel, which has the following characteristics: density 7.8 $\mathrm{g/mm^3}$, elastic modulus 2.08×10^5 $\mathrm{N/mm^2}$, Poisson's ratio 0.3 and Hardness 700 HV10. The bearing parameters are shown in Table 1.

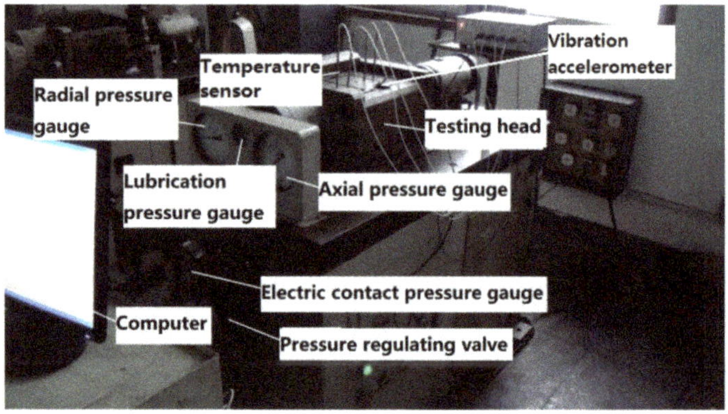

Figure 2. Physical drawing of testing machine.

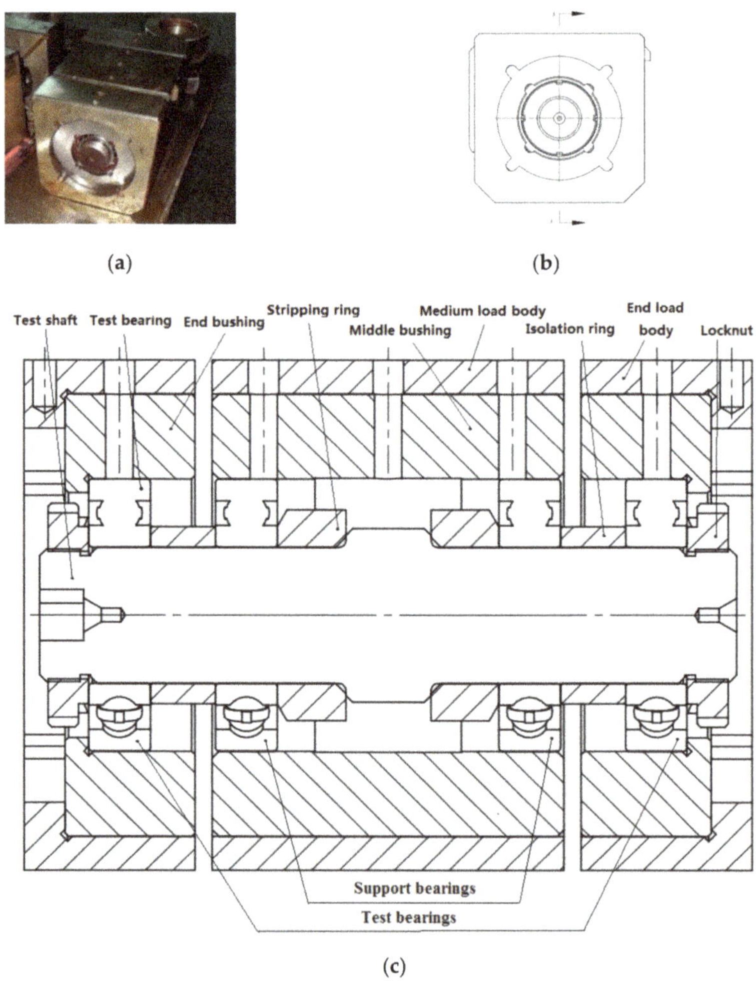

Figure 3. Testing head. (**a**) Physical drawing, (**b**) Main view, (**c**) Section view.

Table 1. Parameters of angular contact ball bearings with type 7008AC.

Parameters	Values	Parameters	Values
Inner diameter d/mm	40	Number of balls	19
Outer diameter D/mm	68	Ball diameter/mm	7.144
Width B/mm	15	Contact angle/°	25

The research is conducted at a motor speed 6000 r/min, an axial load of 4.17 kN, and a radial load of 4.58 kN. The DAQ board type is PCI-1711U, and the data-acquisition rate is 20 KHz. The vibration acceleration sensor used is a YD-1 piezoelectric sensor produced by Far East Vibration System Engineering Technology Co., Ltd. (Beijing, China), with a measuring range of ± 2000 g and a resolution of 0.0001 g. The RMS values of vibration amplitudes are obtained by a 1 min interval. The computer collects the vibration data in the RMS value at this interval with a unit of $m{\cdot}s^{-2}$. If significant variation occurs in the bearing ring or the roller, or even if surface fatigue spalling is noticed, the vibration value of the test machine will obviously increase, and the vibration performance will reduce. If the vibration value reaches 25 $m{\cdot}s^{-2}$, the motor will stop running and the experiment will be terminated. (The vibration threshold limit is set according to the service condition of bearings and experience of experimental operators.) The vibration data are automatically collected by the computer control system, as shown in Figure 4.

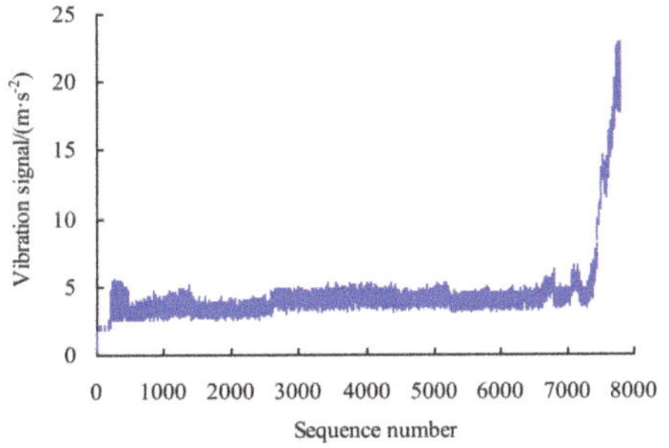

Figure 4. Vibration signals of bearing.

The vibration time series of rolling bearings have obvious nonlinearity, randomness, and uncertainty in Figure 4. Therefore, the vibration performance reliability of bearings should be predicted and evaluated dynamically based on the complex and changeable vibration information. The vibration data from the 1st to the 212th point range from 0.3 to 2.5 $m{\cdot}s^{-2}$. The vibration performance of the bearings appears stable during this period. The vibration data from the 213th to the 472nd point range from 2.5 to 5.5 $m{\cdot}s^{-2}$. During this period, the vibration signals are stronger and highly fluctuating. The vibration data from the 473rd to the 2574th point range from 2.5 to 5.1 $m{\cdot}s^{-2}$. During this period, the vibration performance of the bearings is stable. The values of vibration data from the 2575th to the 6659th point range from 3 to 5.5 $m{\cdot}s^{-2}$. During this period, the vibration signals of bearings are strengthened. The values from the 6660th to the 7446th data points are around 3.4–7.7 $m{\cdot}s^{-2}$. During this period, the vibration signals are stronger and highly fluctuating. The values of the 7447th to 7793rd data points are in the range of 8.2–22.9 $m{\cdot}s^{-2}$. During this period, the vibration signals increase linearly. Therefore, it can be considered that the vibration performance of the bearings is at the initial wear stage during the period corresponding to the 1st to the 472nd data point. The vibration performance is at the optimal vibration performance state during the period corresponding to the 473rd to the 2574th data point, which is regarded as the first time series (intrinsic series). The vibration performance is at the normal wear stage during the period corresponding to the 2575th to the 6659th data point, which is regarded as the second time series. The vibration performance is at the degeneration stage during the period corresponding from the 6660th to the 7446th data point, which is regarded as the third time series. The vibration performance is at the deterioration stage during the period corresponding to the 7447th to the 7793rd data point, which is regarded as the fourth time series.

3.1.1. PDF of Data Samples of Time Series

For the first time series, various order origin moments can be obtained using the maximum entropy method as $[m_{11}, m_{21}, m_{31}, m_{41}, m_{51}] = [-0.7710, 1.1829, -1.6809, 3.0058, -5.0425]$; the Lagrange multipliers $[c_{01}, c_{11}, c_{21}, c_{31}, c_{41}, c_{51}] = [-0.526176, -1.0192, -0.8837, -0.3003, 0.0325, 0.0628]$; the mapping parameters $a_1 = 1.8819$ and $b_1 = -7.1511$. Set the significance level α is 0.01; that is, the confidence level $P = 99\%$. The maximum entropy estimated interval is [2.5507, 4.4738] m·s^{-2} for the first time series. According to Equation (10), the probability density estimated function $f_1(x)$ is calculated as shown in Figure 5.

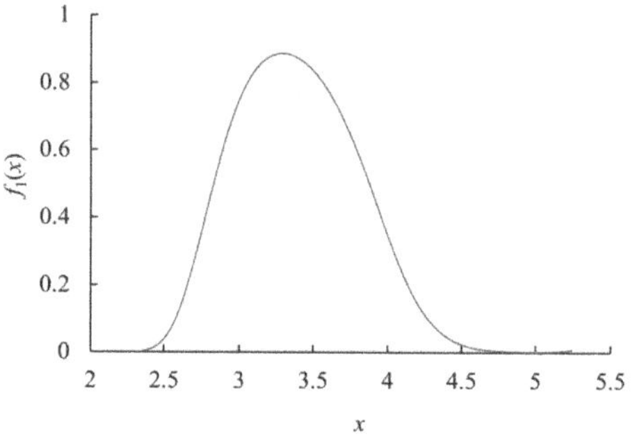

Figure 5. PDF of data sample of the intrinsic sequence.

For the second time series, various order origin moments can be obtained using the maximum entropy method as $[m_{12}, m_{22}, m_{32}, m_{42}, m_{52}] = [-0.4015, 0.8709, -0.7686, 1.6745, -1.9542]$; the Lagrange multipliers $[c_{02}, c_{12}, c_{22}, c_{32}, c_{42}, c_{52}] = [-0.3629, -0.3569, -0.2062, -0.1813, -0.1369, 0.0291]$; the mapping parameters $a_2 = 1.9572$ and $b_2 = -8.3179$. The significance level α is set as 0.01; that is, the confidence level $P = 99\%$. The maximum entropy estimated interval is [3.1823, 5.1008] m·s^{-2} for the second time series.

For the third time series, various order origin moments can be obtained using the maximum entropy method as $[m_{13}, m_{23}, m_{33}, m_{43}, m_{53}] = [-0.8840, 1.5477, -2.2030, 4.5372, -7.4059]$; the Lagrange multipliers $[c_{03}, c_{13}, c_{23}, c_{33}, c_{43}, c_{53}] = [-1.4292, -1.1189, -0.2384, 0.0274, -0.0268, 0.0145]$; the mapping parameters $a_3 = 1.1379$ and $b_3 = -6.3153$. The significance level α is set as 0.01; that is, the confidence level $P = 99\%$. The maximum entropy estimated interval is [3.3405, 7.3435] m·s^{-2} for the third time series.

For the fourth time series, various order origin moments can be obtained using the maximum entropy method as $[m_{14}, m_{24}, m_{34}, m_{44}, m_{54}] = [-0.1634, 1.2181, -0.2573, 3.2716, -0.7843]$; the Lagrange multipliers $[c_{04}, c_{14}, c_{24}, c_{34}, c_{44}, c_{54}] = [-2.3395, -0.5070, -0.0421, 0.2472, -0.0660, -0.0299]$; the mapping parameters $a_4 = 0.3328$ and $b_4 = -5.1758$. The significance level α is set as 0.01; that is, the confidence level $P = 99\%$. The maximum entropy estimated interval is [8.1599, 22.5150]m·s^{-2} for the fourth time series.

The PDF $f_w(x)$ of four time series are shown in Figure 6.

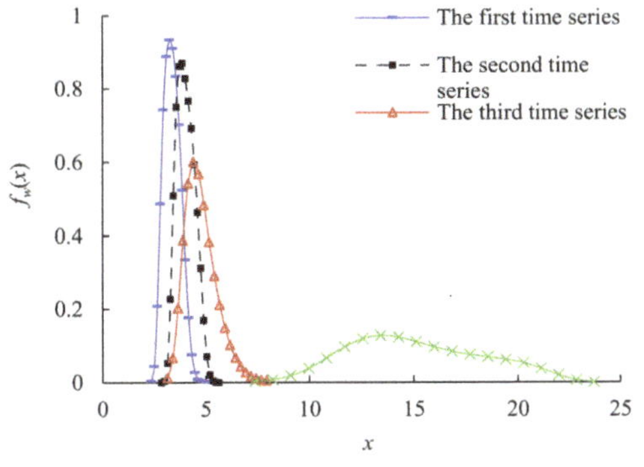

Figure 6. PDF of data samples of time series.

3.1.2. PPDF of Data Samples of Time Series

Intersection interval $[x_{w1}, x_{w2}]$ and intersection area η_w are calculated for the PDF of the wth time series and the intrinsic series, as shown in Table 2. The similarity degrees between time series η_w, μ_w, β_{1w}, β_{2w} and β_{3w} are calculated using the overlapping area method, membership degree method, Hamming approach degree method, Euclidean approach degree method, and cardinal approach degree method, as shown in Table 3.

Table 2. Intersection intervals and overlapped areas of PDF.

Sequence Number w	Intersection Interval $[x_{w1}, x_{w2}]/(\mathrm{m \cdot s^{-2}})$	Intersection Area η_w
1	[2.5000, 5.1000]	1
2	[3.6312, /]	0.3590
3	[3.9596, /]	0.0730
4	[/, /]	0

Table 3. Similarity degrees calculated using different methods.

Sequence Number w	Similarity Degrees			
	Membership Degree μ_w	Euclidean Approach Degree β_{1w}	Hamming Approach Degree β_{2w}	Cardinal Approach Degree β_{3w}
1	1	1	1	1
2	0.7897	0.7538	0.7897	0.8825
3	0.6094	0.6036	0.6094	0.7573
4	0.0023	0.0017	0.0023	0.0045

The experimental data in Figure 3 show that the vibration performance has obvious different uncertainty and nonlinearity for different fault diameters, which belongs to the poor-information problem with an unknown trend. In order to study the variation trend of PMR for the vibration performance, the product functions $f_1(x)f_w(x)$ and posterior probability density functions $hyf_w(x)$ of the four time series are constructed using the Bayesian principle, as shown in Figures 7 and 8.

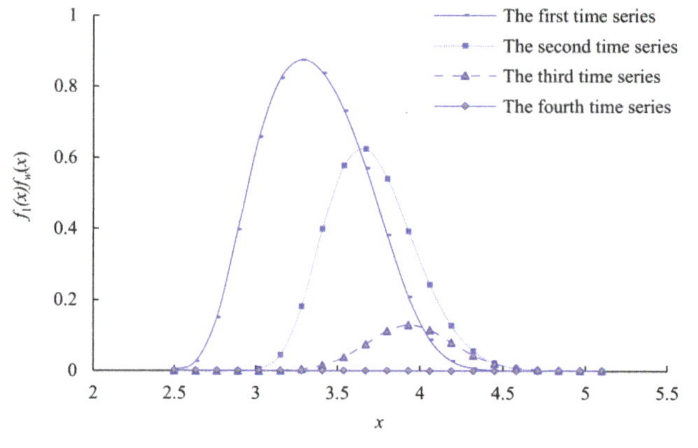

Figure 7. Product $f_1(x)f_w(x)$ of PDF.

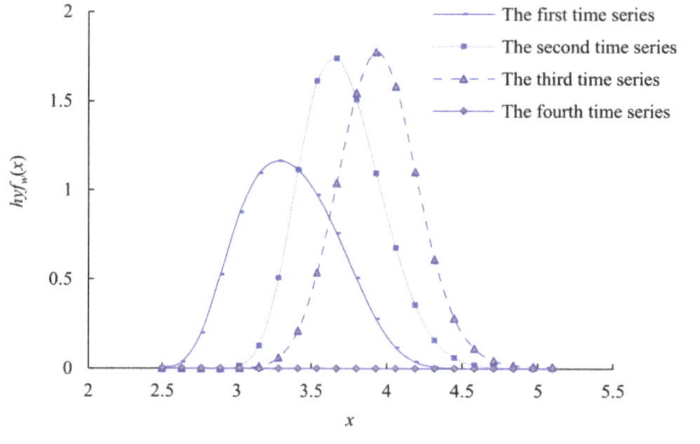

Figure 8. PPDF of 4 time series.

3.1.3. PMR of Time Series

The intersection interval $[x_{1w}, x_{2w}]$ and intersection area S_w are calculated for the PPDF of the wth time series and the intrinsic series, as shown in Table 4. The values of PMR, $R_w(1)$, $R_w(2)$, $R_w(3)$, $R_w(4)$, and $R_w(5)$ are calculated using overlapping area method, membership degree method, Hamming approach degree method, Euclidean approach degree method, and cardinal approach degree method for different time series, as shown in Table 5. The PMR of each time series decreases gradually, and the slope of the curve decreases gradually, as shown in Figure 9.

Table 4. Intersection intervals and overlapped areas of PPDF.

Sequence Number w	Intersection Interval $[x_{w1}, x_{w2}]/(\mathrm{m{\cdot}s^{-2}})$	Intersection Area η_w
1	[2.5000, 5.1000]	1
2	[3.3579, /]	0.6341
3	[3.5821, /]	0.4107
4	[/, /]	0

Table 5. PMR calculated using different methods for different time series.

Sequence Number w	Values of PMR/%				
	Overlapping Area Method $R_w(1)$	Membership Degree Method $R_w(2)$	Hamming Approach Degree Method $R_w(3)$	Euclidean Approach Degree Method $R_w(4)$	Cardinal Approach Degree Method $R_w(5)$
1	100.00	100.00	100.00	100.00	100.00
2	59.82	61.35	60.99	61.35	62.26
3	36.71	38.18	38.14	38.18	39.27
4	0.00	0.00	0.00	0.00	0.00

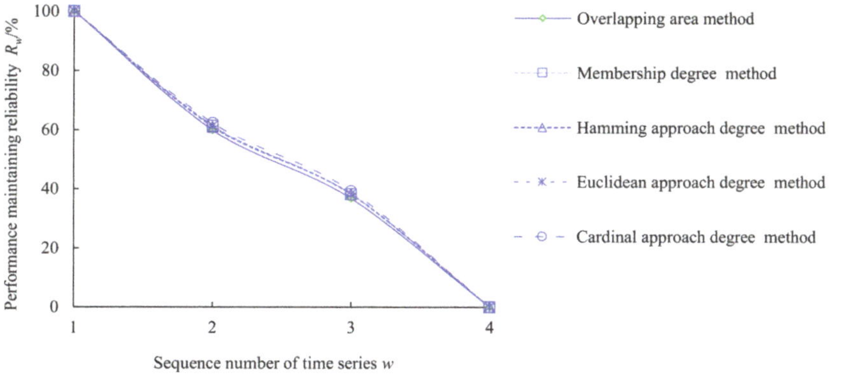

Figure 9. PMR of time series.

3.1.4. PMRR of Time Series

The values of PMRR, $d_w(1)$, $d_w(2)$, $d_w(3)$, $d_w(4)$, and $d_w(5)$ are calculated using the overlapping area method, membership degree method, Hamming approach degree method, Euclidean approach degree method, and cardinal approach degree method, as shown in Table 6.

Table 6. PMRR calculated using different methods.

Sequence Number w	Values of PMRR/%				
	Overlapping Area Method $d_w(1)$	Membership Degree Method $d_w(2)$	Hamming Approach Degree Method $d_w(3)$	Euclidean Approach Degree Method $d_w(4)$	Cardinal Approach Degree Method $d_w(5)$
1	0	0	0	0	0
2	−40.18	−38.65	−39.01	−38.65	−37.74
3	−63.29	−61.82	−61.86	−61.82	−60.73
4	−100	−100	−100	−100	−100

3.1.5. Fusion Results of Multiple Weighting Methods

In the process of bootstrap generation, take the sample data of PMR of the second time series as an example. Let the sampling number $q = 5$, the times for the bootstrap re-sampling $B = 10{,}000$, and the significance level $\alpha = 0$. $R_2 = (R_2(1), R_2(2), R_2(3), R_2(4), R_2(5)) = (59.82\%, 61.35\%, 60.99\%, 61.35\%, 62.26\%)$. The bootstrap re-sampling samples $R_{2\text{bootstrap}}$ can be obtained by an equiprobable sampling, as shown in Figure 10.

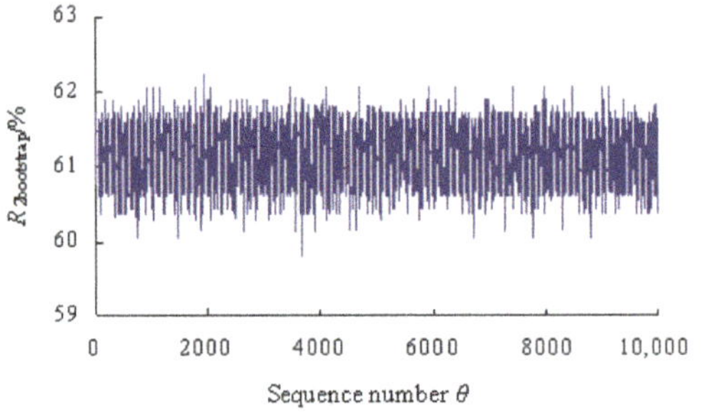

Figure 10. Generated data of PMR of second time series.

The PDF of the PCR data sample can be calculated for the second time series, as shown in Figure 11. Thus, the estimated true value of PMR R_{20} = 61.18% and the maximum entropy estimated interval $[R_{2L}, R_{2U}]$ = [59.68%, 62.39%] are obtained for the sample data of PMR of the second time series. Similarly, the estimated true values and the estimated intervals can be obtained for the sample data of PMR of the other three time series. The results are shown in Table 7.

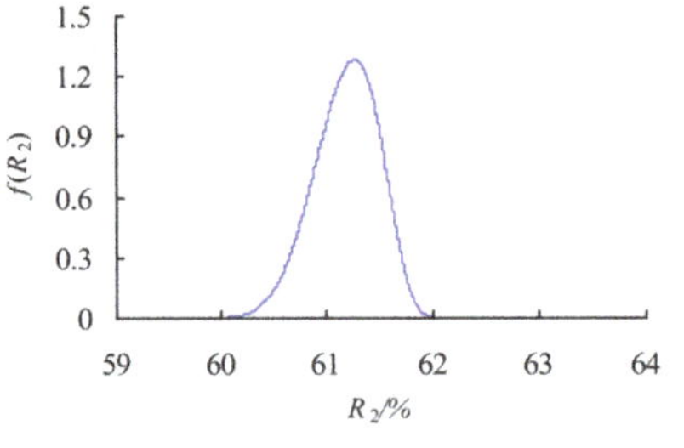

Figure 11. PDF of PMR of the second time series.

Table 7. Estimated true values and estimated intervals for different time series.

Sequence Number w	Estimated True Value R_{w0}/%	Estimated Intervals $[R_{2L}, R_{2U}]$/%
1	100	/
2	61.18	[59.68, 62.39]
3	38.13	[36.71, 39.55]
4	0	/

The calculated results of the estimated truth curve and the upper- and lower-bound curve are shown for different time series in Figure 12. The estimated true value, lower-bound value and upper-bound value of PMR decrease gradually and have trends of nonlinear reduction over time. From the first time series to the second time series, the estimated true value of PMR is reduced from 100% to 61.18% very rapidly; from the second time series to the third time series, the estimated true value of PMR is reduced from 61.18% to 38.13% slowly; from the third time series to fourth time series, the estimated true value of PMR is reduced from 38.13% to 0 rapidly, relatively.

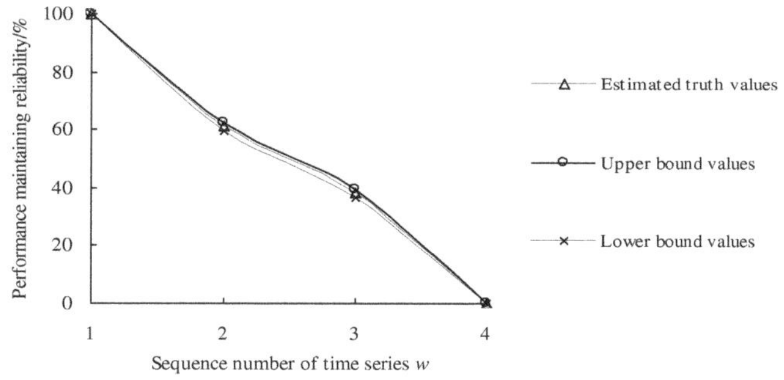

Figure 12. Maximum entropy estimated results of PMR.

The PMRR for vibration performance of rolling bearings are calculated for different time series, as shown in Table 8 and Figure 13. The estimated true value, lower-bound value and upper-bound value of PMRR decrease gradually and have trends of nonlinear reduction. From the first time series to second time series, the estimated true value of PCRR is very rapidly reduced from 0% to −38.82%; from the second time series to the third time series, the estimated true value of PCRR is slowly reduced from −38.82% to −61.87%; from the third time series to the fourth time series, the estimated true value of PCRR is reduced from −61.87% to −100% relatively rapidly.

Table 8. Estimated true values and estimated intervals of PMRR.

Sequence Number w	Estimated True Value d_{w0}/%	Estimated Intervals $[d_{2L}, d_{2U}]$/%
1	0	[/, /]
2	−38.82	[−40.32, −37.61]
3	−61.87	[−63.29, −60.45]
4	−100	[/, /]

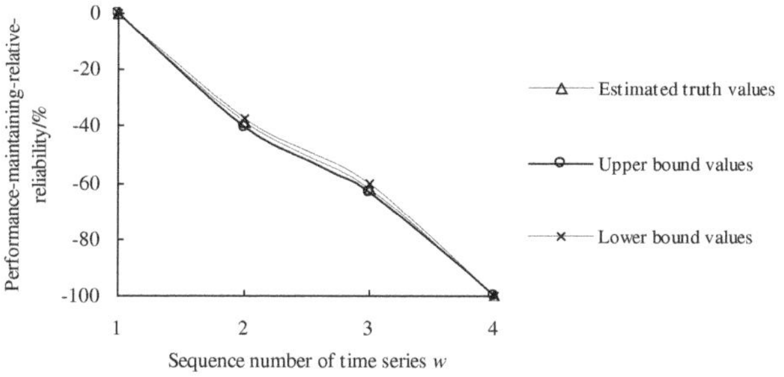

Figure 13. Maximum entropy estimated results of PMRR.

This analysis result is more in line with engineering practice, because the prior information and current sample information of the vibration performance degradation information of bearings in service were comprehensively considered by the proposed method. In addition, the HMEBM also considers the similarity between time series, which contains the hidden information of the degradation process of optimal vibration performance state.

3.2. Case 2

The test machine and the bearing used in this case are exactly the same as those of Case 1. The vibration data are shown in Figure 14 by changing the test conditions of the motor to a speed of 4000 r/min, an axial load of 4.17 kN, and a radial load of 4.58 kN.

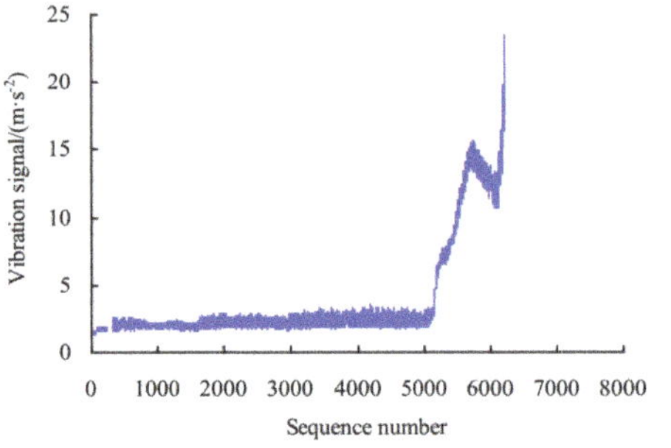

Figure 14. Vibration signals of bearing (Case 2).

3.2.1. PDF of Data Samples of Time Series (Case 2)

For the first time series, various order origin moments can be obtained using the maximum entropy method as $[m_{11}, m_{21}, m_{31}, m_{41}, m_{51}] = [-0.0463, 0.8915, -0.2456, 2.1047, -0.7743]$; the Lagrange multipliers $[c_{01}, c_{11}, c_{21}, c_{31}, c_{41}, c_{51}] = [0.6246, 0.3796, -0.4400, -0.3470, -0.0294, 0.0463]$; the mapping parameters $a_1 = 4.4481$ and $b_1 = -9.1186$. The significance level α is set as 0.01; that is, the confidence level $P = 99\%$. The maximum entropy estimated interval is [1.5354, 2.5003] $m{\cdot}s^{-2}$ for the first time series.

The PDF $f_w(x)$ of four time series are shown in Figure 15.

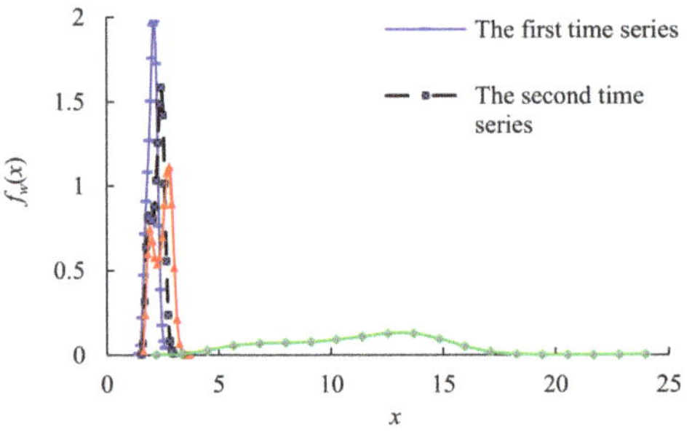

Figure 15. PDF of data samples of time series (Case 2).

3.2.2. PPDF of Data Samples of Time Series (Case 2)

The intersection interval $[x_{w1}, x_{w2}]$ and intersection area η_w are calculated for the PDF of the wth time series and the intrinsic series as shown in Table 9. The similarity degrees between time series η_w, μ_w, β_{1w}, β_{2w}, and β_{3w} are calculated using the overlapping area method, membership degree method, Hamming approach degree method, Euclidean approach degree method, and cardinal approach degree method, as shown in Table 10.

Table 9. Intersection intervals and overlapped areas of PDF (Case 2).

Sequence Number w	Intersection Interval $[x_{w1}, x_{w2}]/(m{\cdot}s^{-2})$	Intersection Area η_w
1	[1.5000, 2.6000]	1
2	[2.2776, /]	0.6272
3	[2.3655, /]	0.4404
4	[/, /]	0

Table 10. Similarity degrees calculated using different methods (Case 2).

Sequence Number w	Similarity degrees			
	Membership Degree μ_{w}	Euclidean Approach Degree β_{1w}	Hamming Approach Degree β_{2w}	Cardinal Approach Degree β_{3w}
1	1	1	1	1
2	0.9053	0.8718	0.9053	0.9503
3	0.7138	0.6722	0.7138	0.8329
4	0.0066	0.0049	0.0066	0.0131

The experimental data in Figure 14 show that the vibration performance has obvious different uncertainty and nonlinearity for different fault diameters, which belongs to the poor-information problem with an unknown trend. In order to study the variation trend of PMR for the vibration performance, the product functions $f_1(x)f_w(x)$ and posterior probability density functions $hyf_w(x)$ of the four time series are constructed using the Bayesian principle, as shown in Figures 16 and 17.

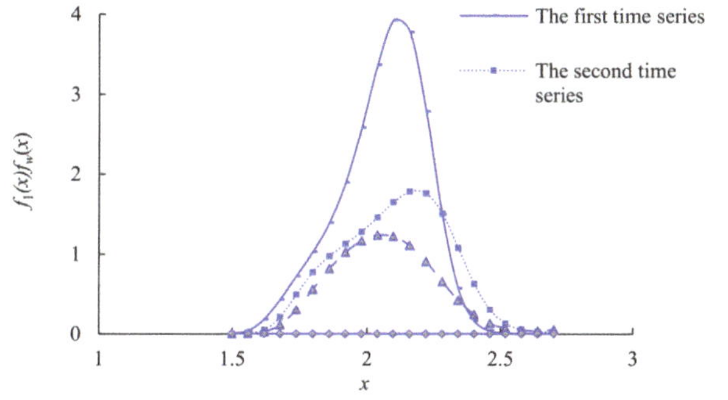

Figure 16. Product $f_1(x)f_w(x)$ of PDF (Case 2).

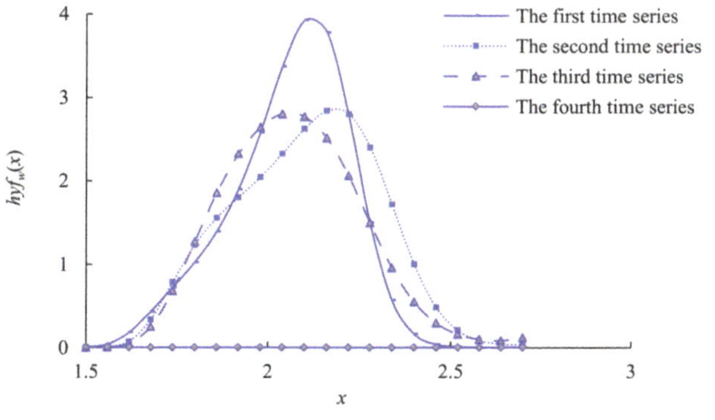

Figure 17. PPDF of 4 time series (Case 2).

3.2.3. PMR of Time Series (Case 2)

The intersection interval $[x_{1w}, x_{2w}]$ and intersection area S_w are calculated for the PPDF of the wth time series and the intrinsic series, as shown in Table 11. The values of PMR, $R_w(1)$, $R_w(2)$, $R_w(3)$, $R_w(4)$, and $R_w(5)$ are calculated using the overlapping area method, membership degree method, Hamming approach degree method, Euclidean approach degree method, and cardinal approach degree method for different time series, as shown in Table 12. The PMR of each time series decreases gradually, and the slope of the curve decreases gradually, as shown in Figure 18.

Table 11. Intersection intervals and overlapped areas of PPDF (Case 2).

Sequence Number w	Intersection Interval $[x_{w1}, x_{w2}]/(\mathrm{m \cdot s^{-2}})$	Intersection Area η_w
1	[1.5000, 2.6000]	1
2	[1.7124, 1.9161, 2.2126]	0.7473
3	[1.7431, 1.9981, 2.2740]	0.6751
4	[/, /]	0

Table 12. PMR calculated using different methods for different time series (Case 2).

Sequence Number w	Values of PMR/%				
	Overlapping Area Method $R_w(1)$	Membership Degree Method $R_w(2)$	Hamming Approach Degree Method $R_w(3)$	Euclidean Approach Degree Method $R_w(4)$	Cardinal Approach Degree Method $R_w(5)$
1	100.00	100.00	100.00	100.00	100.00
2	64.30	72.08	71.14	72.08	73.34
3	57.54	62.41	61.67	62.41	64.53
4	0.00	0.00	0.00	0.00	0.00

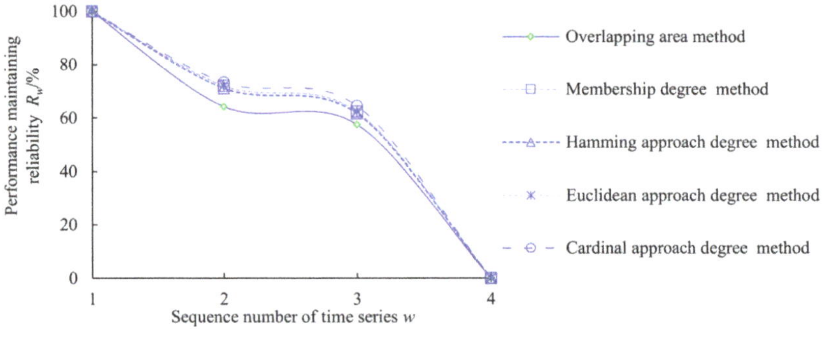

Figure 18. PMR of time series (Case 2).

3.2.4. PMRR of Time Series (Case 2)

The values of PMRR, $d_w(1)$, $d_w(2)$, $d_w(3)$, $d_w(4)$, and $d_w(5)$ are calculated using the overlapping area method, membership degree method, Hamming approach degree method, Euclidean approach degree method, and cardinal approach degree method, as shown in Table 13.

Table 13. PMRR calculated using different methods (Case 2).

Sequence Number w	Values of PMRR/%				
	Overlapping Area Method $d_w(1)$	Membership Degree Method $d_w(2)$	Hamming Approach Degree Method $d_w(3)$	Euclidean Approach Degree Method $d_w(4)$	Cardinal Approach Degree Method $d_w(5)$
1	0	0	0	0	0
2	−35.7	−27.92	−28.86	−27.92	−26.66
3	−42.46	−37.59	−38.33	−37.59	−35.47
4	−100	−100	−100	−100	−100

3.2.5. Fusion Results of Multiple Weighting Methods (Case 2)

In the process of bootstrap generation, take the sample data of PMR of the second time series as an example. Let the sampling number $q = 5$, the times for the bootstrap re-sampling $B = 10{,}000$, and the significant level $\alpha = 0$. $R_2 = (R_2(1), R_2(2), R_2(3), R_2(4), R_2(5)) = (59.82\%, 61.35\%, 60.99\%, 61.35\%, 62.26\%)$. The bootstrap re-sampling samples $R_{2\text{bootstrap}}$ can be obtained by an equiprobable sampling, as shown in Figure 19.

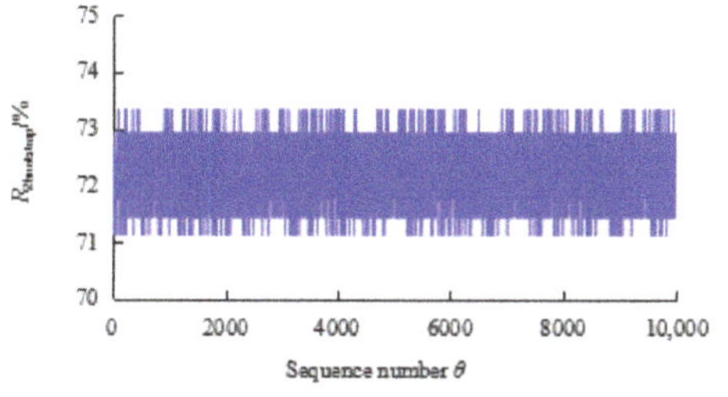

Figure 19. Generated data of PMR of second time series (Case 2).

The PDF of the PCR data sample can be calculated for the second time series, as shown in Figure 20. Thus, the estimated true value of PMR $R_{20} = 61.18\%$ and the maximum entropy estimated interval $[R_{2L}, R_{2U}] = [59.68\%, 62.39\%]$ are obtained for the sample data of PMR of the second time series. Similarly, the estimated true values and the estimated intervals can be obtained for the sample data of PMR of the other three time series. The results are shown in Table 14.

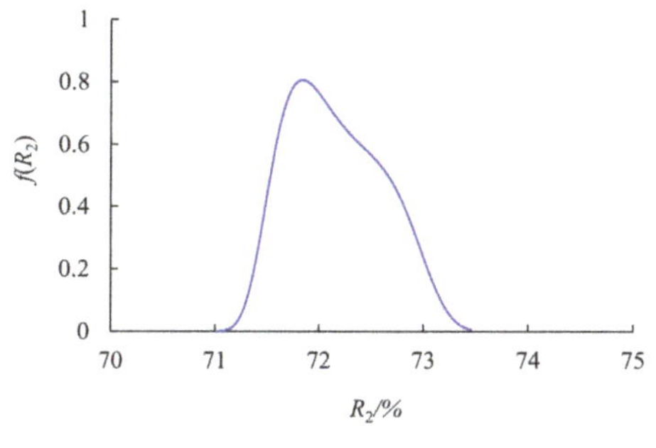

Figure 20. PDF of PMR of the second time series (Case 2).

Table 14. Estimated true values and estimated intervals for different time series (Case 2).

Sequence Number w	Estimated True Value R_{w0}/%	Estimated Intervals $[R_{2L}, R_{2U}]$/%
1	100	/
2	72.16	[71.02, 73.46]
3	62.72	[61.51, 64.69]
4	0	/

The calculated results of the estimated truth curve and the upper and lower bound curve are shown for different time series in Figure 21. The estimated true value, lower-bound value and upper-bound value of PMR decrease gradually and have trends of nonlinear reduction over time. From the first time series to the second time series, the estimated true value of PMR is rapidly reduced from 100% to 72.16%; from the second time series to the third time series, the estimated true value of PMR is slowly reduced from 72.16% to 62.72%; from the third time series to the fourth time series, the estimated true value of PMR is reduced from 62.72% to 0 very rapidly.

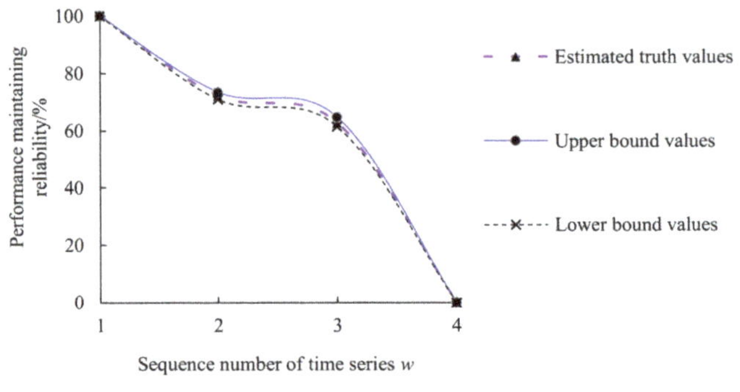

Figure 21. Maximum entropy estimated results of PMR (Case 2).

The PMRR for vibration performance of rolling bearings are calculated for different time series, as shown in Table 15 and Figure 22. The estimated true value, lower-bound value and upper-bound value of PMRR decrease gradually and have trends of nonlinear reduction. From the first time series to the second time series, the estimated true value of PCRR is rapidly reduced from 0% to −27.84%; from the second time series to the third time series, the estimated true value of PCRR is slowly reduced from −27.84% to −37.28%; from the third time series to the fourth time series, the estimated true value of PCRR is reduced from −37.28% to −100% very rapidly.

Table 15. Estimated true values and estimated intervals of PMRR (Case 2).

Sequence Number w	Estimated True Value d_{w0}/%	Estimated Intervals $[d_{2L}, d_{2U}]$/%
1	0	[/, /]
2	−27.84	[−28.98, −26.54]
3	−37.28	[−38.49, −35.31]
4	−100	[/, /]

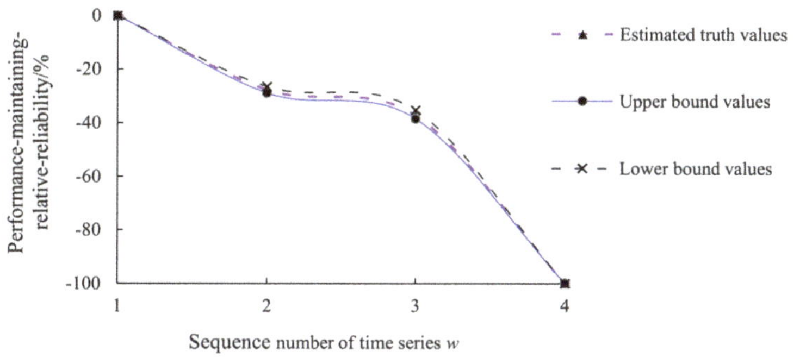

Figure 22. Maximum entropy estimated results of PMRR (Case 2).

This analysis result is more in line with engineering practice because the prior information and current sample information of the vibration performance degradation information of bearings in service were comprehensively considered by the proposed method. In addition, the HMEBM also considers the similarity between time series, which contains hidden information regarding the degradation process of optimal vibration performance state.

4. Conclusions

Based on the HMEBM, the reliability model is established to evaluate the operation performance status of rolling bearings, which has no requirements for the priori information of data samples, types, and components of bearings.

1. Considering the operation status of rolling bearing, the variation degree of the optimal vibration performance status can be calculated more accurately to ensure effective maintenance of the system, reduce faults, and improve quality under the condition of traditional probability statistics.
2. The similarities between time series are obtained using the overlapping area method, membership degree method, Hamming approach degree method, Euclidean approach degree method, and cardinal approach degree method.
3. The maximum entropy method, Bayesian theory, and bootstrap method are fused fully to discover more information in time series of bearing vibration performance. The estimated true values and maximum entropy estimated intervals of PMR and PMRR are calculated to dynamically monitor the health status of rolling bearings online.

The performance degradation law of rolling bearing will be further studied under a different load and speed. In order to verify the universality of the proposed method, HMEBM will be used to evaluate the performance degradation process of other type of bearings.

Author Contributions: Conceptualization, L.Y.; methodology, L.Y.; validation, L.Y. and Y.H.; formal analysis, W.Z.; investigation, L.Y.; resources, L.Y.; data curation, L.Y.; writing—original draft preparation, L.Y.; writing—review and editing, W.Z.; visualization, Y.C.; supervision, J.X.; project administration, S.D.; funding acquisition, Y.H., W.Z. and Y.C. All authors have read and agreed to the published version of the manuscript.

Funding: This research was funded by [Opening Foundation of State Key Laboratory of Air-Conditioning Equipment and System Energy Conservation] grant number [ACSKL2019KT02], [National Natural Science Foundation of China] grant number [U1804145], And The APC was funded by [Yusheng Hu], [Youth Programs of the National Natural Science Foundation of China] grant number [51905152 and 52005158], And The APC was funded by [Yusheng Hu, Xiaoqiang Wang, Wenhu Zhang and Yongcun Cui].

Data Availability Statement: The data used to support the findings of this study are available from the corresponding author upon request.

Conflicts of Interest: The authors declare that there is no conflict of interest regarding the publication of this paper.

Nomenclature

w	Order number of time series.
r	Number of time series.
N	Number of original data.
i	Order number of origin moment.
j	Highest order number of origin moment.
$x_w(k)$	kth performance data in time series.
$\mathbf{X}_w$	wth time series.
$f_w(x)$	Probability density function of time series.
c_{iw}	Lagrange multiplier.
$H_w(x)$	Information entropy of time series.
Ω_w	Feasible domain for the data sample of time series.
$lnf_w(x)$	Logarithmic value of $f_w(x)$.
m_{iw}	Order origin moment.
$a_w;b_w$	Mapping parameters.
$hyf_w(x)$	PPDF of the wth time series $\mathbf{X}_w$.
Ω_{1w}	Intersection of feasible regions of data samples.
R_1	PMR of intrinsic series.
η_w	Overlapping area of PDF.
S_{1w}	Overlapping area of PPDF.
d_{pk}	Minkowski distance.
β_{3w}	Cardinal approach degree.
x_{minw}; x_{maxw}	Lower bound value and upper bound value in the time series.
x_{Lw}; x_{Uw}	Lower boundary value and upper boundary value of estimated interval.
x_{1L}; x_{1U}	Lower and upper bound values of confidence intervals for the PPDF of intrinsic series.
x^*_{1L}; x^*_{1U}	Lower and upper bound values of confidence intervals for the PDF of intrinsic series.
x_{1w}; x_{2w}	Abscissa values of the intersections for the PPDF of the wth time series and Intrinsic series.
x^*_{1w}; x^*_{2w}	Abscissa values of the intersections for the PDF of the wth time series and the intrinsic series.
$R_w(1)$	PMR calculated by using overlapping area method.
$R_w(2)$	PMR calculated by using membership degree method.
$R_w(3)$	PMR calculated by using Hamming approach degree method.
$R_w(4)$	PMR calculated by using Euclidean approach degree method.
$R_w(5)$	PMR calculated by using Cardinal approach degree method.
R_w	Data sample of PMR for the wth time series.
$R_w(\gamma)$	γth data in the PMR data sample for the wth time series.
$R_{w\theta}$	θth bootstrap re-sampling sample.
B	Times of bootstrap re-sampling and number of bootstrap samples.
$R_{w\theta}(\Theta)$	Θth data in the θth bootstrap re-sampling sample of PMR
$R_{wbootstrap}$	Generated sample.
R_{wL}; R_{wU}	Lower-bound value and upper-bound value of PMR.
d_{w0}; d_{wL}; d_{wU}	Estimated true value, lower-bound value, and upper-bound value of PMRR.
PDF	Probability density function.
PMR	Performance maintaining reliability.
PMRR	Performance maintaining relative reliability.
PPDF	Posterior probability density function.
HMEBM	Hierarchical maximum entropy Bayesian method.

References

1. Wang, H.C.; Chen, J. Performance degradation assessment of rolling bearing based on bispectrum and support vector data description. *J. Vib. Control.* **2014**, *13*, 2032–2041. [CrossRef]
2. Liu, J. A dynamic modelling method of a rotor-roller bearing-housing system with a localized fault including the additional excitation zone. *J. Sound. Vib.* **2020**, *469*, 115144. [CrossRef]
3. Ma, S.; Zhang, X.; Yan, K.; Zhu, Y.; Hong, J. A study on bearing dynamic features under the condition of multiball-cage collision. *Lubricants* **2022**, *10*, 9. [CrossRef]
4. Ai, Y.T.; Guan, J.Y.; Fei, C.W.; Tian, J.; Zhang, F.L. Fusion information entropy method of rolling bearing fault diagnosis based on n-dimensional characteristic parameter distance. *Mech. Syst. Signal Pract.* **2017**, *88*, 123–136. [CrossRef]
5. He, Q.B.; Wu, E.H.; Pan, Y.Y. Multi-scale stochastic resonance spectrogram for fault diagnosis of rolling element bearings. *J. Sound. Vib.* **2018**, *420*, 174–184. [CrossRef]
6. Moshrefzadeh, A.; Fasana, A. The autogram: An effective approach for selecting the optimal demodulation band in rolling element bearings diagnosis. *Mech. Syst. Signal Pract.* **2018**, *105*, 294–318. [CrossRef]
7. Ye, L.; Xia, X.T.; Chang, Z. Dynamic prediction for accuracy maintaining reliability of superprecision rolling bearing in service. *Shock. Vib.* **2018**, *2018*, 7396293. [CrossRef]
8. Zio, E. Reliability engineering: Old problems and new challenges. *Reliab. Eng. Syst. Safe* **2009**, *94*, 125–141. [CrossRef]
9. Li, H.L.; Guo, C.H. Piecewise cloud approximation for time series mining. *Knowl. Based Syst.* **2011**, *24*, 492–500. [CrossRef]
10. Xia, X.T.; Liu, B.; Li, Y.F.; Chen, X.; Zhu, W. Research on quality achieving reliability of bearing based on fuzzy weight. *J. Aerosp. Power* **2018**, *33*, 3013–3021.
11. Bhaduri, M.; Zhan, J. Using empirical recurrence rates ratio for time series data similarity. *IEEE Access* **2018**, *6*, 30855–30864. [CrossRef]
12. Shimizu, S. Weibull distribution function application to static strength and fatigue life of materials. *Tribol. Trans.* **2012**, *55*, 267–277. [CrossRef]
13. Raje, N.; Sadeghi, F. Statistical numerical modelling of sub-surface initiated spalling in bearing contacts. *Proc. Inst. Mech. Eng. Part J J. Eng. Tribol.* **2009**, *223*, 849–858. [CrossRef]
14. Zaretsky, E.V.; Branzai, E.V. Rolling bearing service life based on probable cause for removal-A tutorial. *Tribol. Trans.* **2017**, *60*, 300–312. [CrossRef]
15. Zhang, N.; Wu, L.; Wang, Z.; Guan, Y. Bearing remaining useful life prediction based on naive Bayes and Weibull distributions. *Entropy* **2018**, *20*, 944. [CrossRef]
16. Xia, X.T.; Lv, T.M. Dynamic prediction model for rolling bearing friction torque using grey bootstrap fusion method and chaos theory. *Adv. Mater. Res.* **2012**, *443-444*, 87–96. [CrossRef]
17. Das, D.; Zhou, S.Y. Statistical process monitoring based on maximum entropy density approximation and level set principle. *IIE Trans.* **2015**, *3*, 215–229. [CrossRef]
18. Chatterjee, A.; Mukherjee, S.; Kar, S. Poverty level of households: A multidimensional approach based on fuzzy mathematics. *Fuzzy Inf. Eng.* **2014**, *4*, 463–487. [CrossRef]
19. Xia, X.T.; Lv, T.M.; Meng, F.N. Gray chaos evaluation model for prediction of rolling bearing friction torque. *J. Test Eval.* **2010**, *38*, 291–300.
20. Xia, X.T. Reliability analysis of zero-failure data with poor information. *Qual. Reliab. Eng. Int.* **2012**, *28*, 981–990. [CrossRef]
21. Xiao, W.; Cheng, W.; Zi, Y.; Zhao, C.; Sun, C.; Liu, Z.; Chen, J.; He, Z. Support evidence statistics for operation reliability assessment using running state information and its application to rolling bearing. *Mech. Syst. Signal Pract.* **2015**, *60–61*, 344–357. [CrossRef]
22. Ghosh, S.; Polansky, A.M. Smoothed and iterated bootstrap confidence regions for parameter vectors. *J. Multivar. Anal.* **2014**, *132*, 171–182. [CrossRef]
23. Wan, J.P.; Zhang, K.S.; Chen, H. The bootstrap and Bayesian bootstrap method in assessing bioequivalence. *Chaos Solitons Fract.* **2009**, *41*, 2246–2249.
24. Wang, Y.; Zhou, W.; Dong, D.; Wang, Z. Estimation of random vibration signals with small samples using bootstrap maximum entropy method. *Measurement* **2017**, *105*, 45–55. [CrossRef]
25. Chang, J.Y.; Hall, P. Double-bootstrap methods that use a single double-bootstrap simulation. *Biometrika* **2015**, *102*, 203–214. [CrossRef]
26. Wu, F.X.; Wen, W.D. Scatter factor confidence interval estimate of least square maximum entropy quantile function for small samples. *Chin. J. Aeronaut.* **2016**, *29*, 1285–1293. [CrossRef]
27. Edwin, A.B.; Angel, K.M. A clustering method based on the maximum entropy principle. *Entropy* **2015**, *1*, 151–180.
28. Kwon, Y. Design of Bayesian zero-failure reliability demonstration test for products with Weibull lifetime distribution. *J. Appl. Reliab.* **2014**, *14*, 220–224.
29. Jha, D.K.; Virani, N.; Reimann, J.; Srivastav, A.; Ray, A. Symbolic analysis-based reduced order Markov modeling of time series data. *Signal Process* **2018**, *149*, 68–81. [CrossRef]
30. Xiao, N.C.; Li, Y.F.; Wang, Z.; Peng, W.; Huang, H.Z. Bayesian reliability estimation for deteriorating systems with limited samples using the maximum entropy approach. *Entropy* **2013**, *12*, 5492–5509. [CrossRef]

lubricants

MDPI

Article

SACGNet: A Remaining Useful Life Prediction of Bearing with Self-Attention Augmented Convolution GRU Network

Juan Xu [1], Shiyu Duan [1], Weiwei Chen [2], Dongfeng Wang [3] and Yuqi Fan [4,*]

1 Key Laboratory of Knowledge Engineering with Big Data, School of Computer Science and Information Engineering, Hefei University of Technology, Hefei 230009, China; xujuan@hfut.edu.cn (J.X.); 2020171145@mail.hfut.edu.cn (S.D.)
2 Shanghai Aerospace Control Technology Institute, Shanghai 201109, China; youthjiang@126.com
3 Luoyang Bearing Research Institute Co., Ltd., Luoyang 471033, China; zyswdf@163.com
4 School of Computer and Information, Hefei University of Technology, Hefei 230009, China
* Correspondence: yuqi.fan@hfut.edu.cn

Abstract: In recent years, the development of deep learning-based remaining useful life (RUL) prediction methods of bearings has flourished because of their high accuracy, easy implementation, and lack of reliance on a priori knowledge. However, there are two challenging issues concerning the prediction accuracy of existing methods. The run-to-failure sequential data and its RUL labels are almost inaccessible in real-world scenarios. Meanwhile, the existing models usually capture the general degradation trend of bearings while ignoring the local information, which restricts the model performance. To tackle the aforementioned problems, we propose a novel health indicator derived from the original vibration signals by combining principal components analysis with Euclidean distance metric, which was motivated by the desire to resolve the dependency on RUL labels. Then, we design a novel self-attention augmented convolution GRU network (SACGNet) to predict the RUL. Combining a self-attention mechanism with a convolution framework can both adaptively assign greater weights to more important information and focus on local information. Furthermore, Gated Recurrent Units are used to parse the long-term dependencies in weighted features such that SACGNet can utilize the important weighted features and focus on local features to improve the prognostic accuracy. The experimental results on the PHM 2012 Challenge dataset and the XJTU-SY bearing dataset have demonstrated that our proposed method is superior to the state of the art.

Keywords: self-attention; gated neural network; remaining useful life prediction; health indicator

Citation: Xu, J.; Duan, S.; Chen, W.; Wang, D.; Fan, Y. SACGNet: A Remaining Useful Life Prediction of Bearing with Self-Attention Augmented Convolution GRU Network. *Lubricants* **2022**, *10*, 21. https://doi.org/10.3390/lubricants10020021

Received: 11 January 2022
Accepted: 1 February 2022
Published: 3 February 2022

1. Introduction

Bearings are one of the key components in a rotating machinery system. The remaining useful life (RUL) of a bearing is often defined as the length of a bearing from the current time to failure [1]. If the damage time or the trend of the vibration signal can be predicted from the collected vibration signal of a bearing, it is beneficial for identifying the adverse running condition in time to avoid the sudden danger of bearings. Thus, the RUL of a bearing is essential for the maintenance and management of mechanical systems [1,2].

In general, the RUL prediction of bearings can be sorted into two different directions: physics-based methods and data-driven methods. Physics-based methods focus on physical and mathematical models, e.g., partial differential equations and state-space models, which require extensive prior knowledge [3–6].

Data-driven RUL methods directly use historical data to model the degradation process of bearings without any prior knowledge.

Deep learning is a popular approach among data-driven methods, which can directly build a deep neural network to model the degradation process as a functional relationship between health states and original sensory data [7].

Deep learning-based RUL approaches typically include the steps of data acquisition, health indicator (HI) construction, and remaining useful life prediction [7].

Data acquisition is to collect the run-to-failure signals from different sensors that can reflect the degradation process of bearings. The complete lifecycle data of a bearing are usually high-dimension and nonlinear. Therefore, a suitable processing method to retain the degradation features (i.e., HI) is necessary.

Health indicators such as the RUL labels are constructed by selecting the appropriate characteristics from the original sensory signal. Since the damage extent of bearings cannot be directly observed, the RUL labels are almost inaccessible in real-world scenarios. Thereby, the critical information from the original data are extracted as HI to train the prediction model, which is a crucial issue to the model [8].

Afterwards, several deep neural networks are designed to extract deep features from the original sensory data and then predict the RUL. The prevalent models include RNN, LSTM, CNN, etc. RNN are often used for predicting the time-series data. The prevalent RNNs are mainly LSTM, GRU, and their variants, which can learn the general degradation trend of the input data. However, they often overlook the local features in input data. With respect to the sequence vibration signal of the bearing, these vibration data have merely small fluctuations in long time series. Until the end of the bearing's life, the vibration fluctuates dramatically, which is often difficult to predict; hence, the existing models cannot obtain satisfactory prediction results [9].

To tackle the aforementioned issues, we propose a novel health indicator-based remaining useful life prediction approach of bearings. The main contributions of this paper are summarized as follows:

1. We combine the PCA with Euclidean distance metric methods to construct a health indicator to tackle the problem of lack of RUL labels. Facing the high-dimensional and long-term series data, PCA can reduce the data dimensionality while retaining sufficient useful features. The Euclidean distance is to measure the similarity between data to distinguish the different degradation stages. Compared with the existing linear RUL labels, our HI is not only capable of representing the general degradation trend of bearings, but it also can retain more local features from the original vibration signal, which benefit the corresponding model's learning and calculations.
2. We design a novel self-attention augmented convolution GRU network (SACGNet) to predict the RUL. Combining the self-attention mechanism with a convolution framework can both adaptively assign greater weights to more important information and focus on local information. Furthermore, Gated Recurrent Units (GRU) are used to parse the long-term dependencies in weighted features so that SACGNet can utilize the important weighted features and focus on local features to improve the prognostic accuracy.
3. Based on the designed HI and SACGNet, a novel remaining useful life prediction approach is proposed. We conduct ablation experiments and different comparison experiments on the PHM 2012 Challenge dataset and XJTU-SY bearing dataset. The experimental results prove the superiority of our proposed method.

The remaining part of this paper is organized as follows: In Section 2, we introduce related works in the field of RUL prediction. We describe our proposed method in detail in Section 3. The experimental results are discussed in Section 4. Finally, we conclude the paper in Section 5.

2. Related Works

2.1. Health Indicator Construction

In deep learning-based RUL methods, HI construction currently has two branches in general. One approach extracts simple physical fault characterization from the original vibration signals as HI, using statistical methods or signal processing methods. For instance, the root mean square (RMS) of the original vibration signal [10] or the percentage of useful life (the current life divided by the total useful life) [11–14]. However, such HIs cannot represent enough useful degradation information of the original data. Thereby, using such HIs as model input makes the model fail to accurately capture the degradation trend for RUL prediction.

The other branch constructs the virtual HI by fusing multiple physical characteristics or multi-sensor signals. These HI can filter out abnormal trends in the early degradation stages, which is more suitable for model learning [15–18]. Guo et al. selected six related-similarity features and combined eight time-frequency features so as to form an original feature set that contains rich degradation signatures of bearings. Then, the selected features are fused into an HI through an RNN [19]. Li et al. used KPCA to integrate multiple features and introduced the EWMA to reduce the fluctuations for the constructed HI [20]. Li et al. designed the generative adversarial network to learn the data distribution in the health states of machine, using the output of the discriminator as HI [21]. Liang et al. proposed a novel index by calculating offset distance and offset angle between the current state and normal state of devices [22].

In summary, the existing HI can only represent the global degradation process of the vibration signal, but it fails to retain more local features. In order to extract more representative features from the vibration signal and facilitate for the model learning, a more effective HI construction method is proposed in this paper.

2.2. Prediction Model

With respect to regression model design, LSTM, GRU, CNN, and the attention mechanism have been successively introduced into the field of RUL prediction.

LSTM uses input gates, forgetting gates, and output gates to regulate the information of the input sequence, which enables the network to learn the long-term dependence of the data and gain favorable results. Hinchi et al. used convolutional layers to directly extract local features from sensor data, combined them with LSTM layers to capture the degradation process of the bearing, and finally output the prediction values [23]. Whereas LSTM solves the problem of gradient disappearance of traditional RNN to some extent, the deliberate design of LSTM for RUL prediction is very time consuming [24,25].

GRU is a variant of LSTM, whose structure is further simplified to show better performance than LSTM in smaller datasets [26,27]. Cao et al. use the BiGRU model to solve the problem of distribution discrepancy [28]. However, with regard to LSTM and GRU, they only use the features learned in the previous time step for regression prediction and often do not pay attention to local features in the long time series [29].

CNN can extract features with less computational effort because of the sparsity of parameter sharing of the convolutional kernel and inter-layer connectivity. More importantly, CNN focuses on the local features in the original vibration signal, which is suitable for RUL prediction [30–33]. Wang et al. proposed a multi-scale convolutional network to improve the domain adaptation capability of the RUL prediction model [34].

It is noted that the original vibration signal often contains different features with different levels of importance. The features that contain more important information should be paid more attention. Hence, an attention mechanism is introduced to RUL prediction of bearing to adaptively extract input features [35]. The self-attention mechanism aims to correlate different states of sequences, which reduces the dependence on external information and is more suitable for capturing the internal relevance of data or features [36,37]. Chen et al. constructed an encoder–decoder model based on the attention mechanism to mine useful degradation information from a long historical vibration signal [38]. Chen et al. proposed an attention-based deep learning framework for RUL prediction, which adopted LSTM to extract features, and then combined with the attention layer to fusion the features, LSTM extracted and manually extracted features [39].

3. Proposed Method

Without loss of generality, given a bearing, vibration signals $V = \{v_1, v_2, \ldots v_m\}$, input V to the health indicator construction module and get the $HI = \{h_1, h_2, \ldots h_m\}$. We expect the model to predict the h_{t+1} value after input $h_1, h_2, \ldots h_t$.

Then, our proposed SACGNet learns the deep features in HI:

$$F(h_1, h_2, \ldots, h_t, \theta) : h \rightarrow h_{t+1} \tag{1}$$

where θ is the parameter on the model.

There is also a testing dataset of the vibration signal $V' = \{v'_1, v'_2, \ldots v'_m\}$, which after HI construction obtains $H' = \{h'_1, h'_2, \ldots h'_n\}$.

Finally, inputing the H' to the trained SACGNet, the model will predict the correct value: $\hat{y}'_t = F(h'_1, h'_2, \ldots, h'_t)$, where $\hat{y}'_t$ is the model's prediction value.

The whole structure of SACGNet is shown in Figure 1, including the health indicator construction module and remaining useful life prediction module. First, input the original signal of the bearings to the health indicator construction module to obtain the HI. After data normalization and sliding window processing, input it to SACGNet for training. In the testing stage, the predicted values are output by autoregression.

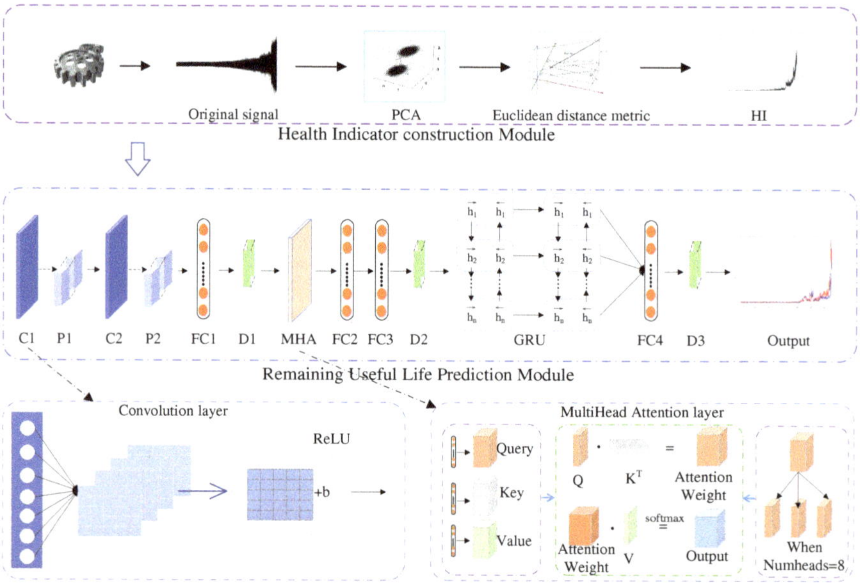

Figure 1. Proposed complete model.

3.1. Health Indicator Construction Module

If the dimension of the original signal is set to d, the matrix form of the original vibration signal $V = \{v_1, v_2, \ldots v_m\}$, where v_i denoted an acquired vibration data that can be written as:

$$V = \begin{pmatrix} v_{11} & \cdots & v_{1d} \\ \vdots & \ddots & \vdots \\ v_{m1} & \cdots & v_{md} \end{pmatrix}. \tag{2}$$

Principal components analysis (PCA) linearly transforms the data into a new coordinate system such that the first major variance of any data projection is at the first coordinate (called the first principal component), the second major variance is at the second coordinate, and so on.

V^T is the de-averaged data. The singular value decomposition of V is:

$$V = W \sum H^T \tag{3}$$

where the matrix W is the eigenvector matrix of VV^T, $\sum$ is a non-negative rectangular diagonal matrix, and H is the eigenvector matrix of $V^T V$.

Assuming zero empirical means, the principal component $w(1)$ of the dataset V can be defined as:

$$w(1) = arg\ max_{\|w\|=1} Var\{W^T V\}. \tag{4}$$

To obtain the k-th principal component, the previous k-1 principal components must first be subtracted from V.

$$\hat{V_{k-1}} = V - \sum_{i-1}^{k-1} w_i w_i^T V \tag{5}$$

Then, the k-th principal component is obtained to update a new dataset and continue to search for principal components.

$$w_k = arg\ max_{\|w\|=1} E\{(w^T \hat{V_{k-1}})^2\} \tag{6}$$

Through PCA, we can reduce the original vibration signal's dimensionality from d to k. We retain the k principal components of the original signals; thus, the dimension of V_{pca} is k, which can be abbreviated as $\{v_{pca1}, v_{pca2}, v_{pca3}, \ldots, v_{pcan}\}$.

$$V_{pca} = pca(V) = \begin{pmatrix} w_{11} & \cdots & w_{1k} \\ \vdots & \ddots & \vdots \\ w_{m1} & \cdots & w_{mk} \end{pmatrix} = \{v_{pca1}, v_{pca2}, v_{pca3} \ldots, v_{pcam}\} \tag{7}$$

Using standard PCA to reduce the dimensionality of the original vibration data, we can only retain the principal components of the data. In this paper, based on PCA, to reduce the dimensionality of vibration data, we use Euclidean distance to calculate the distance between the low-dimensional data to construct HI. The metric Euclidean distance can obtain the similarity between one data in the time series and the neighboring points, which can better reflect the trend of the neighboring data in the original vibration signal, which means "capturing local features" we mentioned.

By calculating the average of the Euclidean distance from each point in V_{pca} to the sequential neighboring points, we can obtain the HI corresponding to each point. HI = $\{h_1, h_2, \ldots h_m\}$. The calculation process of h_i is as follows:

$$h_i = \frac{1}{2}\left(\sqrt{\sum_{j=1}^{k}(v_{pcai_j} - v_{pca(i+1)_j})^2} + \sqrt{\sum_{j=1}^{k}(v_{pcai_j} - v_{pca(i-1)_j})^2}\right). \tag{8}$$

HI will be input to the constructed SACGNet to make the model learn the relationship between them. In order to make HI meet the dimensionality requirements of the model input, a sliding window is set to process the data into the shape required by the model, with a sliding window size of 20. Then, we obtain the $X = \{x_1, x_2, \ldots x_n\}$.

3.2. Remaining Useful Life Prediction Module

In this section, we describe our SACGNet in detail, as shown in the Table 1. We combine a 1D convolution (Conv1d) block with self-attention mechanisms to extract deep features from the input data. The Conv1d block focuses more on local features, and the self-attention mechanism can extract global features of the data. GRU can identify long-term features in the input data, which is beneficial to adapt the bearings under different operating conditions, thereby improving the prediction accuracy of our model [40,41].

Table 1. The architecture of SACGNet.

No	Symobl	Operator	Kernel Size	Dimension
1	Input	Input signal	/	(None, 20, 2)
2	C1	Convolution	4 × 4	(None, 20, 80)
3	P1	Average pooling	1 × 1	(None, 20, 80)
4	C2	Convolution	4 × 4	(None, 20, 80)
5	P2	Average pooling	1 × 1	(None, 20, 80)
6	FC1	Fully connected	80 × 1	(None, 20, 80)
7	D1	Dropout	/	(None, 20, 80)
8	MHA	Multi-Head Attention	/	(None, 20, 80)
9	FC2	Fully connected	80 × 1	(None, 20, 80)
10	FC3	Fully connected	80 × 1	(None, 20, 80)
11	D2	Dropout	/	(None, 20, 80)
12	GRU	Gated recurrent units	/	(None, 80)
13	D3	Dropout	/	(None, 80)
14	FC4	Fully connected	1 × 1	(None, 1)

For convenience, we use C, P, FC, D, MHA, and GRU to denote the Conv1d layer, the pooling layer, the fully connected layer, the dropout layer, Multi-Head attention layer, and the GRU layer, respectively.

In the convolution layer, the calculation of the input data can be written as follows:

$$X_c = ReLU(X \odot f_i + b_i) \tag{9}$$

where $\odot$ represents convolution operation. f_i represents the ith convolution filter, and b_i is the bias. The convolution layers used $ReLU$ as the activation function. Compared to images, the vibration signal is time-series data; hence, the one-dimensional convolution (Conv1d) neural network can be used to perform convolutional operations. The filters of the Conv1d layer are set to 80, the kernel size is set to 4, the stride is set to 1. In our paper, we select the average pooling layer.

The specific calculation process of the self-attention mechanism can be summarized into two processes: calculation the weight coefficients based on the Query and Key, and summation of the weight values based on the weight coefficients. The first process can further include the following: first, calculate the similarity or relevance between Query and Key, and then normalize the found relevance. Its attention function can be described as mapping a Query and a pair of key-value pairs to an output, where Queries, Keys, and values are vectors and the output is computed as a weighted sum of values, where the weight assigned to each value is computed by the compatibility function of the Query with the corresponding Key.

In order to learn the expression of multiple meanings, the input data will be transformed; W_Q, W_K, W_V is the matrix of assigned weights. Self-attention represents a focus on itself, so the equation can be denoted as follows:

$$\begin{cases} Q = X_c W_Q = Linear(X_c) \\ K = X_c W_K = Linear(X_c) \\ V = X_c W_V = Linear(X_c). \end{cases} \tag{10}$$

The output matrix of self-attention is expressed as:

$$Attention(Q, K, V) = softmax(\frac{QK^T}{\sqrt{d_k}})V \tag{11}$$

where d_k is the dimension of K, and the use of $\sqrt{d_k}$ is to change the attention matrix into a standard normal distribution.

Multi-head attention can make the model pay attention to the information from different representational subspaces; the output of the self-attention mechanism layer is three-dimensional vectors, which are written as X_a:

$$X_a = Multi - Head(Q, K, V) = Concat(head_1, \ldots head_h)W. \tag{12}$$

In this paper, we choose h = 8, $d_k = d_q = d_v = 80, W \in R^{hd_v \times d_{model}}$, which are the empirical values selected in the experiment.

Gated Recurrent Units (GRUs) are a gating mechanism in recurrent neural networks. The calculation of the GRU can be written as follows:

$$\begin{cases} X_a = \{a_1, a_2, \ldots a_n\} \\ z_t = \sigma_g(W_z a_t + U_z h_{t-1} + b_z) \\ r_t = \sigma_g(W_r a_t + U_r h_{t-1} + b_r) \\ \hat{h}_t = \phi_h(W_h x_t + U_h * (r_t * h_{t-1}) + b_h) \\ h_t = (1 - z_t) * h_{t-1} + z_t * \hat{h}_t. \end{cases} \tag{13}$$

Among them, a_t is input vector X_a at time t, h_t is output vector, $\hat{h}_t$ is the candidate activation vector, z_t is the update gate vector, r_t is reset gate vector, W, U, and b are the parameter matrices, vector σ_g is a sigmoid function, and ϕ_h is a hyperbolic tangent. The GRU layer receives the features extracted from the Conv1d layer and the Multi-Head attention layer and then outputs the prediction value. The units of GRU are set to 80.

Finally, after the fully connected layer, the final output is obtained:

$$\hat{y}_t = FCN(h_t). \tag{14}$$

SACGNet is trained using the error back-propagation algorithm and gradient descent method. The loss function of the training process is the mean square error function:

$$\text{MSE} = \frac{1}{n}\sum_{i=1}^{n}(y_i - \hat{y}_i)^2 \tag{15}$$

where y_i is the true value, $\hat{y}_i$ is the prediction value, and n is the total number of samples.

In addition, Adam is chosen as the optimizer of this paper, and the learning rate is set to 10^{-3} [42].

Dropout is added to our SACGNet with the parameter set to 0.5 in order to reduce overfitting by preventing complex co-adaptations on training data. The algorithm pseudocode is shown in as follow Algorithm 1.

Algorithm 1: Proposed SACGNet.

1. The SACGNet algorithm for training is defined as follows:
Input: Hyper-parameters of model (batch size, epoch, dropout rate, learning rate, etc.), original signal $V = \{v_1, v_2, v_3, \ldots v_m\}$
2. $HI = \{h_1, h_2, \ldots h_m\}$
3. By sliding window processing:
$HI = \{h_1, h_2, \ldots h_m\} \rightarrow X = \{x_1, x_2, \ldots x_n\}$
Each x represents a batch h, the number of a batch is i
4. $Y = \{y_1, y_2, \ldots y_n\} = \{h_{i+1}, h_{i+2}, \ldots h_m\}$
5. For $i = 1, 2, \ldots n$ do:
$X = \frac{x_i - Min(X)}{Max(X) - Min(X)}, Y = \frac{y_i - Min(Y)}{Max(Y) - Min(Y)}$
end
6. Build SACGNet model
7. w (parameters of the SACGNet) and b (biases) are initialized to zeros
8. Input X and Y to train SACGNet
$X_c = Conv(X)$
$Q = X_c W_Q = Linear(X_c) \quad K = X_c W_K = Linear(X_c) \quad V = X_c W_V = Linear(X_c)$
$Attention(Q, K, V) = softmax(\frac{QK^T}{\sqrt{d_k}})V$
$X_a = Multi - Head(Q, K, V)$
$Output = GRU(X_a)$
$Compute\ MSE\ by$ (15)
$w \leftarrow Adam(MSE, w)$
$b \leftarrow Adam(MSE, b)$
end
Output: Trained SACGNet model for prediction
END

4. Experiments and Results

In this section, we use the IEEE PHM Challenge 2012 bearing dataset and the XJTU-SY Bearing dataset to validate the effectiveness of our method.

4.1. Dataset Description

The IEEE PHM 2012 Challenge dataset was collected from the PRONOSTIA testbed, as shown in Figure 2.

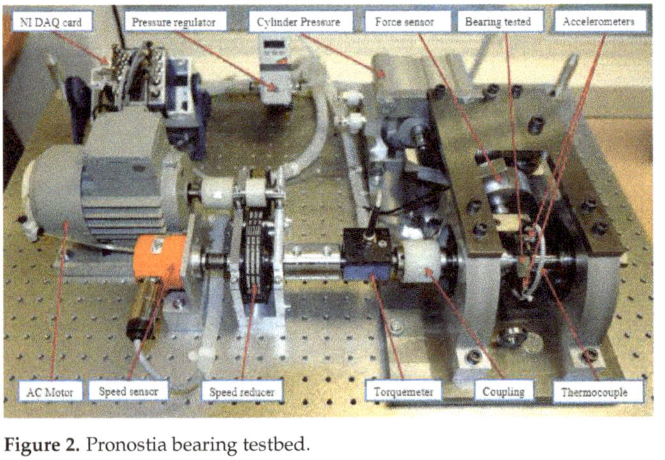

Figure 2. Pronostia bearing testbed.

The PRONOSTIA test platform contains a rotating part, load part, and data collection part. The motor power of the rotating part is 250 W. The power is transferred to the bearing by the axis of rotation. The load part provides a load of 4000 N to make the bearing degrade quickly. The acceleration sensor is placed on a bearing seat in horizontal and vertical directions to select the vibration signals. The sampling frequency of the acceleration sensor is 25.6 kHz. When the test platform starts to work, the vibration signal is recorded every 10 s, and the sampling time is 0.1 s [43].

The data provided by IEEE PHM Challenge 2012 include three different operating conditions. Seven bearings (bearings 1-1 to 1-7) work in the first condition, the motor speed is 1800 rpm, and the load is 4000 N. Seven bearings (bearings 2-1 to 2-7) work in the second condition, the motor speed is 1650 rpm, and the load is 4200 N. Three bearings (bearings 3-1 to 3-3) work in the third condition, the motor speed is 1500 rpm, and the load is 5000 N. Table 2 illustrates the details of the PHM 2012 dataset.

In this paper, the vibration dada of bearings 1-1, 1-2, 2-1, 2-2, and 3-1 are selected, respectively, as the training set, while the rest of the bearings are selected as the testing set. Table 2 illustrates the details of the PHM dataset.

Table 2. The detail of the PHM dataset.

Working Condition	Rotation Speed	Load	Dataset	Sample Number	Bearing Lifetime	Division
			Bearing1-1	2803	7 h 47 m	training
			Bearing1-2	871	2 h 25 m	training
			Bearing1-3	1802	5 h 10 s	testing
1	1800 rpm	4000 N	Bearing1-4	1139	3 h 9 m 40 s	testing
			Bearing1-5	2302	6 h 23 m 30 s	testing
			Bearing1-6	2302	6 h 23 m 29 s	testing
			Bearing1-7	1502	4 h 10 m 11 s	testing
			Bearing2-1	911	2 h 31 m 40 s	training
			Bearing2-2	797	2 h 12 m 40 s	training
			Bearing2-3	1202	3 h 20 m 10 s	testing
2	1650 rpm	4200 N	Bearing2-4	612	1 h 41 m 50 s	testing
			Bearing2-5	2002	5 h 33 m 30 s	testing
			Bearing2-6	572	1 h 35 m 10 s	testing
			Bearing2-7	172	28 m 30 s	testing
			Bearing3-1	515	1 h 25 m 40 s	training
3	1500 rpm	5000 N	Bearing3-2	1637	4 h 32 m 40 s	training
			Bearing3-3	352	58 m 30 s	testing

The XJTU-SY bearing dataset is provided by the Institute of Design Science and Fundamental Research of Xi'an Jiaotong University and contains the run-to-failure vibration data from 15 rolling bearings [44].

As shown in Figure 3, the bearing testbed is composed of an alternating current (AC) induction motor, a motor speed controller, a support shaft, two support bearings (heavy duty roller bearings), and a hydraulic loading system. This testbed is designed to conduct the accelerated degradation tests of the testing bearings under different operating conditions (i.e., different radial force and rotating speed). The radial force is generated by the hydraulic loading system and applied to the housing of tested bearings, and the rotating speed is set and kept by the speed controller of the AC induction motor [44].

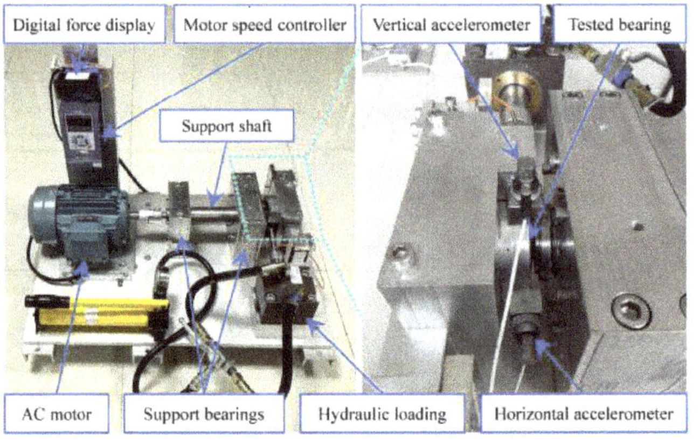

Figure 3. XJTU-SY bearing testbed.

Three different operating conditions are set in the accelerated degradation experiments, and five bearings are used under each operating condition. The sampling frequency is 25.6 kHz, and the sampling period is 1 min. Table 3 illustrates the details of the XJTU-SY dataset.

Table 3. The details of the XJTU-SY dataset.

Working Condition	Rotation Speed	Load	Dataset	Sample Number	Bearing Lifetime	Division
1	2100 rpm	12,000 N	Bearing1-1	123	2 h 3 m	training
			Bearing1-2	161	2 h 41 m	training
			Bearing1-3	158	2 h 38 m	testing
			Bearing1-4	122	2 h 2 m	testing
			Bearing1-5	52	52 m	testing
2	2250rpm	11,000 N	Bearing2-1	491	8 h 11 m	training
			Bearing2-2	161	2 h 41 m	training
			Bearing2-3	533	8 h 53 m	testing
			Bearing2-4	42	42 m	testing
			Bearing2-5	339	5 h 39 m	testing
3	2400 rpm	10,000 N	Bearing3-1	2538	42 h 18 m	training
			Bearing3-2	2496	41 h 36 m	training
			Bearing3-3	371	6 h 11 m	testing
			Bearing3-4	1515	25 h 15 m	testing
			Bearing3-5	114	1 h 54 m	testing

In this paper, the mean square error (MSE), root mean square error (RMSE), mean absolute error (MAE), and mean absolute percentage error (MAPE) are used to evaluate the prediction accuracy. They are respectively computed as follows:

$$\mathrm{MSE} = \frac{1}{n}\sum_{i=1}^{n}(y_i - \hat{y}_i)^2 \tag{16}$$

$$\mathrm{RMSE} = \sqrt{\mathrm{MSE}} = \sqrt{\frac{1}{n}\sum_{i=1}^{n}(y_i - \hat{y}_i)^2} \tag{17}$$

$$\mathrm{MAE} = \frac{1}{n}\sum_{i=1}^{n}|y_i - \hat{y}_i| \tag{18}$$

$$\mathrm{MAPE} = \frac{1}{n}\sum_{i=1}^{n}|\frac{y_i - \hat{y}_i}{y_i}|. \tag{19}$$

In Equations (18)–(21), y_i is the label, $\hat{y}_i$ is the model's prediction value, and n is the total number of samples.

4.2. Different HIs Results

In this section, we compare different HI construction methods to validate the superiority of our proposed HI construction method.

Figure 4 shows the results of different HI construction methods on the bearing 1-3 of the PHM dataset. It can be clearly observed from Figure 4a that the original vibration signal of the bearing 1-3 is in a very smooth state with little fluctuation when the bearing has just begun to work. In the degradation state, the vibration signal usually fluctuates slightly, whereas the overall trend is upward. The signal fluctuation will increase sharply when the bearing finally completely degrades.

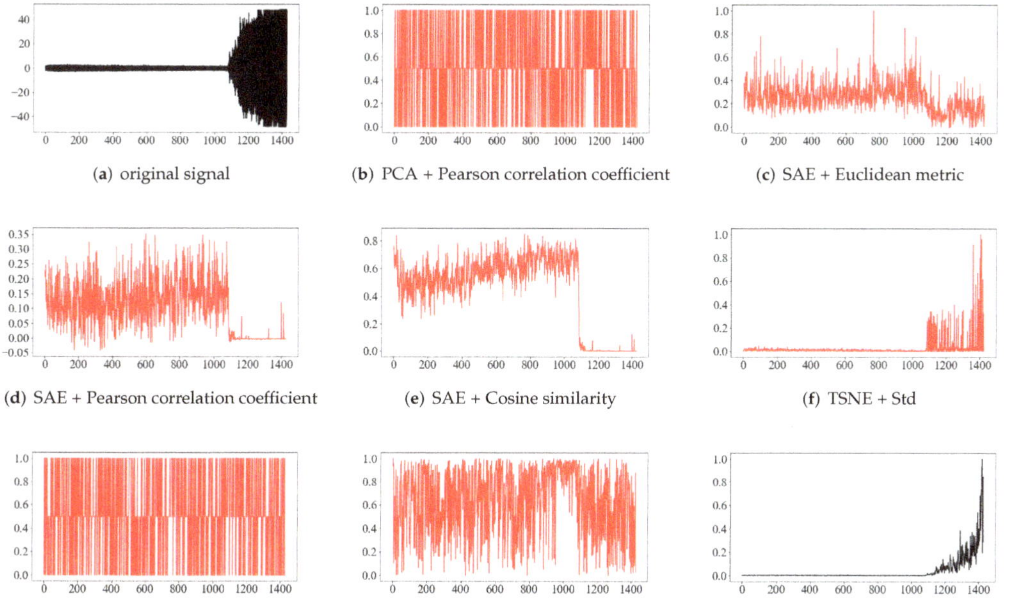

(**a**) original signal; (**b**) PCA + Pearson correlation coefficient; (**c**) SAE + Euclidean metric; (**d**) SAE + Pearson correlation coefficient; (**e**) SAE + Cosine similarity; (**f**) TSNE + Std; (**g**) TSNE + Pearson correlation coefficient; (**h**) TSNE + Cosine similarity; (**i**) Our method

Figure 4. Different HI construction methods on the PHM dataset.

As shown in Figure 4b–h, the HIs constructed by the combination of TSNE and other distance metrics basically do not have any regular change trend. Meanwhile, the methods of SAE and Euclidean distance metric can retain the change trend of the original vibration signal, but the early degradation and complete degradation stage of the bearing cannot be completely distinguished. For the vibration signal of the bearing, PCA is a linear transformation method for each of its principal components. Specifically, the linearity of each point is calculated to obtain the principal components and then downscaled; thus, the global trend of the original signal can be retained. Meanwhile, SAE is a nonlinear learning model that requires a lot of training data to get a satisfactory performance. In contrast, our proposed method, as shown in Figure 4i, is more suitable to reflect the change trend of the original signal and can distinguish the early degradation and complete degradation stage of the bearing, which is beneficial for SACGNet to improve its prediction accuracy.

In order to further illustrate the superiority of our HI construct method, we use the percentage of the use life as the HI; then, we compare the RUL prediction results with our proposed methods on the PHM dataset. The red lines are true values of HI, and the blue lines are the prediction values of the model. Figure 5a–c is the RUL results of comparison HI, while Figure 5d–f is the RUL results of our methods.

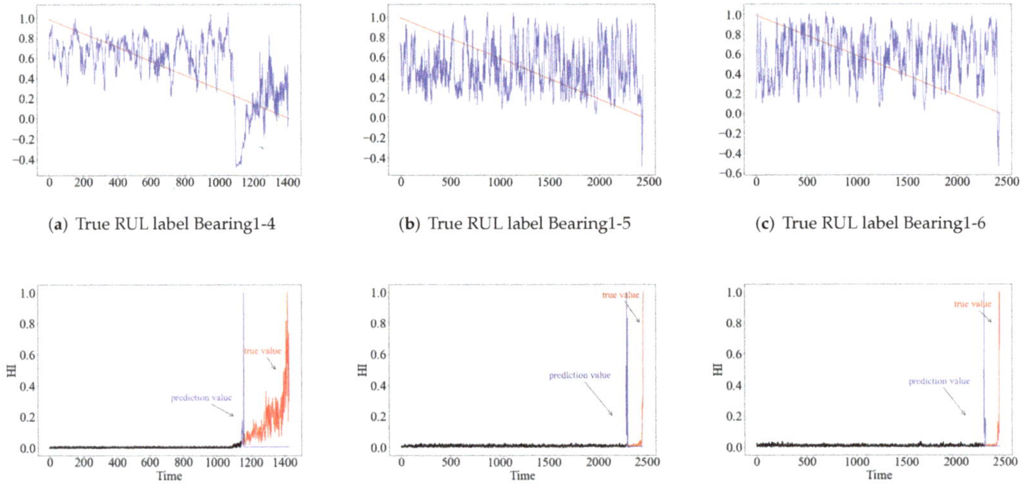

(**a**) True RUL label Bearing1-4 (**b**) True RUL label Bearing1-5 (**c**) True RUL label Bearing1-6

(**d**) HI label Bearing1-4 (**e**) HI label Bearing1-5 (**f**) HI label Bearing1-6

Figure 5. The remaining useful life prediction using the True RUL label and our method as the label.

It can be seen that when the bearing true remaining useful life percentage is used as the HI label, the model fails to learn the degradation trend of the bearing vibration signal, and the final prediction results are not well-fitted. In contrast, using our proposed HI for prediction, the degradation trend of the vibrtion data is depicted more accurately, and the prediction accuracy of the model is improved. Note that our model can properly predict RUL in the stages of rapid degradation of bearing operation, which is of great value for the actual industrial scenarios.

4.3. Ablation Experiments

In order to observe the effects of the different layers in the proposed model, we conduct ablation experiments on the PHM dataset and the XJTU-SY dataset. We let the model constructs remain and respectively remove the Conv1d layer and Multi-Head Attention layer for comparison models. We call them the NoAttention model and NoConv1d model, respectively.

As seen in Table 4, with respect to the PHM dataset, our method achieved the best results in 10 of the 11 bearing data for MSE, RMSE, MAE, and MAPE metrics. Using the MSE metric, our model did not achieve the best result for bearing 2-5; the discrepancy with respect to the best results (i.e., NoAttention model) is 0.002. For the RMSE, MAE, and MAPE metrics, our model does not achieve the optimal results for bearing 2-7, the discrepancy with respect to the best results (i.e., the NoAttention model) is 0.148, 0.124, and 177.228, respectively.

Table 4. Ablation experiments in PHM dataset.

Metric		Bearing 1-3	Bearing 1-4	Bearing 1-5	Bearing 1-6	Bearing 1-7	Bearing 2-3	Bearing 2-4	Bearing 2-5	Bearing 2-6	Bearing 2-7	Bearing 3-3
MSE	SACGNet	**0.010**	**0.053**	**0.039**	**0.042**	**0.012**	**0.017**	**0.042**	0.066	**0.042**	**0.061**	**0.078**
	NoAttention	0.203	0.138	0.212	0.193	0.223	0.049	0.063	0.064	0.061	0.062	0.162
	NoConv1d	0.055	0.099	0.160	0.060	0.069	0.162	0.110	0.161	0.129	0.076	0.095
RMSE	SACGNet	**0.101**	**0.230**	**0.197**	**0.205**	**0.108**	**0.131**	**0.204**	**0.202**	**0.205**	0.397	**0.280**
	NoAttention	0.451	0.372	0.461	0.439	0.472	0.220	0.250	0.253	0.246	0.249	0.403
	NoConv1d	0.236	0.314	0.401	0.245	0.263	0.403	0.332	0.402	0.359	0.276	0.309
MAE	SACGNet	**0.041**	**0.157**	**0.077**	**0.079**	**0.022**	**0.033**	**0.081**	**0.071**	**0.083**	0.220	**0.161**
	NoAttention	0.373	0.304	0.382	0.359	0.394	0.201	0.148	0.225	0.167	0.096	0.368
	NoConv1d	0.216	0.215	0.273	0.203	0.256	0.387	0.271	0.375	0.303	0.178	0.205
MAPE	SACGNet	**1.300**	**1.461**	**5.800**	**2.707**	**2.526**	**13.290**	**14.128**	**15.778**	**48.944**	188.952	**11.879**
	NoAttention	26.616	3.542	64.021	38.050	86.298	89.249	33.654	47.493	83.542	11.724	22.317
	NoConv1d	16.809	2.627	33.060	19.339	47.315	175.061	62.932	77.439	140.799	78.596	15.225

Furthermore, we also conducted the ablation experiment on the XJTU-SY dataset, and the results are shown in Table 5. Using the MAE and MAPE metrics, our model did not achieve the best results for bearings 1-5, 2-3, and 2-4 but only a discrepancy of 1.9% from the best results. Using the MSE and RMSE metrics, our model did not achieve the best results for bearings 2-3 and 2-4, but the discrepancy with the best results (i.e., NoAttention model) was only 3.03%. Except for the afore-mentioned results, the performance of our model is superior to the comparison models.

Table 5. Ablation experiments in the XJTU-SY dataset.

Metric		Bearing 1-3	Bearing 1-4	Bearing 1-5	Bearing 2-3	Bearing 2-4	Bearing 2-5	Bearing 3-3	Bearing 3-4	Bearing 3-5
MSE	SACGNet	**0.022**	**0.028**	**0.129**	0.102	0.261	**0.116**	**0.134**	**0.037**	**0.249**
	NoAttention	0.071	0.045	0.151	0.099	0.238	0.189	0.150	0.050	0.252
	NoConv1d	0.282	0.141	0.150	0.262	0.292	0.122	0.147	0.093	0.254
RMSE	SACGNet	**0.147**	**0.166**	**0.360**	0.320	0.511	**0.341**	**0.369**	**0.193**	**0.500**
	NoAttention	0.266	0.212	0.388	0.315	0.488	0.435	0.387	0.223	0.502
	NoConv1d	0.531	0.376	0.387	0.512	0.540	0.350	0.383	0.304	0.504
MAE	SACGNet	**0.117**	**0.088**	0.206	0.307	0.428	**0.249**	**0.256**	**0.069**	**0.447**
	NoAttention	0.229	0.137	0.198	0.301	0.400	0.333	0.276	0.098	0.450
	NoConv1d	0.447	0.274	0.194	0.479	0.462	0.297	0.294	0.211	0.447
MAPE	SACGNet	**12.904**	**0.714**	3.217	**12.090**	0.862	8.714	**17.105**	**31.240**	**1.251**
	NoAttention	39.583	1.903	0.798	12.442	0.750	8.626	19.341	53.358	1.286
	NoConv1d	84.217	4.435	0.568	19.283	0.998	9.188	33.756	172.834	1.314

The results of the ablation experiments conducted on both bearing datasets prove that our proposed model achieves the best results on the largest number of testing bearings. In a comprehensive analysis, the degradation features of different bearings with different operating conditions are different, Conv1d can extract the local features of the original vibration signals, and the Self-Attention mechanism focuses on the global features, which can be integrated to achieve more excellent results.s

4.4. Results of Different Models

In this section, we compare the overall prediction accuracy of our proposed model with state-of-the-art methods on the PHM dataset and XJTU-SY dataset. The compared models include CNN, RNN, LSTM, and GRU.

As shown in Table 6, with respect to the PHM dataset, our model achieved the best results in nine out of 11 bearings for MSE and RMSE metrics, and in nine out of 11 bearings for MAE and MAPE metrics. Using the MSE and RMSE metrics, our model did not achieve the best results for bearings 2-5 and 2-7. Using the MAE and MAPE metrics, our model did not achieve the best values for bearings 2-7.

Table 6. Different models' result in the PHM dataset.

Metric		Bearing 1-3	Bearing 1-4	Bearing 1-5	Bearing 1-6	Bearing 1-7	Bearing 2-3	Bearing 2-4	Bearing 2-5	Bearing 2-6	Bearing 2-7	Bearing 3-3
MSE	SACGNet	**0.010**	**0.053**	**0.039**	**0.042**	**0.012**	**0.017**	**0.042**	0.066	**0.042**	0.061	**0.078**
	CNN	0.276	0.236	0.361	0.339	0.401	0.017	0.044	0.033	0.049	0.063	0.159
	RNN	0.087	0.079	0.191	0.167	0.089	0.098	0.107	0.102	0.106	0.052	0.080
	LSTM	0.051	0.154	0.099	0.088	0.100	0.134	0.126	0.080	0.129	0.153	0.232
	GRU	0.156	0.130	0.220	0.212	0.122	0.047	0.173	0.074	0.150	0.621	0.255
RMSE	SACGNet	**0.101**	**0.230**	**0.197**	**0.205**	**0.108**	**0.131**	**0.204**	0.202	**0.205**	0.397	**0.280**
	CNN	0.526	0.486	0.601	0.583	0.633	0.132	0.209	0.182	0.221	0.250	0.399
	RNN	0.295	0.282	0.437	0.409	0.299	0.313	0.327	0.319	0.326	0.229	0.282
	LSTM	0.227	0.393	0.315	0.296	0.317	0.366	0.354	0.283	0.360	0.392	0.482
	GRU	0.395	0.360	0.469	0.461	0.350	0.216	0.416	0.272	0.387	0.788	0.505
MAE	SACGNet	**0.041**	**0.157**	**0.077**	**0.079**	**0.022**	**0.033**	**0.081**	**0.071**	**0.083**	0.220	**0.161**
	CNN	0.431	0.401	0.492	0.473	0.529	0.079	0.121	0.127	0.129	0.094	0.361
	RNN	0.272	0.230	0.405	0.371	0.277	0.305	0.297	0.307	0.302	0.137	0.208
	LSTM	0.082	0.270	0.294	0.275	0.308	0.352	0.185	0.219	0.215	0.369	0.376
	GRU	0.378	0.315	0.449	0.282	0.337	0.163	0.250	0.167	0.220	0.744	0.433
MAPE	SACGNet	**1.300**	**1.461**	**5.800**	**2.707**	**2.526**	**13.290**	**14.128**	**15.778**	**48.944**	188.952	**11.879**
	CNN	34.128	4.702	84.794	68.833	116.331	34.082	26.482	26.338	66.653	10.532	21.550
	RNN	20.391	2.999	64.366	51.722	58.369	135.458	77.727	63.910	132.203	33.092	14.056
	LSTM	4.084	3.941	46.745	34.372	64.918	157.411	43.524	45.461	107.898	160.438	25.883
	GRU	28.107	4.765	52.206	52.417	70.271	72.181	63.668	35.833	110.955	364.585	28.784

In order to indicate the superior performance of our model, we add the prediction results of all the comparison models in the PHM dataset. Without loss of generality, we visualize the prediction results for bearing 1-4, 1-5, and 1-6 in order to compare them with the previous prediction results, as shown in Figure 6.

From Figure 6a,d,g,j,m, it can be seen that CNN has the worst fitting results on the testing data. The difference between the CNN predicted values and the original signal is very obvious, because the single CNN models are unsuitable for processing the time-series data. The prediction results of RNN and GRU are slightly superior to those of CNN, but there is still a visible difference from the original vibration signal. Furthermore, the RNN performance is significantly inferior to LSTM on long-term series.

The prediction results of SACGNet and LSTM are similar on bearing 1-4, and the fitting results of SACGNet are more favorable when combined with the four evaluation metrics in Table 6. From Figure 6b,c,e,f,h,i,k,l,n,o, it can be seen that on bearing 1-5 and 1-6, SACGNet has superior prediction results, which is consistent with the comparison results of the four evaluation metrics in Table 6.

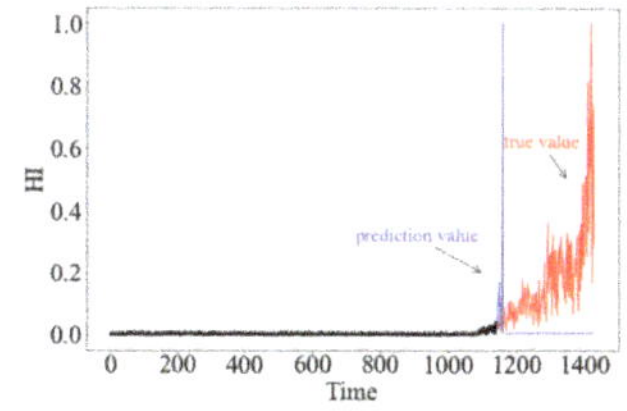

(**a**) SACGNet prediction result of bearing1-4

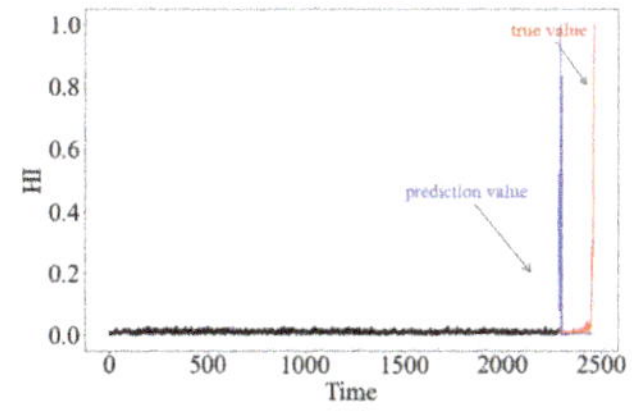

(**b**) SACGNet prediction result of bearing1-5

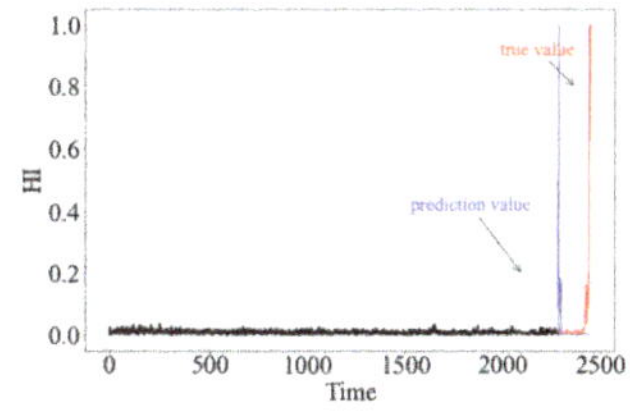

(**c**) SACGNet prediction result of bearing1-6

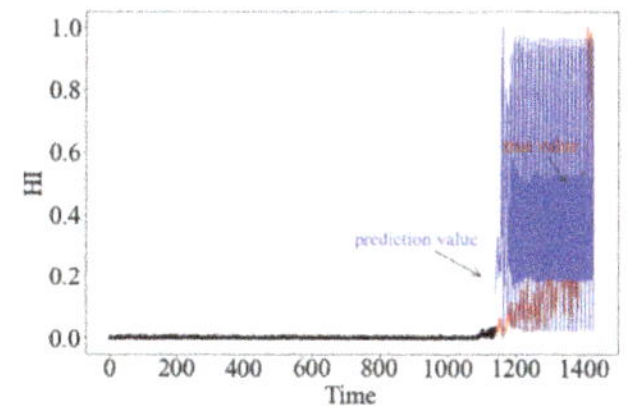

(**d**) CNN prediction result of bearing1-4

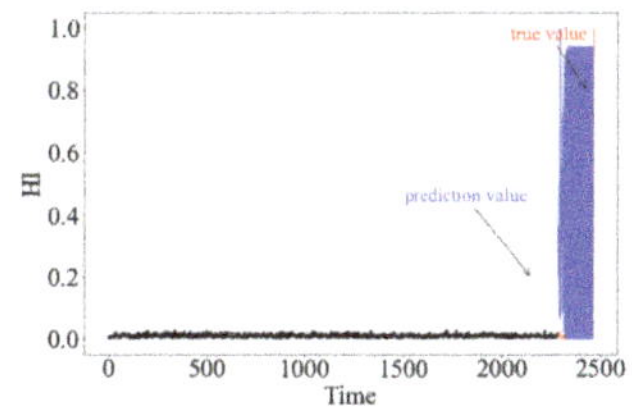

(**e**) CNN prediction result of bearing1-5

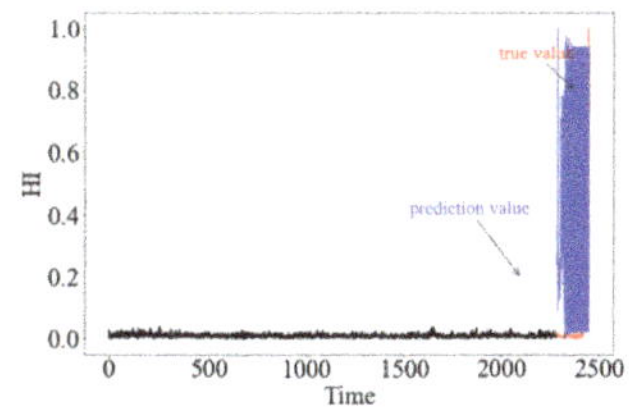

(**f**) CNN prediction result of bearing1-6

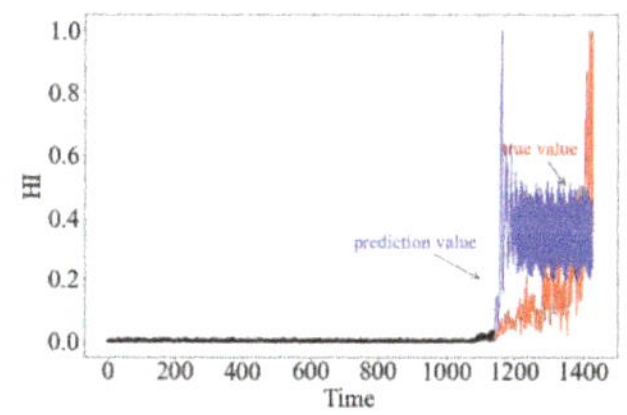

(**g**) RNN prediction result of bearing1-4

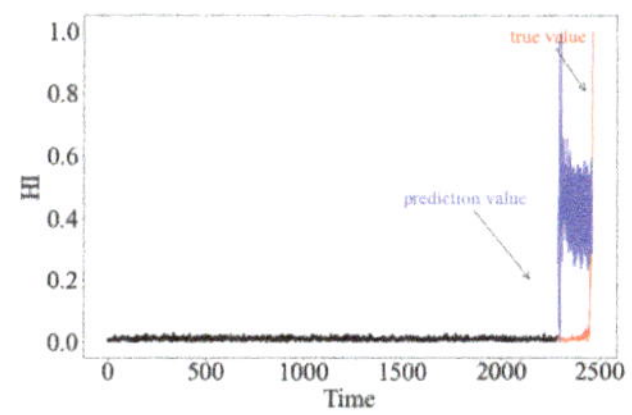

(**h**) RNN prediction result of bearing1-5

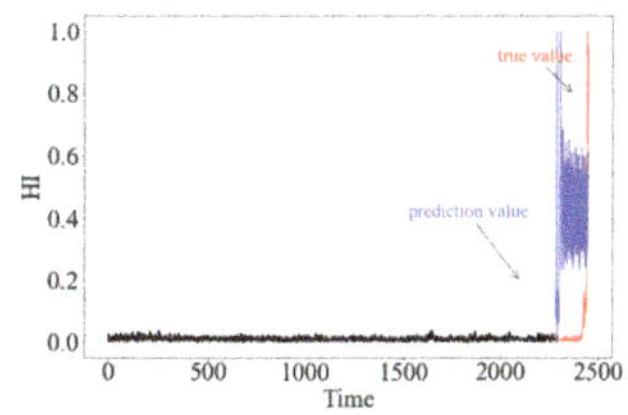

(**i**) RNN prediction result of bearing1-6

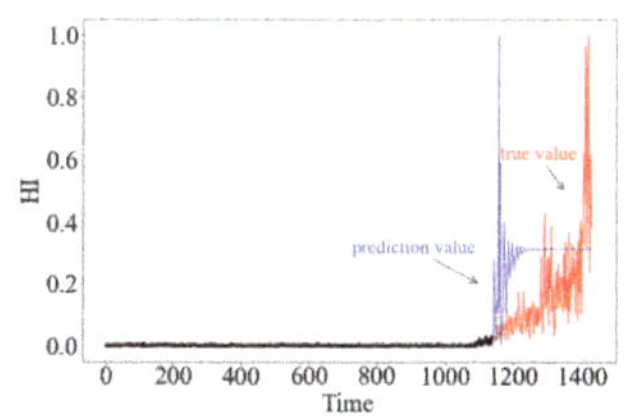

(**j**) LSTM prediction result of bearing1-4

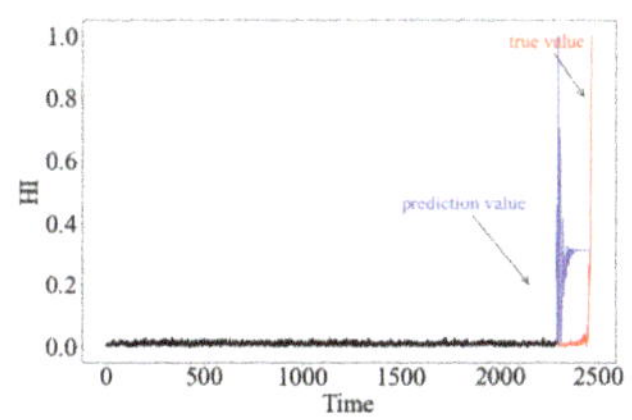

(**k**) LSTM prediction result of bearing1-5

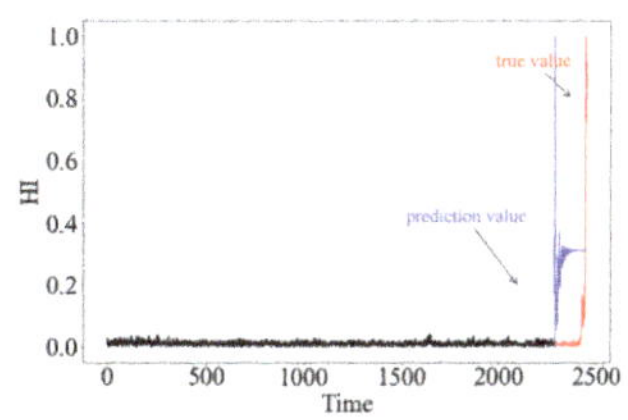

(**l**) LSTM prediction result of bearing1-6

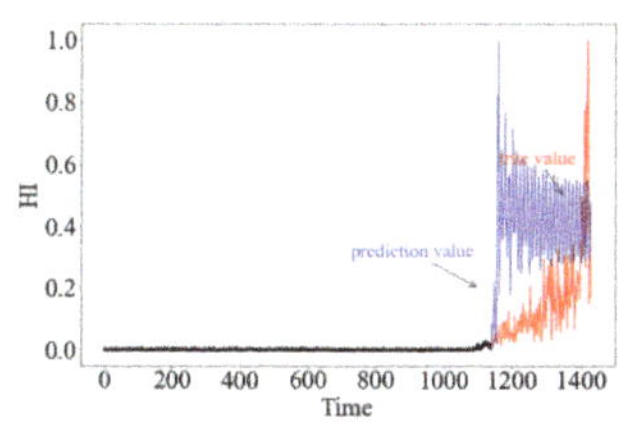

(**m**) GRU prediction result of bearing1-4

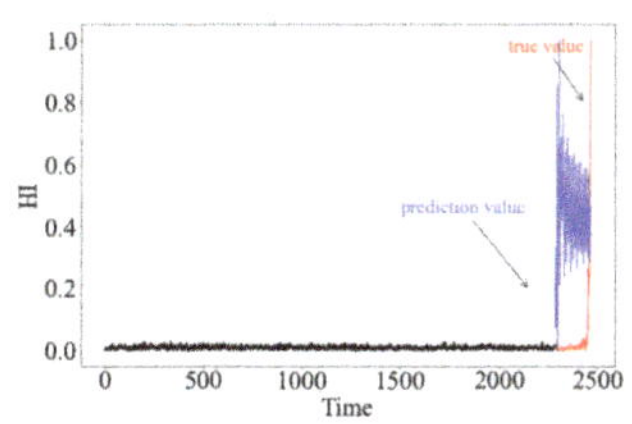

(**n**) GRU prediction result of bearing1-5

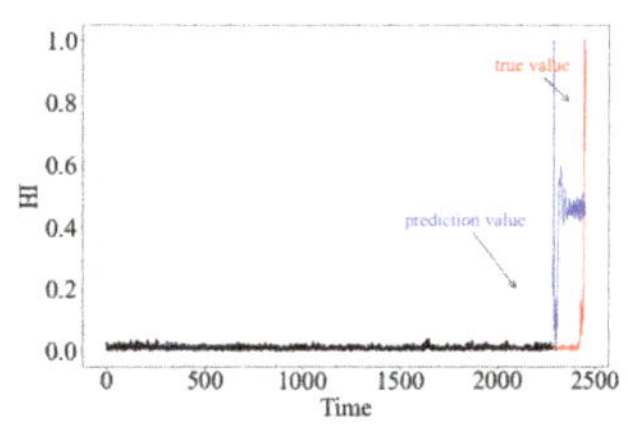

(**o**) GRU prediction result of bearing1-6

Figure 6. The prediction results of different models in the PHM dataset.

Upon observation of the original vibration signal of bearing 2-7, the early degradation states of bearing 2-7 show very sharp fluctuations, and the amplitude difference between the early degradation and the complete degradation stage is very small. The vibration fluctuation of the early degradation is even more severe than that of the complete degradation stage. That may be the reason why our model did not achieve optimal results on bearing 2-7.

As shown in Table 7, with respect to the XJTU-SY dataset, it can be seen that our model has achieved the best resluts on most of the testing bearings, expect for the MSE of bearings 2-3 and 3-5, RMSE of bearings 2-3, 2-5 and 3-5, MAE of bearings 1-5, 2-3, 2-5, and 3-5, and MAPE of bearings 1-5, 2-3, and 2-5, respectively.

Table 7. Different models' results in the XJTU-SY dataset.

Metric		Bearing 1-3	Bearing 1-4	Bearing 1-5	Bearing 2-3	Bearing 2-4	Bearing 2-5	Bearing 3-3	Bearing 3-4	Bearing 3-5
MSE	SACGNet	**0.022**	**0.028**	**0.129**	0.102	**0.261**	**0.116**	**0.134**	**0.037**	0.249
	CNN	0.024	0.036	0.151	0.060	0.276	0.190	0.161	0.041	0.255
	RNN	0.297	0.122	0.139	0.127	0.292	0.984	0.141	0.084	0.225
	LSTM	0.276	0.136	0.150	0.276	0.292	0.260	0.615	0.097	0.147
	GRU	0.276	0.144	0.150	0.123	0.292	0.107	0.170	0.331	0.236
RMSE	SACGNet	**0.147**	**0.166**	**0.360**	0.320	**0.511**	0.341	**0.369**	**0.193**	0.500
	CNN	0.154	0.191	0.389	0.244	0.525	0.436	0.401	0.203	0.505
	RNN	0.545	0.349	0.373	0.357	0.540	0.314	0.375	0.290	0.474
	LSTM	0.525	0.368	0.387	0.525	0.540	0.510	0.784	0.312	0.384
	GRU	0.526	0.380	0.387	0.351	0.540	0.331	0.413	0.575	0.486
MAE	SACGNet	**0.117**	**0.088**	0.206	0.307	**0.428**	0.249	**0.256**	**0.069**	0.447
	CNN	0.134	0.093	0.200	0.228	0.444	0.333	0.297	0.077	0.454
	RNN	0.469	0.249	0.194	0.332	0.462	0.231	0.311	0.252	0.421
	LSTM	0.442	0.280	0.194	0.520	0.462	0.457	0.730	0.135	0.297
	GRU	0.446	0.290	0.194	0.334	0.462	0.246	0.368	0.563	0.433
MAPE	SACGNet	**12.904**	**0.714**	3.217	12.090	**0.862**	8.714	**17.105**	**31.240**	**1.251**
	CNN	17.219	0.762	0.922	8.611	0.929	8.883	21.782	34.925	1.292
	RNN	85.029	4.089	1.628	14.217	0.998	8.480	29.475	184.134	1.257
	LSTM	84.945	4.599	0.568	19.232	0.998	15.022	90.756	77.418	1.546
	GRU	85.115	4.736	0.568	14.007	0.998	8.644	42.677	425.427	1.258

The reason for such results may be due to the fact that the experimental environment of 2-3, 2-5, and 3-5 is slightly different from the conditions of the bearing dataset we chose as the training set, and these datasets produce fluctuations in the degradation stage that are equal to or even higher than the final complete damage. Specifically, the CNN model achieved the best results on the bearing 2-3 dataset for four metrics, which may be because the degradation process of bearing 2-3 is filled with many small-scale local fluctuations, allowing the CNN to fit this process more directly. As for bearings 2-5 and 3-5, they do not show much sharp fluctuations and also due to the small amount of data compared to the other datasets, RNN and LSTM are able to parse their long-term serial relationships on these two datasets.

In summary, from the comparison experiments on the two datasets, in most of the cases, our model achieves the best prediction accuracy compared to the other existing models, which demonstrates that our proposed model is applicable for RUL prediction.

5. Conclusions

In this paper, we explored a health indicator-based remaining useful life prediction method. First, we combine principal component analysis (PCA) with Euclidean distance measurement to construct the health indicator for tackling the dependency on RUL labels. Then, we design a self-attention augmented convolution GRU network (SACGNet) to predict the RUL task, which utilizes the globe features as well as local important features to improve the prognostic accuracy. To verify the effectiveness of the model, we conducted extensive experiments on two bearing datasets, respectively, and the results demonstrate that SACGNet is superior to these existing models under several evaluation criteria. Meanwhile, the model has excellent generalization performance for multiple bearings.

Author Contributions: J.X. contributed to the conception of the study; S.D. performed the experiment and wrote the manuscript; W.C. contributed significantly to analysis; D.W. performed the data analyses; Y.F. helped perform the analysis with constructive discussions. All authors have read and agreed to the published version of the manuscript.

Funding: This work was supported in part by the National Key Research And Development Plan under Grant 2018YFB2000505, in part by the Key Research and Development Plan of Anhui Province under Grant 202104a04020003, and in part by the Fundamental Research Funds for the Central Universities under Grant PA2021KCPY0045.

Institutional Review Board Statement: Not applicable.

Informed Consent Statement: Not applicable.

Data Availability Statement: Public datasets used in our paper: https://github.com/wkzs111/phm-ieee-2012-data-challenge-dataset (accessed on 10 December 2021), https://biaowang.tech/xjtu-sy-bearing-datasets/ (accessed on 10 December 2021).

Conflicts of Interest: The authors declare no conflict of interest.

References

1. Uckun, S.; Goebel, K.; Lucas, P.J. Standardizing research methods for prognostics. In Proceedings of the International Conference on Prognostics and Health Management, Denver, CO, USA, 6–9 October 2008; pp. 1–10.
2. Glowacz, A. Acoustic fault analysis of three commutator motors. *Mech. Syst. Signal Process.* **2019**, *133*, 106226. [CrossRef]
3. Zarei, J.; Tajeddini, M.A.; Karimi, H.R. Vibration analysis for bearing fault detection and classification using an intelligent filter. *Mechatronics* **2014**, *24*, 151–157. [CrossRef]
4. Glowacz, A.; Glowacz, W.; Kozik, J.; Piech, K.; Gutten, M.; Caesarendra, W.; Liu, H.; Brumercik, F.; Irfan, M.; Khan, Z.F. Detection of deterioration of three-phase induction motor using vibration signals. *Meas. Sci. Rev.* **2019**, *19*, 241–249. [CrossRef]
5. Ordóñez, C.; Lasheras, F.S.; Roca-Pardinas, J.; de Cos Juez, F.J. A hybrid ARIMA–SVM model for the study of the remaining useful life of aircraft engines. *J. Comput. Appl. Math.* **2019**, *346*, 184–191. [CrossRef]
6. Ma, S.; Zhang, X.; Yan, K.; Zhu, Y.; Hong, J. A Study on Bearing Dynamic Features under the Condition of Multiball—Cage Collision. *Lubricants* **2022**, *10*, 9. [CrossRef]
7. Lei, Y.; Li, N.; Guo, L.; Li, N.; Yan, T.; Lin, J. Machinery health prognostics: A systematic review from data acquisition to RUL prediction. *Mech. Syst. Signal Process.* **2018**, *104*, 799–834. [CrossRef]
8. Singleton, R.K.; Strangas, E.G.; Aviyente, S. Extended Kalman filtering for remaining-useful-life estimation of bearings. *IEEE Trans. Ind. Electron.* **2014**, *62*, 1781–1790. [CrossRef]
9. Saidi, L.; Benbouzid, M. Prognostics and Health Management of Renewable Energy Systems: State of the Art Review, Challenges, and Trends. *Electronics* **2021**, *10*, 2732. [CrossRef]
10. Zhang, N.; Wu, L.; Wang, Z.; Guan, Y. Bearing remaining useful life prediction based on Naive Bayes and Weibull distributions. *Entropy* **2018**, *20*, 944. [CrossRef]
11. Malhi, A.; Yan, R.; Gao, R.X. Prognosis of defect propagation based on recurrent neural networks. *IEEE Trans. Instrum. Meas.* **2011**, *60*, 703–711. [CrossRef]
12. Liao, H.; Tian, Z. A framework for predicting the remaining useful life of a single unit under time-varying operating conditions. *IEEE Trans.* **2013**, *45*, 964–980. [CrossRef]
13. Hu, L.; Hu, N.Q.; Fan, B.; Gu, F.S.; Zhang, X.Y. Modeling the relationship between vibration features and condition parameters using relevance vector machines for health monitoring of rolling element bearings under varying operation conditions. *Math. Probl. Eng.* **2015**, *2015*, 123730. [CrossRef]
14. Zhang, Z.X.; Si, X.S.; Hu, C.H. An age-and state-dependent nonlinear prognostic model for degrading systems. *IEEE Trans. Reliab.* **2015**, *64*, 1214–1228. [CrossRef]
15. Hu, C.; Youn, B.D.; Wang, P.; Yoon, J.T. Ensemble of data-driven prognostic algorithms for robust prediction of remaining useful life. *Reliab. Eng. Syst. Saf.* **2012**, *103*, 120–135. [CrossRef]
16. Wang, Y.; Peng, Y.; Zi, Y.; Jin, X.; Tsui, K.L. A two-stage data-driven-based prognostic approach for bearing degradation problem. *IEEE Trans. Ind. Inform.* **2016**, *12*, 924–932. [CrossRef]

17. Giantomassi, A.; Ferracuti, F.; Benini, A.; Ippoliti, G.; Longhi, S.; Petrucci, A. Hidden Markov model for health estimation and prognosis of turbofan engines. In Proceedings of the International Design Engineering Technical Conferences and Computers and Information in Engineering Conference, Washington, DC, USA, 28–31 August 2011; Volume 54808, pp. 681–689.
18. Kumar, H.; Pai, S.P.; Sriram, N.; Vijay, G. Rolling element bearing fault diagnostics: Development of health index. *Proc. Inst. Mech. Eng. Part C J. Mech. Eng. Sci.* **2017**, *231*, 3923–3939. [CrossRef]
19. Guo, L.; Li, N.; Jia, F.; Lei, Y.; Lin, J. A recurrent neural network based health indicator for remaining useful life prediction of bearings. *Neurocomputing* **2017**, *240*, 98–109. [CrossRef]
20. Li, X.; Jiang, H.; Xiong, X.; Shao, H. Rolling bearing health prognosis using a modified health index based hierarchical gated recurrent unit network. *Mech. Mach. Theory* **2019**, *133*, 229–249. [CrossRef]
21. Li, X.; Zhang, W.; Ma, H.; Luo, Z.; Li, X. Data alignments in machinery remaining useful life prediction using deep adversarial neural networks. *Knowl.-Based Syst.* **2020**, *197*, 105843. [CrossRef]
22. Zeming, L.; Jianmin, G.; Hongquan, J.; Xu, G.; Zhiyong, G.; Rongxi, W. A similarity-based method for remaining useful life prediction based on operational reliability. *Appl. Intell.* **2018**, *48*, 2983–2995. [CrossRef]
23. Hinchi, A.Z.; Tkiouat, M. Rolling element bearing remaining useful life estimation based on a convolutional long-short-term memory network. *Procedia Comput. Sci.* **2018**, *127*, 123–132. [CrossRef]
24. Wang, F.; Liu, X.; Deng, G.; Yu, X.; Li, H.; Han, Q. Remaining life prediction method for rolling bearing based on the long short-term memory network. *Neural Process. Lett.* **2019**, *50*, 2437–2454. [CrossRef]
25. Ragab, M.; Chen, Z.; Wu, M.; Foo, C.S.; Kwoh, C.K.; Yan, R.; Li, X. Contrastive adversarial domain adaptation for machine remaining useful life prediction. *IEEE Trans. Ind. Inform.* **2020**, *17*, 5239–5249. [CrossRef]
26. Cheng, Y.; Hu, K.; Wu, J.; Zhu, H.; Lee, C.K. A deep learning-based two-stage prognostic approach for remaining useful life of rolling bearing. *Appl. Intell.* **2021**, 1–16. [CrossRef]
27. Erdenebayar, U.; Kim, Y.; Park, J.U.; Lee, S.; Lee, K.J. Automatic Classification of Sleep Stage from an ECG Signal Using a Gated-Recurrent Unit. *Int. J. Fuzzy Log. Intell. Syst.* **2020**, *20*, 181–187. [CrossRef]
28. Cao, Y.; Jia, M.; Ding, P.; Ding, Y. Transfer learning for remaining useful life prediction of multi-conditions bearings based on bidirectional-GRU network. *Measurement* **2021**, *178*, 109287. [CrossRef]
29. Liu, H.; Liu, Z.; Jia, W.; Lin, X. Remaining useful life prediction using a novel feature-attention-based end-to-end approach. *IEEE Trans. Ind. Inform.* **2020**, *17*, 1197–1207. [CrossRef]
30. Ma, M.; Mao, Z. Deep-convolution-based LSTM network for remaining useful life prediction. *IEEE Trans. Ind. Inform.* **2020**, *17*, 1658–1667. [CrossRef]
31. Liu, R.; Yang, B.; Hauptmann, A.G. Simultaneous bearing fault recognition and remaining useful life prediction using joint-loss convolutional neural network. *IEEE Trans. Ind. Inform.* **2019**, *16*, 87–96. [CrossRef]
32. Zhu, J.; Chen, N.; Peng, W. Estimation of bearing remaining useful life based on multiscale convolutional neural network. *IEEE Trans. Ind. Electron.* **2018**, *66*, 3208–3216. [CrossRef]
33. Cheng, C.; Ma, G.; Zhang, Y.; Sun, M.; Teng, F.; Ding, H.; Yuan, Y. Online bearing remaining useful life prediction based on a novel degradation indicator and convolutional neural networks. *arXiv* **2018**, arXiv:1812.03315.
34. Ge, Y.; Liu, J.; Ma, J. Remaining Useful Life Prediction Using Deep Multi-scale Convolution Neural Networks. *IOP Conf. Ser. Mater. Sci. Eng.* **2021**, *1043*, 032011. [CrossRef]
35. Luo, J.; Zhang, X. Convolutional neural network based on attention mechanism and Bi-LSTM for bearing remaining life prediction. *Appl. Intell.* **2021**, *52*, 1–16. [CrossRef]
36. Vaswani, A.; Shazeer, N.; Parmar, N.; Uszkoreit, J.; Jones, L.; Gomez, A.N.; Kaiser, Ł.; Polosukhin, I. Attention is all you need. In *Advances in Neural Information Processing Systems*; MIT Press: Cambridge, MA, USA, 2017; pp. 5998–6008.
37. Mo, Y.; Wu, Q.; Li, X.; Huang, B. Remaining useful life estimation via transformer encoder enhanced by a gated convolutional unit. *J. Intell. Manuf.* **2021**, *32*, 1–10. [CrossRef]
38. Chen, Y.; Peng, G.; Zhu, Z.; Li, S. A novel deep learning method based on attention mechanism for bearing remaining useful life prediction. *Appl. Soft Comput.* **2020**, *86*, 105919. [CrossRef]
39. Chen, Z.; Wu, M.; Zhao, R.; Guretno, F.; Yan, R.; Li, X. Machine remaining useful life prediction via an attention-based deep learning approach. *IEEE Trans. Ind. Electron.* **2020**, *68*, 2521–2531. [CrossRef]
40. Bello, I.; Zoph, B.; Vaswani, A.; Shlens, J.; Le, Q.V. Attention augmented convolutional networks. In Proceedings of the IEEE/CVF International Conference on Computer Vision, Seoul, Korea, 27–28 October 2019; pp. 3286–3295.
41. Woo, S.; Park, J.; Lee, J.Y.; Kweon, I.S. Cbam: Convolutional block attention module. In Proceedings of the European Conference on Computer Vision (ECCV), Munich, Germany, 8–14 September 2018; pp. 3–19.
42. Kingma, D.P.; Ba, J. Adam: A method for stochastic optimization. *arXiv* **2014**, arXiv:1412.6980.
43. Nectoux, P.; Gouriveau, R.; Medjaher, K.; Ramasso, E.; Chebel-Morello, B.; Zerhouni, N.; Varnier, C. PRONOSTIA: An experimental platform for bearings accelerated degradation tests. In Proceedings of the IEEE International Conference on Prognostics and Health Management, PHM'12, London, UK, 23–27 September 2012; pp. 1–8.
44. Wang, B.; Lei, Y.; Li, N.; Li, N. A hybrid prognostics approach for estimating remaining useful life of rolling element bearings. *IEEE Trans. Reliab.* **2018**, *69*, 401–412. [CrossRef]

lubricants

MDPI

Article

The Influence of Bearing Ring Inclination on Precision Ball Bearing Contact and Heat Generation Performance

Bei Yan [1,*], Xiaohong Zhang [2] and Zanfei Zhu [1]

[1] School of Mechanical Engineering, Chang'an University, Xi'an 710049, China
[2] Key Laboratory of Education Ministry for Modern Design and Rotor-Bearing System, Xi'an Jiaotong University, Xi'an 710049, China
* Correspondence: yanbei@chd.edu.cn

Abstract: The assembly quality between bearing components will lead to an increase in the load imbalance of each rolling element, which becomes a factor that affects the heat generation of precision bearings. The motion and heat generation of angular contact ball bearing (ACBB) under different assembly states were studied, and the quasi-static model of ACBB with outer ring tilting was established. The contact angle, contact load and motion posture of rolling elements were obtained. Then, the relative imbalance of the outer ring was experimentally simulated based on the outer spacer ring with poor end-parallelism, and the thermal characteristics of the spindle system were monitored. Through theoretical and experimental comparison, the influence of different skew degrees of rings on bearing heat generation and spindle rotation accuracy was discussed.

Keywords: bearing heat generation; outer ring inclination; bearing assembly quality

Citation: Yan, B.; Zhang, X.; Zhu, Z. The Influence of Bearing Ring Inclination on Precision Ball Bearing Contact and Heat Generation Performance. *Lubricants* **2022**, *10*, 206. https://doi.org/10.3390/lubricants10090206

Received: 22 May 2022
Accepted: 10 August 2022
Published: 29 August 2022

1. Introduction

Precision bearing-spindle systems are the core support rotary parts of CNC machine tools, and their service performance directly affects the processing accuracy and efficiency of the equipment, in which the errors caused by thermal problems, such as thermal deformation and uneven heating of the bearing-spindle system, are dominant [1–4]. The general mechanism of thermal error generation is as follows: under the joint influence of frictional heat sources, such as rolling bearings and sealing structures, various parts of the shaft system will generate different degrees of heat transfer and accumulation. With the increasing demands on the thermal characteristics of the bearing-spindle system in the precision machining process, the thermal characteristics' variation in the precision bearing-spindle system, triggered by different assembly processes and assembly quality, has become an important link that cannot be ignored in the study of thermal problems of the spindle system in precision machine tools.

Based on bearing force analysis, researchers mainly used the quasi-static analysis model to consider the bearing force characteristics in its rotating state, and studied the force relationship between the raceway and the ball under different external loads, focusing on the influence of the centrifugal load and gyroscopic moment caused by the ball rotation on the force between inside components [5–9]. In order to simplify the calculation, Jones [10] assumed that the bearing fixed ring and rolling element maintained a pure rolling state, namely raceway control assumption. Based on this, combined with the Hertz elastic contact theory, a quasi-static analysis model was constructed, considering the combined effects of centrifugal force and gyroscopic moment of the steel ball. Based on Jones' model, Harris [11] introduced elastohydrodynamic lubrication theory to the quasi-static analysis of bearings, and the calculation of quasi-static analysis of bearings considering the dynamic pressure lubrication effect of lubricant was successfully realized. On this basis, for the purpose of overcoming the limitations of the raceway control assumption in Jones' model, a research team from Xi'an Jiaotong University [12] proposed a new method for calculating the attitude angle of the rolling element by considering the relative sliding action between the raceway and the rolling element, which further improved the quasi-static model of rolling bearing. Furthermore, Wang et al. at Beijing University of Technology [7] introduced a quasi-static analysis model of bearing without raceway control, which truly realized the quasi-static analysis of bearing, considering the real kinematic features of raceways and rolling bodies. Through the analysis of the above research work, it can be found that the current scholars have made a more systematic analysis of the internal contact state of precision bearings based on quasi-static analysis model, as well as the above analysis is mainly carried out under the conditions of correct bearing installation, that is, the above analysis models do not take into account the actual assembly factors, such as abnormal relative position of inner and outer rings and inclination. Nevertheless, in the actual assembly process, especially for the ultra-precision bearing-spindle system, extremely small assembly errors may lead to changes in the internal contact and force state of the bearing, such as the outer ring tilting, due to the misalignment of the spacer ring. At present, there are relatively few bearing mechanics models that consider the bearing ring skew and other undesirable operating conditions. For example, in the classic monograph in the bearing analysis presented by Harris [13], the static analysis model with inner ring tilting is introduced, but it is limited by the model assumptions and cannot carry out the study of contact characteristics under the operating condition of the bearing. Li et al. [14] of Shanghai University performed a numerical simulation analysis of the contact stress in the tilting condition of roller bearings. Li et al. [15] of Xi'an Jiaotong University studied a 5-DOF model of angular contact ball bearings and analyzed the effect of non-uniformly distributed preload force on the contact state of ball bearings during spindle assembly. In addition, Zhang et al. [16] explored the changes in dynamic characteristics due to the outer ring tilting, based on a 3-DOF quasi-static model of rolling bearing, and numerically analyzed the effect of poor parallelism of the outer spacer ring on the vibration characteristics of some bearings. In addition, some scholars have carried out research on bearing-spindle system modeling and dynamic behavior. Jedrzejewski [17] presents modelling of moving sleeve and spindle tip displacements in spindle bearing systems equipped with angular contact ball bearings. The proposed model has been verified and can be used for the compensation of spindle tip displacements. Evidently, the above work further enriches the research on the characteristics of bearing systems. However, for precise analysis and prediction of thermal characteristics of the precision bearing-spindle system, it is necessary to establish a quasi-static model of bearing under the condition of outer ring tilting, which considers the influence of centrifugal force and gyroscopic moment of the rolling elements at the same time. In addition, limited by the bearing assembly process, experimental design and other factors, the current operating characteristics research of bearings under abnormal assembly and skewed rings, etc., mainly relies on theoretical analysis and prediction, and lacks systematic experimental analysis and verification.

Considering the real service environment, the assembly quality between components and the axial thermal deformation of components in the spindle system will lead to an increase in load imbalance of each rolling element, which becomes a factor that affects the heat generation of precision bearings that cannot be ignored. In this paper, based on the quasi-static theory, the motion and heat generation of ACBB under different working conditions and different assembly states were studied, and the quasi-static model of ACBB with outer ring tilting was established. Additionally, the contact angle, contact load and motion posture of the rolling elements were obtained, and the heat generation law of the bearing under complex conditions was analyzed. Furthermore, in order to verify the influence of the heat generation difference caused by the ring skew on the spindle system, the relative imbalance of the outer ring during the bearing service was experimentally simulated, based on the outer spacer ring with poor end-parallelism, and the thermal characteristics of the spindle system were monitored by relying on Xi'an Jiaotong University's NSK precision spindle system experimental platform and the spindle rotation error analysis under different working conditions was experimentally studied. Through theoretical and experimental comparison, the influence of different skew degrees of rings on bearing heat generation and spindle rotation accuracy was discussed.

2. Static Analysis Model of Ball Bearing

This paper analyzes the position relationship of ACBB in the rectangular coordinate system. As shown in Figure 1a, the origin *o* of the bearing coordinate system *xyz* is on the center of the inner ring groove curvature center trajectory circle, the *x*-axis coincides with the bearing centerline, the *y*-axis is radial, and the *z*-axis is determined by the right-hand rule. Simultaneously, the ball azimuth angle is defined as the clockwise angle with the *y*-axis in Figure 1b, and the azimuth angle of the *j*th ball is given.

$$\varphi_j = 2\pi(j-1)/Z \tag{1}$$

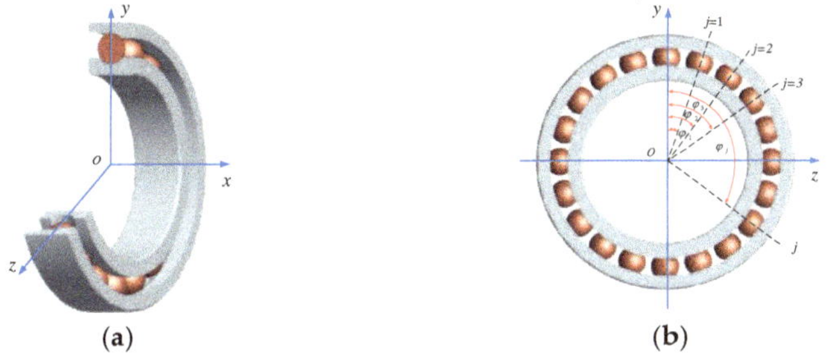

Figure 1. (**a**) The coordinate system of ACBB; (**b**) the position angle φ_j of each ball.

When the bearing operates without being externally loaded, the initial distance between the inner ring's raceway groove curvature center and the outer ring's is A_o, as presented in Figure 2a, which can be calculated as follows:

$$A_o = (f_i + f_o - 1) \times D_b \tag{2}$$

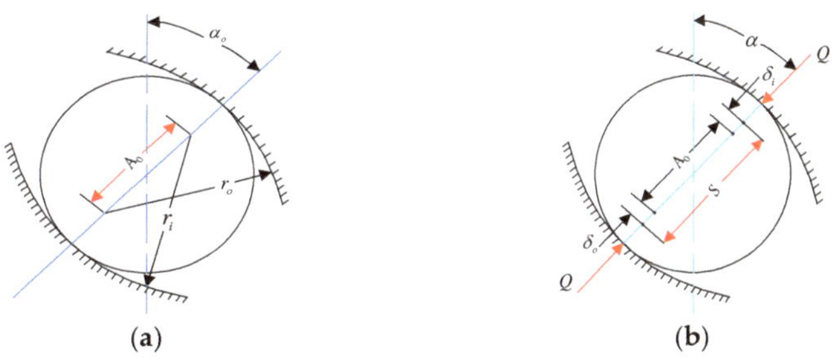

Figure 2. The schematic diagram of ball-raceway contacts relationship: (**a**) the bearing is not externally loaded in the initial state; (**b**) the bearing is externally loaded by the normal force.

Since the groove curvature center is relatively fixed, the ball is externally loaded by the normal force *Q* and the normal limit value of the raceway groove curvature center distance will change accordingly. The relationship formula of displacement and deformation is obtained from Figure 2b.

$$S = A_o + \delta_i + \delta_o \tag{3}$$

$$\delta_n = \delta_i + \delta_o = S - A_o \tag{4}$$

The inner ring and outer ring are misaligned due to the slightly poor assembly quality in the actual process of the bearing-spindle assembly, and the outer ring inclines relative to the inner ring (inclination angle is θ). Assuming that the inner ring is fixed in space, the outer ring moves respectively in axial and radial directions (the axial displacement is δ_a, the radial displacement is δ_r) in the radial plane of bearing, under known external combined axial and radial loads (the axial force is F_a, the radial force is F_r), and the trajectory of the bearing groove curvature center changes after being external loaded, as shown in Figure 3.

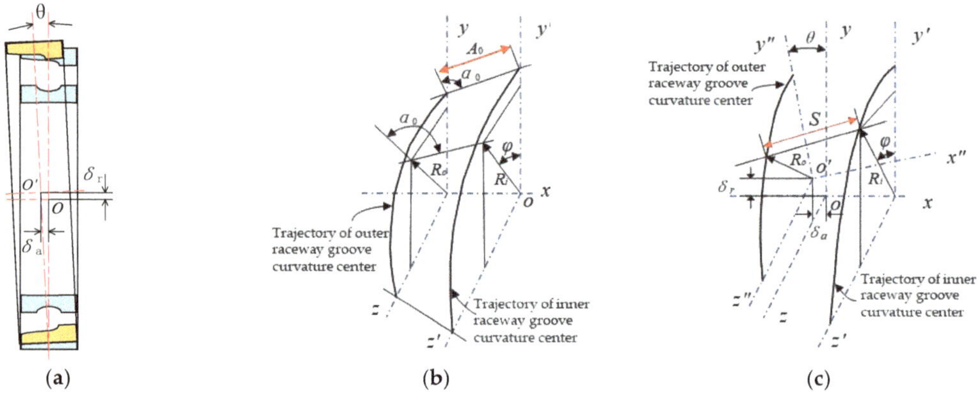

Figure 3. Change in the trajectory of raceway groove curvature center: (**a**) the diagram of outer ring tilt; (**b**) the raceway groove curvature centers under initial conditions; (**c**) the raceway groove curvature centers after loading.

The distance S_j between the ball at any position angle in the circumferential direction and the raceway groove curvature center of the bearing is calculated as follows:

$$S_j = \left[(A_o \sin\alpha_o + \delta_a + \theta R_o \cos\varphi_j)^2 + (A_o \cos\alpha_o - \delta_r \cos\varphi_j)^2\right]^{1/2} \quad (5)$$

where R_o is the groove curvature radius of the outer raceway.

$$R_o = \frac{d_m}{2} - (f_o - 0.5)D_b \cos\alpha_o \quad (6)$$

According to Equation (4), the total contact deformation between the ball and the raceways under the load is given as the following equation:

$$\delta_{nj} = S_j - A_o = A_o\left\{\left[\left(\sin\alpha_o + \overline{\delta_a} + R_o\overline{\theta}\cos\varphi_j\right)^2 + \left(\cos\alpha_o - \overline{\delta_r}\cos\varphi_j\right)^2\right]^{1/2} - 1\right\} \quad (7)$$

where $\overline{\delta}_a = \delta_a / A_o, \overline{\delta}_r = \delta_r / A_o, \overline{\theta} = \theta / A_o$.

According to the load–displacement relationship between the ball and the raceway,

$$Q = K_n \delta^{3/2} \quad (8)$$

Combining Equations (7) and (8), the contact force between the ball at any position angle φ_j and the raceway can be obtained by

$$Q_j = K_n {A_o}^{1.5}\left\{\left[\left(\sin\alpha_o + \overline{\delta_a} + R_o\overline{\theta}\cos\varphi_j\right)^2 + \left(\cos\alpha_o - \overline{\delta_r}\cos\varphi_j\right)^2\right]^{1/2} - 1\right\}^{1.5} \quad (9)$$

The contact angle α_j between the ball at any position angle φ_j and the raceway under static load can be described as follows:

$$\sin\alpha_j = \frac{\sin\alpha_o + \overline{\delta_a} + R_o\overline{\theta}\cos\varphi_j}{\left[\left(\sin\alpha_o + \overline{\delta_a} + R_o\overline{\theta}\cos\varphi_j\right)^2 + \left(\cos\alpha_o - \overline{\delta_r}\cos\varphi_j\right)^2\right]^{1/2}} \quad (10)$$

The normal load Q_j in Equation (9) applied to the raceway is decomposed into two components along the axial and radial, respectively, which are balanced with the external axial load F_a and radial load F_r, and the static equilibrium equations can be expressed as follows:

$$\begin{cases} F_a = \sum\limits_{j=0}^{Z} Q_j \sin\alpha_j \\ F_r = \sum\limits_{j=0}^{Z} Q_j \cos\varphi_j \cos\alpha_j \end{cases} \quad (11)$$

By substituting Equations (9) and (10) into the above equilibrium equations, for the bearing model considering outer ring inclination due to the misalignment between the inner and outer rings, the axial and radial displacements can be solved by the iterative method. After obtaining the displacement results, the ball-raceway contact load and contact angle at any position angle can be calculated accordingly, and the results will be used as the initial value of the quasi-static analysis.

3. Quasi-Static Analysis Model of Ball Bearing

The distance between the inner and outer raceway groove curvature centers is A_0, as shown in Figure 2a, in the absence of external load, while the distance is $A_0 + \delta_i + \delta_o$ under external loads and the raceway groove curvature centers are still collinear with the ball center, as shown in Figure 2b. However, as shown in Figure 4a, when the bearing works at high speed, the ball is subjected to complex loading combinations, including centrifugal force F_c caused by a high level of rotation, gyroscopic moment M_g induced by the spin of the ball, and ball-raceway normal contact forces Q_i and Q_o. Because of the centrifugal effect, the ball-raceway contact angle varies; consequently, the inner raceway groove curvature center moves and the raceway groove curvature centers will not be colinear with the ball center anymore, as shown in Figure 4bIn this article, the model has been developed imposing the inner ring is fixed while the outer ring displaces in the global coordinate system. Figure 4 shows the displacement relationship and as the ball center moves due to the centrifugal effect, the outer and inner raceway contact angles are no longer equal.

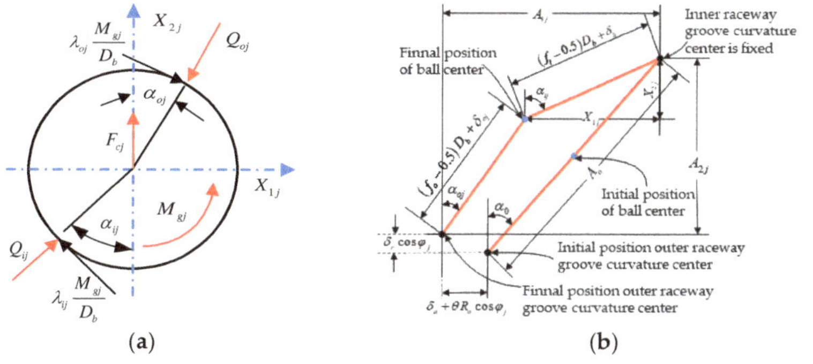

Figure 4. Mechanic and geometric relations: (**a**) the mechanic diagram of ball; (**b**) the geometric relations.

The final distances between the jth ball center and the raceway groove curvature centers, respectively, are described as

$$\Delta_{ij} = (f_i - 0.5)D_b + \delta_{ij} \tag{12}$$

$$\Delta_{oj} = (f_o - 0.5)D_b + \delta_{oj} \tag{13}$$

where δ_{ij} and δ_{oj} are normal contact deformation between the ball and the inner and outer raceways.

After analyzing the relative axial displacement δ_a, relative radial displacement δ_r, and relative inclination angle θ of the inner and outer ring, the axial distance A_{1j} and the radial distance A_{2j} between the inner and outer raceway groove curvature centers at position angle φ_j can be written as

$$A_{1j} = A_o \sin \alpha^\circ + \delta_a + R_o \theta \cos \varphi_j \tag{14}$$

$$A_{2j} = A_o \cos \alpha^\circ - \delta_r \cos \varphi_j \tag{15}$$

In order to facilitate the analysis, the variables X_1 and X_2 are introduced. Combining the Pythagorean theorem, the following equations can be deduced by analyzing the position relations in Figure 4b:

$$\begin{cases} (A_{1j} - X_{1j})^2 + (A_{2j} - X_{2j})^2 - [(f_o - 0.5)D_b + \delta_{oj}]^2 = 0 \\ X_{1j}^2 + X_{2j}^2 - [(f_i - 0.5)D_b + \delta_{ij}]^2 = 0 \end{cases} \tag{16}$$

In addition, it can be observed that at the position of the jth ball,

$$\begin{cases} \cos \alpha_{ij} = \frac{X_{2j}}{(f_i - 0.5)D_b + \delta_{ij}} \\ \sin \alpha_{ij} = \frac{X_{1j}}{(f_i - 0.5)D_b + \delta_{ij}} \\ \cos \alpha_{oj} = \frac{A_{2j} - X_{2j}}{(f_o - 0.5)D_b + \delta_{oj}} \\ \sin \alpha_{oj} = \frac{A_{1j} - X_{1j}}{(f_o - 0.5)D_b + \delta_{oj}} \end{cases} \tag{17}$$

When the centrifugal force F_c, gyroscopic moment M_g, and contact forces Q_i, Q_o act on the ball simultaneously, the forces applied to the ball in both the horizontal and vertical directions satisfy the following equilibrium relations:

$$\begin{cases} Q_{ij} \sin \alpha_{ij} - Q_{oj} \sin \alpha_{oj} - \frac{M_{gj}}{D_b}(\lambda_{ij} \cos \alpha_{ij} - \lambda_{oj} \cos \alpha_{oj}) = 0 \\ Q_{ij} \cos \alpha_{ij} - Q_{oj} \cos \alpha_{oj} + \frac{M_{gj}}{D_b}(\lambda_{ij} \sin \alpha_{ij} - \lambda_{oj} \sin \alpha_{oj}) + F_{cj} = 0 \end{cases} \tag{18}$$

As the ball bearing operates at high speed, the ball-outer race friction increases while the ball-inner race friction decreases, and the characteristics are close to the outer race control assumption, that is, the gyroscopic moment caused by the spin motion of the ball relative to the outer race will be prevented by the friction between the two (they are relatively pure rolling); therefore, $\lambda_{ij} = 0, \lambda_{oj} = 2$.

Then, the centrifugal force and the gyroscopic moment of the jth ball can be expressed as

$$F_{cj} = \frac{1}{2} m d_m \omega^2 \left(\frac{\omega_m}{\omega}\right)_j^2 \tag{19}$$

$$M_{gj} = J\omega^2 \left(\frac{\omega_b}{\omega}\right)_j \left(\frac{\omega_m}{\omega}\right)_j \sin \beta \tag{20}$$

where ω_m and ω_b denote the orbital revolution and spinning angular speeds of the ball, respectively. β is the ball pitch angle.

In order to determine the relations between the bearing displacements δ_a, δ_r and the external loads F_a, F_r, under the condition that the outer inclines relative to the fixed inner ring, the equilibrium equations of outer ring are listed as follows:

$$\begin{cases} F_a - \sum_j^Z \left(Q_{oj} \sin \alpha_{oj} - \lambda_{oj} \frac{M_{gj}}{D_b} \cos \alpha_{oj} \right) = 0 \\ F_r - \sum_j^Z \left(Q_{oj} \cos \alpha_{oj} + \lambda_{oj} \frac{M_{gj}}{D_b} \sin \alpha_{oj} \right) \cos \varphi_j = 0 \end{cases} \quad (21)$$

Meanwhile, the bending moment M of the bearing is calculated as follows:

$$M = \sum_j^Z \left[\left(Q_{ij} \sin \alpha_{ij} - \frac{\lambda_{ij} M_{gj}}{D_b} \cos \alpha_{ij} \right) R_i + \frac{\lambda_{ij} M_{gj}}{D_b} r_i \right] \cos \varphi_j \quad (22)$$

So far, the numerical analysis of quasi-static of the rolling bearing has been completed by solving Equations (16), (18), (21) simultaneously, the relevant characteristics parameters including ball-raceway contact angles and contact loads, axial and radial displacements of outer ring will be obtained.

4. Heat Generation Calculation of Ball Bearing

Taking angular contact ball bearing (NSK7014C) as research object, the specific working conditions that the external axial load F_a = 1400 N, external radial load F_r = 100 N and bearing speed n = 2000–8000 rpm are set. According to the comprehensive sample handbook for rolling bearing products offered by NSK Ltd., for the inclination of the outer ring due to the poor assembly quality, the allowable angle error for the installation tilt between the inner and outer rings of precision bearing is generally within 0.069°. In order to avoid abnormal working conditions, such as the obvious inclination with a large angle of the outer ring, this article focuses on the force and temperature characteristics of precision bearing in the case where the outer ring is slightly inclined. Therefore, the tilt angles of the outer ring are selected as 0°, 0.0052°, 0.0104°, 0.0156°, 0.0208° respectively.

The characteristics of the contact force between the bearing ball and the ring at different speeds are analyzed, and the results show that the variation law of the internal contact force with the bearing speed is very similar. Therefore, the bearing operation speed is set at 8000 rpm to analyze the ball-raceway contact angles and contact forces in the state of different tilt angles of the outer ring.

In order to characterize the ball forces and operating state at any position angle, the ball-raceway contact conditions are plotted in a polar coordinate system, the ball-raceway contact load distribution is shown in Figure 5, and the ball-raceway contact angle distribution is shown in Figure 6. The circumferential direction is the bearing ball number, and the radius direction is the simulation result value of each calculated quantity.

As shown in Figure 5, the polar coordinate circle diagram shifts to the right as a whole, as the outer ring inclination angle increases, that is, the ball-raceway contact forces increase along the radial direction of force applied to the outer ring. At the same time, it can be observed from the local magnification view that in the opposite direction of the radial force applied to the outer ring, the ball-raceway contact forces at first slightly decrease and then slightly increase, as the outer ring inclination angle increases. As shown in Figure 6, when the inclination angle of the outer ring is zero, the ball-raceway contact angles are approximately constant, presenting a regular circle located in the center of the polar coordinate system, while the polar coordinate circle diagram shifts to the right as a whole, as the outer ring inclination angle increases; the ball-raceway contact angles increase along the radial direction of force applied to the outer ring, whereas the contact angles decrease.

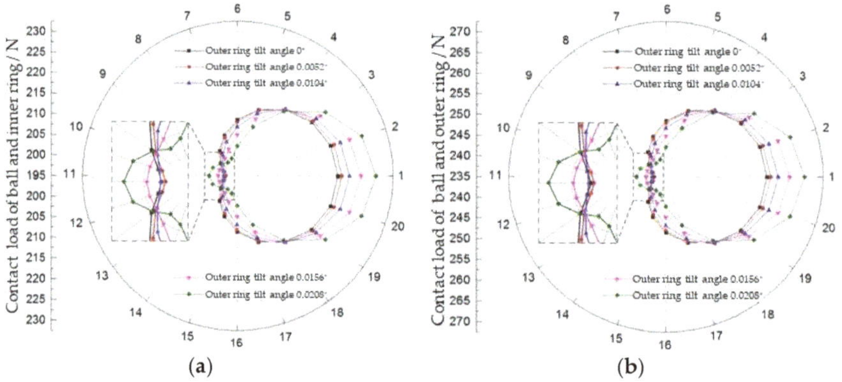

Figure 5. Ball-raceway contact load distribution: (**a**) ball and inner ring; (**b**) ball and outer ring.

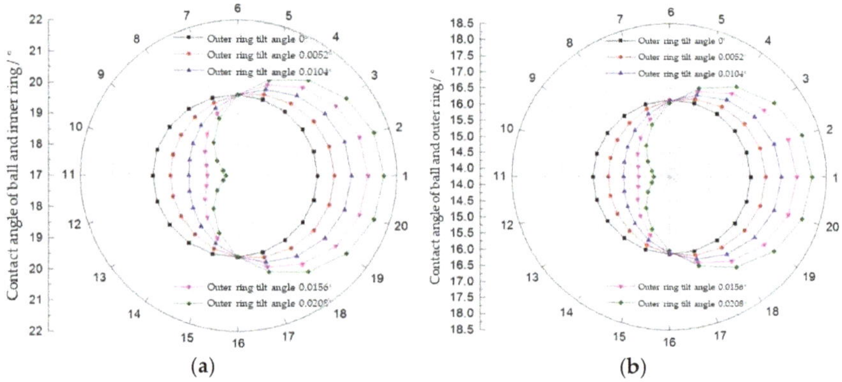

Figure 6. Ball-raceway contact angle distribution: (**a**) ball and inner ring; (**b**) ball and outer ring.

During the bearing operation, the energy loss caused by the internal friction is converted into heat generation. Therefore, the bearing heat generation power is calculated by the friction torque and the rotation speed. Subsequently, the influence factors of bearing friction torque mainly depend on the type, size, load, rotation speed, characteristics and supply of lubricant and so on. Firstly, Palmgren and other scholars performed a large amount of experimental studies, then analyzed the bearing thermal performance under the conditions of different loads with light, medium and heavy states, different lubricant methods and different lubricant media and finally derived the empirical formula for bearing friction torque. Therefore, the empirical formula has a reference function to some extent under the low-speed operation conditions, while the heat generation calculation needs to be improved under the high-speed operating conditions.

The empirical formula for bearing friction torque proposed by Palmgren contains the following two parts: the friction torque M_0 caused by lubricant viscosity and the friction torque M_1 caused by contact loads. However, due to the centrifugal effect during the bearing high-speed operation, the changes in ball-raceway contact characteristics occur, which leads to the ball-raceway contact friction torques' variation. At the same time, the friciton torque differs at different position angles. Therefore, in the calculation process, it is necessary to split the bearing friction torque into the inner and outer rings' components, then transform them into local components of local contact area at each position angle, demonstrated by the following equations:

$$M_{ij} = \frac{1}{Z} \times \frac{D_b}{d_o}\left[5 \times 10^{-11} f_0 (vn)^{2/3} d_m^3 + 5 \times 10^{-4} f_1 \left(\frac{F_{si}}{C_s}\right)^{0.33} F_{\beta i} d_m\right] \quad (23)$$

$$M_{oj} = \frac{1}{Z} \times \frac{D_b}{d_i}\left[5 \times 10^{-11} f_0 (vn)^{2/3} d_m^3 + 5 \times 10^{-4} f_1 \left(\frac{F_{so}}{C_s}\right)^{0.33} F_{\beta o} d_m\right] \quad (24)$$

Similarly, the frictional heat generation caused by the ball spinning motion cannot be negligible during the bearing high-speed operation, and the ball-raceway spin friction moments within the contact regions are given by

$$\begin{cases} M_{si} = \frac{3u_{si} Q_i a_i \varepsilon_i}{8} \\ M_{so} = \frac{3u_{so} Q_o a_o \varepsilon_o}{8} \end{cases} \quad (25)$$

where u_{si} and u_{so} denote the friction coefficient between the balls and inner and outer raceways, a_i and a_o represent the semi-major axes of the elliptical projected area of contact between the balls and inner and outer raceways; ε_i and ε_o are the complete elliptic integrals of the second kind, respectively.

The power loss is related to friction torque and rotation speed, the heat generation within every contact region can be obtained by multiplying friction torque and rotation speed and the total heat generation of bearing is given as

$$H = \sum_{j=1}^{Z} (H_{ij} + H_{oj}) = \sum_{j=1}^{Z} (w_{rolli} M_{ij} + w_{si} M_{si} + w_{rollo} M_{oj} + w_{so} M_{so}) \quad (26)$$

where w_{rolli} and w_{rollo} are the angular velocities of rolling of the ball with respect to the inner and outer race, w_{si} and w_{so} are the angular velocities of spin of the ball with respect to the inner and outer race, respectively.

w_{rolli} and w_{rollo} are expressed as follows:

$$w_{rolli} = w_i \frac{d_m}{D_b},\ w_{rollo} = -w_o \frac{d_m}{D_b} \quad (27)$$

where w_i and w_o are the relative angular velocities of balls with respect to the inner and outer raceways (bearing axis as a reference).

By the establishment of the model for heat generation prediction in angular contact ball bearing, the local and total heat generation of the bearing are obtained with different rotation speeds and diverse outer ring parallelism errors, so as to investigate the influence of outer ring inclination on bearing thermal characteristics. Figure 7 shows the effect of the outer ring inclination on the total heat generation of bearing as the rotation speed varies. That is, the heat generation of the bearing shows a curve change at different speeds. This is because as the speed increases, it can be observed from the empirical formula that the friction torque, caused by the viscosity of the lubricating medium, increases exponentially with the speed change, resulting in a change in the curve in Figure 7. Figure 8 shows the distribution of local heat generation of each ball at 8000 rpm.

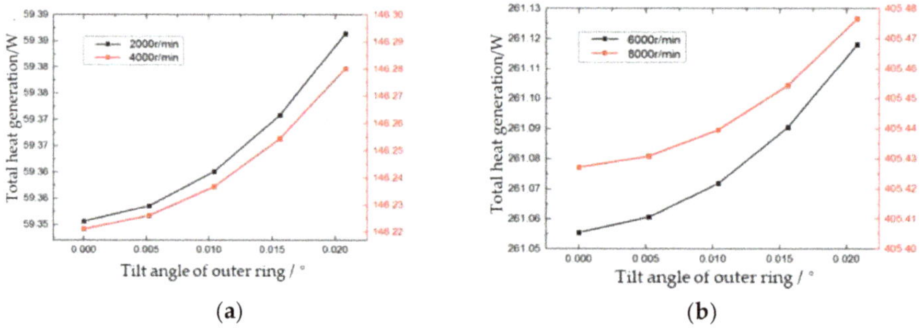

Figure 7. Effect of the rotational speed on the heat generation: (**a**) at bearing rotation speed of 2000 rpm and 4000 rpm; (**b**) at bearing rotation speed of 6000 rpm and 8000 rpm.

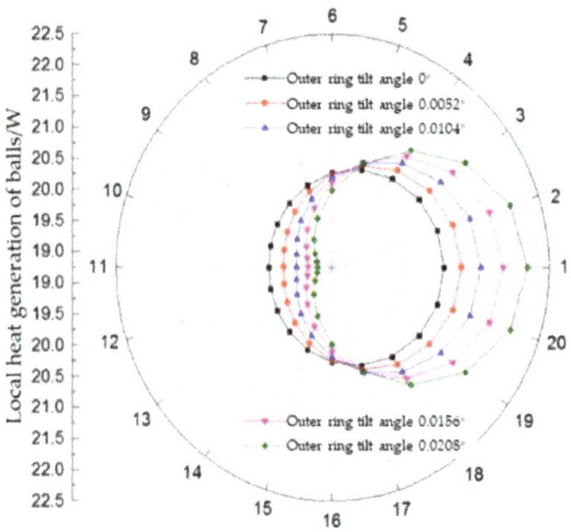

Figure 8. Local heat generation of balls at bearing rotation speed of 8000 rpm.

It can be observed from Figure 8 that the variation tendency of local heat generation of the ball is consistent with the variation tendency of the contact angles, which means that with the increase in the outer ring tilt angle, the heat generation in the radial direction increases. While the heat generation decreases in the opposite direction of the radial force, in the direction perpendicular to the radial force, it is constant. In addition, the heat generation of the angular contact ball bearing shows a slight increase with the increase in the outer ring tilt. The increment of total heat generation is small with the increase in tilt angle at the same speed and is almost unchanged, which indicates that the inclination of the outer ring has little impact on the total heat generation of the bearing. The total heat generation of the bearing is mainly affected by the rotation speed; as the speed increases gradually from 2000 rpm to 8000 rpm, the heat generation tends to increase progressively by double with each 2000 rpm increase, meanwhile, the increase rate slows down gradually.

When the bearing is subjected to external loads, the outer ring inclination will lead to uneven heat generation in the local contact area at each position angle inside the bearing. When the spindle runs at a constant speed of 8000 rpm, the relative tilt angle of the outer ring increases from 0° to 0.0208°, the heat generation at different locations differs by 2.7 times, and the heat generation in the upper half of the bearing is larger than that in the lower half.

5. Experiments and Discussion

It can be observed from the above calculation that, with the impact of outer ring non-parallelism, the heat generation at different positions in the circumferential direction inside the bearing differs significantly, and the difference is more pronounced as the rotation speed increases. However, due to the limit of testing technology for rotating components inside the bearing, it is difficult to verify the effect of the ring non-parallelism on heat generation through the temperature test of the inner ring directly, while it is also difficult for the traditional outer ring temperature measurement to reflect the difference in temperature rise directly, due to the influence of the complex heat transfer process inside the bearing. In this regard, this paper attempts to measure rotary accuracy variation induced by expansion and deformation of the bearing under different heat generation conditions, to indirectly verify the effect of ring non-parallelism on the thermal characteristics of the bearing.

5.1. Description of the Precision Bearing Test-Rig

The experiment platform in this study adopts the WZ15B90-30SE precision spindle provided by NSK Ltd. (Tokyo, Japan), the built-in tool shank specification for the BT40 broaching mechanism and the highest experimental rotation speed can reach 8000 rpm. The bearing type is 7014CTYNSULP4, grease lubricated (during the bearing operation, bearing grease and grease dosage were selected by referring to the bearing manual of NSK), and installed on the spindle in the configuration of DBB. The motor is an SJ-V7.5-03 ZM AC spindle motor manufactured by Mitsubishi Group, and the spindle is driven by the belt. The spindle thermal displacement test experiment selects the spindle error analyzer produced by Lion Precision, Oakdale, MN, USA, and the analyzer obtains the thermal displacement data of the front end of the spindle by measuring the displacement change in the standard ball (bar) fixed on the spindle front through the capacitive displacement sensor. The standard ball is 25.4 mm in diameter, 50 nm in maximum roundness error, 20 mm in diameter at the clamping end, clamped and fixed in the tapered hole at the front end of the spindle using a BT40 × ER32-70 tool holder. The capacitive displacement sensor test heads are respectively mounted in the −X, −Y, −Z directions in Figure 9, and the three directions indicate the horizontal radial, vertical radial, and axial coordinate directions, with the standard ball center as the coordinate center, which is consistent with the coordinate system setting of the previous theoretical analysis in this paper. The C8-2.0 capacitive displacement transducer is set to the standard measurement range (250), the bandwidth is set to 1 kHz (resolution 1.4 nm), and the sampling frequency is 360 Hz. Before the formal test, the standard ball is clamped to the front end of the spindle by the tool holder, and then the displacement transducer is installed at the corresponding measurement position and adjusted to the center on the standard ball using the fine adjustment mechanism.

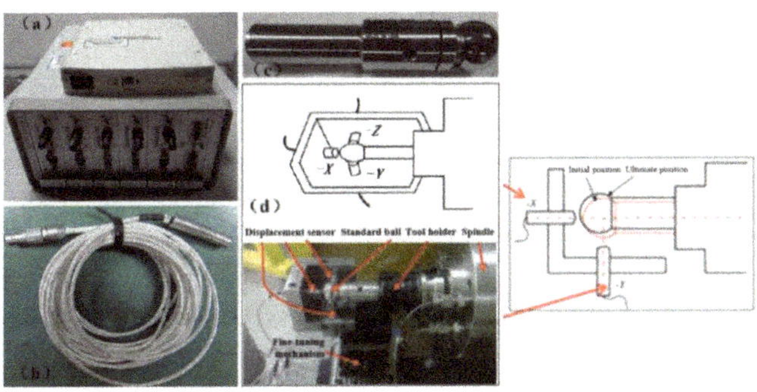

Figure 9. Spindle error analyzer equipment and the sensor installation: (**a**) spindle error analyzer; (**b**) capacitance displacement sensor; (**c**) standard ball (stick); (**d**) displacement sensor installation diagram.

5.2. Analysis of Trajectory and Accuracy Error of Spindle Rotation

The test simulates the different outer ring balance degrees of the bearing by adjusting the end face parallelism error of the short outer spacer inside the mechanical spindle system, and then investigates its effect on the performance of the shaft system. In the experimental simulation, bearing spacers with five tilting angles of 0°, 0.0052°, 0.0104°, 0.0156°, 0.0208° were machined, respectively.

After the spindle rotates and the temperature data reach a stable value, the radial X and Y channel displacement signals are collected and recorded, and the spindle rotation trajectory is plotted at each speed, as shown in Figure 10.

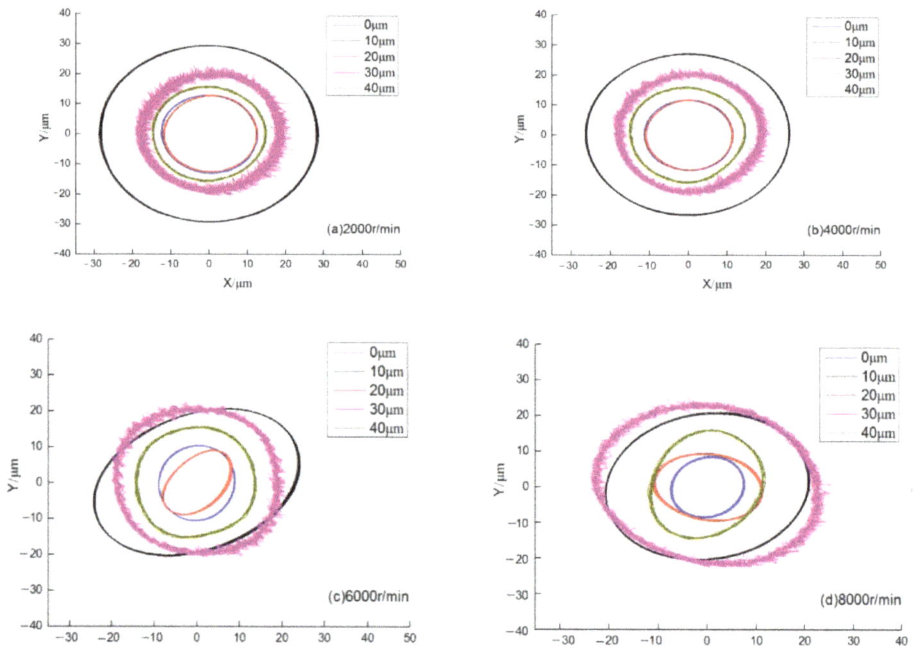

Figure 10. Effect of the spacer ring balance on the spindle rotation trajectory at different bearing rotational speeds: (**a**) 2000 rpm; (**b**) 4000 rpm; (**c**) 6000 rpm; (**d**) 8000 rpm.

As can be observed from Figure 10, the spindle rotation trajectory is in the best condition when the outer ring is not tilted. When the outer spacer parallelism is poor, the spindle rotary trajectory becomes worse, both in terms of roundness and radial runout. The TIR (total indicated runout) value of the spindle front end in both radial directions increases, the rotary trajectory expands, and the width of the spindle front end runout trajectory becomes larger accordingly, which shows that the outer spacer parallelism error causes uneven heat generation inside the bearing and tilt deformation, and immediately results in an increase in the NRRO (non-repeatable runout) value at the spindle front end. However, it can be observed that the slewing accuracy does not completely increase with the increase in the outer ring unbalance, and the slewing accuracy shows the phenomenon of interval runout. In addition, it can be observed from the above analysis that in the heat generation calculation of different inclinations of the bearing outer ring, the uneven heat generation of the bearing is more obvious as the inclination increases, which shows a certain linear relationship. However, in the experiment of spindle rotation accuracy under different outer ring inclinations, it is found that the rotation accuracy of the spindle is the best when the outer ring is not inclined. However, with the increase in outer ring inclination, the rotation accuracy does not have a similar linear relationship. This is because, for a precision spindle, its rotation accuracy is jointly affected by many factors, such as thermal deformation of the bearing, shaft structure, and modal frequency, etc.

The quantitative analysis results of the spindle-related motion errors can also be obtained from the rotary error calculation, as shown in Figure 11 below, which compares the spindle motion error values under each spacer accuracy type.

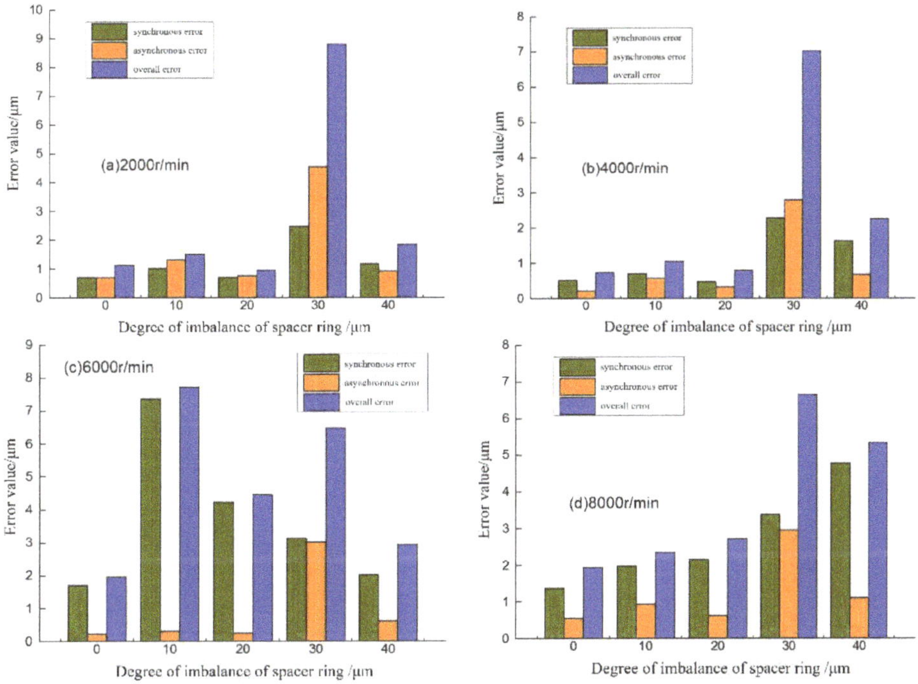

Figure 11. Effect of the spacer ring balance on the spindle rotation accuracy error at different bearing rotational speeds: (**a**) 2000 rpm; (**b**) 4000 rpm; (**c**) 6000 rpm; (**d**) 8000 rpm.

As can be observed from the above figure, the spacer parallelism error causes uneven heat generation in the circumferential direction of bearing and tilt deformation, which will lead to the deterioration of rotational accuracy of the spindle front end, and the impact is relatively small at the rotation speed of 2000 rpm and 4000 rpm, while the impact is great at 6000 rpm and 8000 rpm. In contrast, the outer ring unbalance has a greater impact on the spindle synchronization error, especially at higher speeds. However, further analysis shows that the rotation accuracy of the spindle front end is worse than other working conditions with the 0.0156°(30 microns) inclined spacer and at 6000 rpm, which may be because at this moment, the spindle is close to its natural vibration mode, resulting in inconsistency with other rotational speeds. At the rotation speeds 2000 rpm, 4000 rpm, and 8000 rpm, it can be observed that as the inclination of the bearing outer ring gradually increases from 0 to 0.0052°(10 microns), 0.0104°(20 microns) and 0.0208°(40 microns), the rotation accuracy of the spindle deteriorates, which indirectly verifies the effect of bearing ring inclination on its internal heat generation.

6. Conclusions

In view of the skew of the spindle bearing ring in the machining and assembly process, this paper carried out research work from both theoretical analysis and experimental verification, and the conclusions are as follows:

(1) Under a certain external load, the outer ring's non-parallelism will lead to uneven heat generation in the local contact area at each angle position of bearing. With the increase in the outer ring's non-parallelism, the difference in heat generation at different angular positions will increase;
(2) The experiments show that the parallelism error of the spacer ring causes uneven circumferential heat generation and tilt deformation of the bearing, which will lead to the deterioration of spindle bearing rotational accuracy at high speed and in particular, the synchronization error of the spindle will be more obvious.

Author Contributions: Conceptualization, B.Y., X.Z. and Z.Z.; methodology, B.Y.; software, X.Z.; validation, B.Y., X.Z. and Z.Z.; formal analysis, B.Y. and X.Z.; investigation, B.Y. and Z.Z.; resources, B.Y. and Z.Z.; data curation, X.Z. and B.Y.; writing—original draft preparation, B.Y.; writing—review and editing, X.Z. and Z.Z.; visualization, B.Y.; supervision, B.Y.; project administration, B.Y.; funding acquisition, B.Y. All authors have read and agreed to the published version of the manuscript.

Funding: This research was funded by the National Natural Science Foundation of China, grant number 52105242.

Institutional Review Board Statement: Not applicable.

Informed Consent Statement: Not applicable.

Data Availability Statement: https://doi.org/10.1177/0954406218802324 (accessed on 8 August 2022).

Conflicts of Interest: The authors declare no conflict of interest.

References

1. NSK. Machine Tool Spindle Bearing Selection and Mounting Guide. Available online: http://www.nsk-literature.com/en/machine-tool-spindle-bearings/ (accessed on 1 January 2020).
2. Cui, G.W.; Gao, D.; Yao, Y.X. Thermal error separating and modeling for heavy-duty CNC machine tools. *J. Harbin Inst. Technol.* **2012**, *44*, 51–56.
3. Chen, J.B. Research on Thermal-Mechanical Coupling Modeling and Optimal Design Method of High-Speed Electric Spindle. Ph.D. Thesis, Xi'an University of Technology, Xi'an, China, 2021.
4. Wu, L. Thermal Characteristics Study of CNC Machine Tool Spindle–Bearing System. Ph.D. Thesis, Jilin University, Changchun, China, 2016.
5. Ma, Z.K. Research on Dynamic Behaviors of Ball Bearing Based on Quasi-Static Method. Ph.D. Thesis, Zhejiang University, Hangzhou, China, 2010.
6. Ma, S.J.; Zhang, X.H.; Yan, K.; Zhu, Y.S.; Hong, J. A Study on Bearing Dynamic Features under the Condition of Multiball–Cage Collision. *Lubricants* **2022**, *10*, 9. [CrossRef]
7. Wang, W.Z.; Hu, L.; Zhang, S.G.; Zhao, Z.Q.; Ai, S.Y. Modeling angular contact ball bearing without raceway control hypothesis. *Mech. Mach. Theory* **2014**, *82*, 154–172. [CrossRef]
8. Wang, W.Z.; Hu, L.; Zhang, S.G.; Kong, L.J. Modeling high-speed angular contact ball bearing under the combined radial, axial and moment loads. *Proc. Inst. Mech. Eng. Part C J. Mech. Eng. Sci.* **2014**, *228*, 852–864. [CrossRef]
9. Li, Z.; Guan, X.L.; Zhong, R.; Wang, Q.S. Analysis of dynamic characteristics of angle contact bearings with combined loads. *Chin. J. Mech. Eng.* **2020**, *56*, 116–125.
10. Jones, A.B. A general theory for elasticaly constrained ball and radial roller bearings under arbitrary load and speed conditions. *J. Basic Eng.* **1960**, *82*, 309–320. [CrossRef]
11. Harris, T.A. An analytical method to predict skidding in high speed roller bearings. *ASLE Trans.* **1966**, *9*, 229–241. [CrossRef]
12. Ding, C.A.; Zhou, F.Z.; Zhu, J.; Zhang, L. Raceway control theory and determination of rolling element attitude angle. *Chin. J. Mech. Eng.* **2001**, *37*, 58–61. [CrossRef]
13. Harris, T.A. *Rolling Bearing Analysis*; Wiley: New York, NY, USA, 1984.
14. Li, W.W.; Chen, X.Y.; Shen, X.J.; Zhang, X.L. Contact pressure calculation and analysis of roller bearings in skewing condition. *China Mech. Eng.* **2011**, *22*, 2034–2039.
15. Li, X.; Li, H.; Hong, J.; Zhang, Y. Heat analysis of ball bearing under nonuniform preload based on five degrees of freedom quasi-static model. *Proc. Inst. Mech. Eng. Part J J. Eng. Tribol.* **2015**, *230*, 709–728. [CrossRef]
16. Zhang, H.; Zhang, J.H.; Zhu, Y.S.; Qiu, Z.H.; Wang, Y.W. Numerical analysis for impact of spacing ring nonparallelism on bearing performance. *J. Xi'an Jiaotong Univ.* **2014**, *48*, 86–90.
17. Jedrzejewski, J.; Kwasny, W. Modelling of angular contact ball bearings and axial displacements for high-speed spindles. *CIRP Annals.* **2010**, *59*, 377–382. [CrossRef]

lubricants

MDPI

Article

Thermal Dynamic Exploration of Full-Ceramic Ball Bearings under the Self-Lubrication Condition

Junxing Tian [1], Yuhou Wu [1,2], Jian Sun [1,*], Zhongxian Xia [1], Kexuan Ren [1], He Wang [1], Songhua Li [1] and Jinmei Yao [1]

[1] National-Local Joint Engineering Laboratory of NC Machining Equipment and Technology of High-Grade Stone, Shenyang Jianzhu University, Shenyang 110168, China
[2] School of Mechanical Engineering, Shenyang Jianzhu University, Shenyang 110168, China
* Correspondence: sunjian09@126.com

Abstract: A silicon nitride ceramic bearing has good self-lubricating characteristics. It still has a good operational status under the condition of a lack of oil. However, the temperature distribution of a silicon nitride ceramic bearing during its operation is unclear. To clarify the thermal distribution of a full-ceramic ball silicon nitride ceramic bearing under self-lubricating conditions, the changing trend of the rolling friction temperature between the rolling elements and channels with different accuracies is analyzed using the friction testing machine. The bearing heat generation model based on the silicon nitride material coefficient is established, and the life test machine measures the temperature of the bearing to verify the accuracy of the simulation model. The results show that the friction temperature between the ceramic ball and channel decreases with the increase in ceramic ball level. With an increase in the ceramic ball pressure and temperature, the friction temperature rises. Under self-lubrication, when the bearing bears a heavy load, the influence of the rotating speed on temperature rise tends to decrease. Under the condition of high speed, with the increase in load, the change range of temperature rise shows an upward trend. The important relationship between the bearing's heat and bearing's load and speed is revealed. It provides some theoretical guidance for the thermal analysis of a silicon nitride ceramic ball bearing under the self-lubricating condition to improve the service life and reliability of full-ceramic ball bearings.

Keywords: full-ceramic ball bearing; self-lubricating; ceramic ball; heat generation analysis; conduct heat; friction torque; temperature distribution

Citation: Tian, J.; Wu, Y.; Sun, J.; Xia, Z.; Ren, K.; Wang, H.; Li, S.; Yao, J. Thermal Dynamic Exploration of Full-Ceramic Ball Bearings under the Self-Lubrication Condition. *Lubricants* **2022**, *10*, 213. https://doi.org/10.3390/lubricants10090213

Received: 31 May 2022
Accepted: 30 August 2022
Published: 6 September 2022

1. Introduction

Engineering ceramic materials, such as silicon nitride (Si_3N_4), silicon carbide (SiC) and zirconia (ZrO_2), are widely used in high-precision fields such as the aerospace and military industries because of their excellent characteristics of wear resistance, corrosion resistance, high- and low-temperature resistance, high hardness, high strength, low density, low expansion coefficient and self-lubrication. Silicon nitride (Si_3N_4) stands out in engineering ceramics because of its higher hardness, wear resistance and light weight [1,2]. It is considered to be the best substitute for bearing steel. Full-ceramic bearings made of silicon nitride materials have the characteristics of high hardness, high wear resistance, being lightweight and high thermal stability. Therefore, full-ceramic ball bearings have great advantages in high-temperature, high-speed and acid corrosion environments [3,4]. Ordinary bearings must add lubricating oil and other media to improve their performance. Mohamed G. A. Nassef researched the superiority of graphene in enhancing the load-carrying capacity and damping of grease in rolling bearings [5]. The self-lubricating characteristic of silicon nitride helps to improve the operating characteristics of full-ceramic bearings in a poor oil state. However, due to the root problem of friction heat generation between materials, under the working conditions of a heavy load and high speed, the friction heat generation between various components in a ceramic bearing is the main reason affecting the formation of a thermal field in the working process of the bearing [6]. The study is on the effect of grease level on mechanical vibrations associated with the damaged rolling bearings.

The formation of the thermal field is directly related to the service performance and service life of the bearing. Because the working state of the bearing is complex, the formation of the bearing thermal field is closely related to its running state [7]. The thermal state of the bearing comes from the friction between the moving pairs of the bearing. The load working state amplifies the heat source and the external environment assists in the increase or decrease in the heat source. The formation of this heat has a great impact on the service performance of the bearing [8]. Sun Jian analyzed the factors that affect the surface quality of silicon nitride ceramics. It has been found that between friction depth and speed, the latter factor has a greater impact on friction force and the quality of the surface [9]. Ma S. analyzed the radial load applied to the bearing; the peak impact force is more sensitive to the bearing's speed changes [10]. Kannel et al. studied the bearing's rolling element by taking the infinite surface heat transfer as the premise and obtained the calculation method of the contact surface temperature of the rolling element [11]. Hannon obtained the formula that the corresponding size change of the bearing is determined by the change in the bearing's temperature gradient, and put forward a new friction heat generation model of the rolling bearing by analyzing the relationship between the internal component temperature of the bearing and the friction heat generation power [12,13].

It can be seen from the above that many well-known scholars have conducted in-depth research on bearing heat generation and obtained many results. However, the thermal analysis of bearings is mainly focused on metal bearings, while the thermal analysis of full-ceramic bearings is still relatively rare. Many scholars are studying the dry friction characteristics of silicon nitride ceramics and analyzing the friction changes formed by cage materials and ceramic crystal structures [14,15]. In the application of the bearing, the main reason for its heat generation is the mutual movement between the inner and outer rings of the bearing and between the cage and the rolling element, in which the load and rotation speed have an important impact on its temperature [16]. The surface roughness and roundness of the bearing's rolling element have a certain influence on the running heat of the bearing. The amount of heat produced by this movement depends on several factors, including the degree of roughness and roundness of the ceramic ball's surface and the bearing's inner and outer ring channels [17]. The inclination of the bearing's outer ring also has an important influence on the heat generation of the bearing [18]. The temperature distribution of the entire ceramic bearing will also be impacted by the frictional characteristics between the cage and the rolling body. Based on Palmgren's empirical formula, a thermal dynamic analysis model is established for full-ceramic bearings under self-lubricating conditions [19]. It is difficult to test the bearing under high-speed and heavy-load conditions. At the same time, it is difficult to measure the temperature of each part of the bearing. Thus, the analysis and prediction of the thermal changes between various bearing parts are necessary. Therefore, the analysis of the friction between the rolling element and the groove of the ceramic bearing is the basis of bearing heat generation. The bearing's heat under a high speed and heavy load is analyzed through the heat generation model and the test of the ceramic bearing, which is helpful to predict the service performance of full-ceramic ball bearings.

2. Experimental Section

2.1. Experimental Materials

In order to better reveal the influence of ceramic ball level and friction state on the thermal state of silicon nitride under the condition of self-lubricating of full-ceramic ball bearing, the friction temperature between silicon nitride ceramic ball and inner and outer rings was explored. This experiment used a ceramic ball and ceramic block with a channel for experimental analysis. The friction and wear tests of silicon nitride ceramic balls and silicon nitride ceramic blocks with different specifications were carried out. The characteristics of silicon nitride material are shown in Table 1. The level and specification of the ceramic ball are shown in Table 2.

Table 1. Material properties.

Density ($g.cm^{-3}$)	Young's Modulus (GPa)	Poisson's Ratio	Thermal Expansion Coefficient ($10^{-6}/°C$)	Specific Heat Capacity (J/(kg∗K))	Fracture Toughness ($MPa*m^{1/2}$)	Compressive Strength (MPa)
3.07–3.21	(20 °C) 310	0.27	3.5–3.9 (20–1000 °C)	711.7	7–12	4500

Table 2. Ceramic ball level and specification.

Level	Roughness (μm)	Roundness (μm)
G3	0.009	0.07
G5	0.013	0.1
G10	0.18	0.24

2.2. Characterization Methods

The test was carried out by a friction and wear testing machine (MFT-500, Rtec Instruments, San Jose, CA, USA). The parameters used in the friction test are shown in Table 3. The test process and ball clamp status are shown in Figure 1. In the experiment, the friction and wear tester recorded the friction coefficient; at the same time, a FLIR thermal imager (FLIR Systems, Inc., Wilsonville, OR, USA; Accuracy is ±1%) was used to detect and record the temperature. After the experiment, firstly, the detected temperature value was averaged. Then, the worn spherical surface was observed by a three-dimensional microscope (KEYENCE, Osaka, Japan). Next, the surface morphology of the Si_3N_4 ball after wear was analyzed by scanning electron microscopy (SEM) (FEI, Hillsboro, OR, USA), and the wear mechanism under different loads and rotation speeds was deduced. Then, based on the temperature and coefficient of the ceramic ball in the process of rolling friction, the bearing heat generation model was established. The model was used to predict the temperature rise state of full-ceramic bearings. Finally, the temperature change trend of the bearing is tested by the life test machine (Hangzhou Bearing Testing Center, Hangzhou, China). The test equipment of bearing is shown in Figure 2. The predicted bearing temperature is compared with the test temperature value to verify the correctness and accuracy of the model.

Table 3. Test grouping.

Group	The Precision of Silicon Nitride Ceramic Ball	Speed v (r/min)	Apply Load F(N)	Test Time (s)
N_i (i = 1 ... 8)	G3, G5, G10	500	5, 15, 25, 35, 45, 55, 65, 75	1800
F_i (i = 1 ... 9)	G3, G5, G10	100, 200, 400, 500, 600, 700, 800, 900, 1000	35	1800
G3, G5, G10	G3, G5, G10	500	35	1800

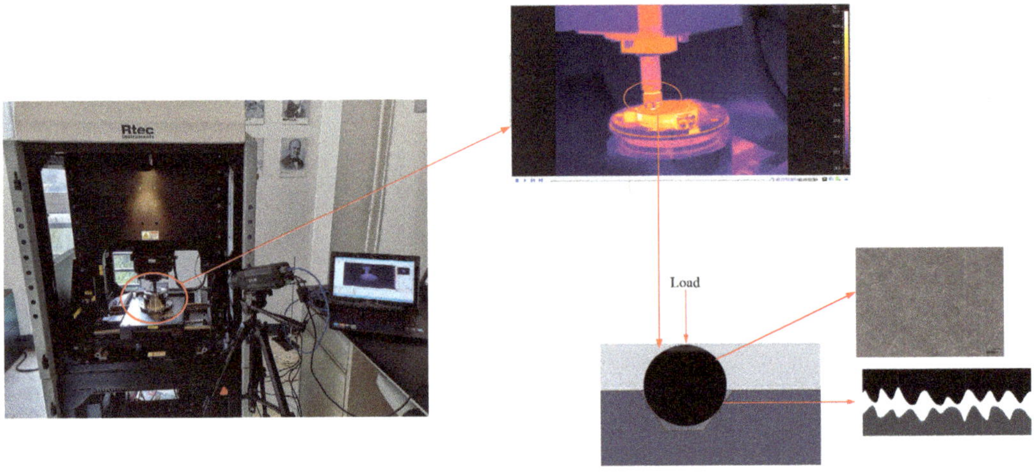

Figure 1. Friction experiment and the geometric contact of the roll with the channel.

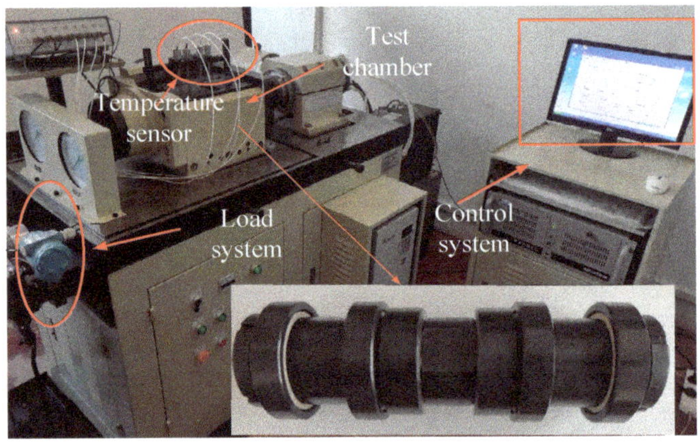

Figure 2. Bearing life strengthening testing machine.

3. Results and Discussion

3.1. The Analysis of Silicon Nitride Ceramic Ball Friction State

It can be seen from Figure 3 that the precision level of the ceramic ball is high and the friction temperature is low. The friction temperature of the G3 ceramic ball changes slightly. As the level of the ceramic ball decreases, the roughness decreases, and the friction temperature increases with the same load environment. This is due to the roughness increase, which leads to the increase in the friction force and temperature. Figure 3 shows the initial friction temperature of the ceramic balls is relatively high in the friction experiment process. With the progress of friction, the surface temperature fluctuates between 5 °C and 10 °C at G1 and G2, but the fluctuation range reaches about 40 °C at G10. Figure 4 shows that the friction coefficient of the three levels of ceramic balls is relatively high at the initial stage of friction. The coefficient of friction of the G3 level ceramic balls is relatively low, and the change is relatively smooth. The friction coefficient of the G10 ceramic balls is high, and the overall change is relatively large.

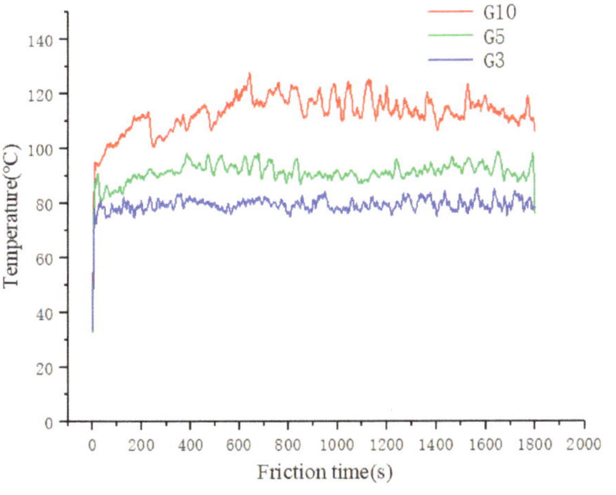

Figure 3. The friction temperature of the different level ceramic ball under the same conditions.

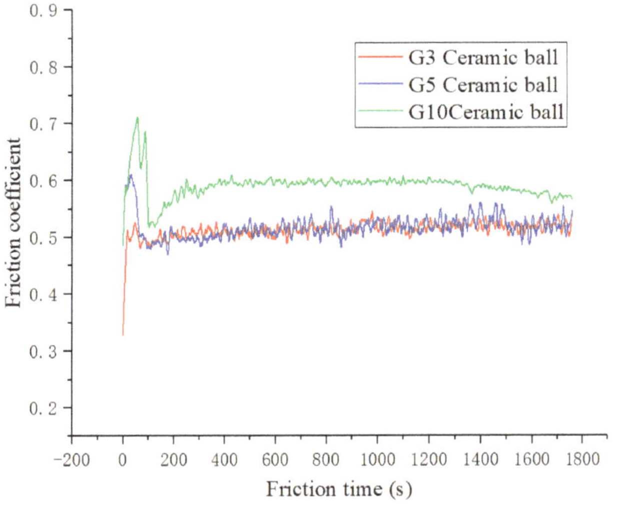

Figure 4. The friction coefficient of the different level ceramic ball under the same conditions.

The friction temperature changes with the transformation in the roundness of the ceramic ball. When the roundness of the ceramic ball is relatively high, the temperature formed by friction decreases, which means that the roundness of the ceramic ball is very high and the temperature rise conversion rate is relatively smooth. However, with the extension in friction time, the friction temperature of the ceramic ball with low roundness is relatively large.

Figure 5 shows that the temperature rise of the bearing in operation will increase with the increase in its load. The temperature rise of the ceramic ball bearings is relatively low, and the effect of load on the temperature rise tends to be linear. However, with the load increase, the temperature of the G10 ceramic ball changes significantly at the beginning, and then the temperature rise changes slowly. This is because the surface roughness of the G3 ceramic ball is relatively low and the surface is smooth. The friction force is small under the load condition and the friction temperature is low. However, the surface roughness of the G10 ceramic ball is high. In the initial stage of bearing load, it is affected by some wave peaks on the surface, which increase the friction force. The decrease in friction force reduces the range of temperature rise.

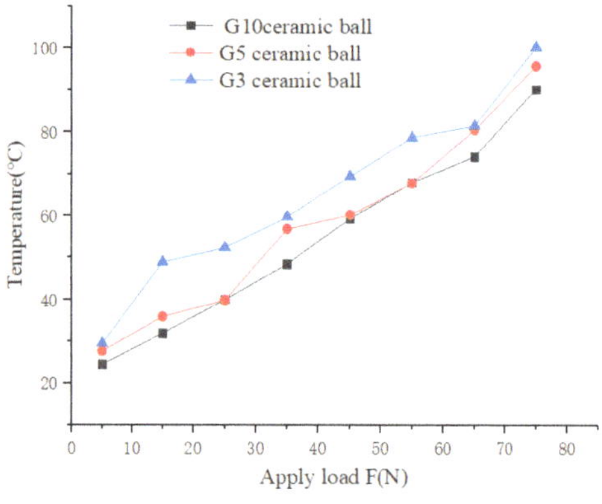

Figure 5. The change in the ceramic ball temperature under different pressures.

When the rotational speed increases, the bearing's temperature rises; however, as the level of the ceramic ball increases, there is less of a temperature shift, as shown in Figure 6. As the speed increases, the temperature of the bearing with G3 ceramic balls steadily rises. However, as the speed increases, the temperature rise of the bearing using G10 ceramic balls changes substantially. When the rotation speed of the bearings with G10 and G5 ceramic balls is 600 r/min, the temperature changes arrive at the turning point. When the speed is greater than 600 r/min, the temperature tends to slow down with the speed increase. This is because when the speed of the bearing's rolling body is low, the contact area between the rolling body and the raceway surface increases under the action of centrifugal force, leading to the aggravation of the friction phenomenon. Under certain load conditions, when the speed is greater than 600 r/min, partial peaks on the contact surface are reduced because of the centrifugal force of the friction force between the bearing's rolling element and the raceway contact surface. Thus the surface friction force is reduced. As a result, the effect of the speed on the temperature is reduced.

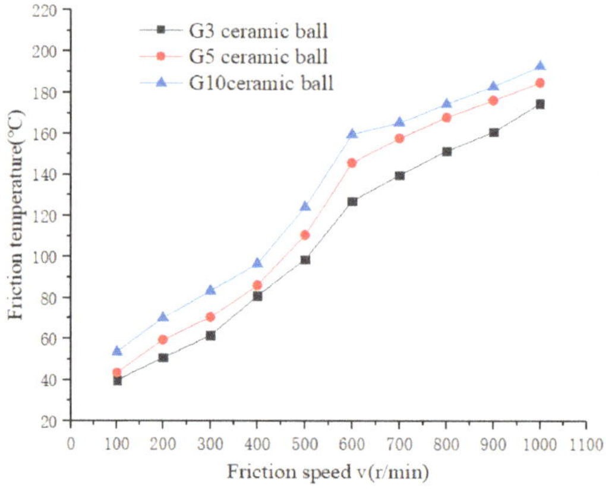

Figure 6. The temperature change of ceramic ball under different rotational speeds.

As shown in Figure 7, the surface roughness of the ceramic ball increases significantly. There are grind marks on the ball's surface after the friction experiment. The appearance of wear marks directly affects the friction and temperature. It can be observed from Figure 8b,c that when the ceramic ball is under the same speed and pressure conditions, the friction marks on the surface of the G3 ceramic ball are relatively shallow and the wear marks of G5 and G10 are deeper. This is because the contact surface of a high-precision ceramic ball is relatively smooth and the friction is relatively small. The surface wear traces and the wear defects of the G10 ceramic ball are increased when the balls are under heavy-load conditions. μm

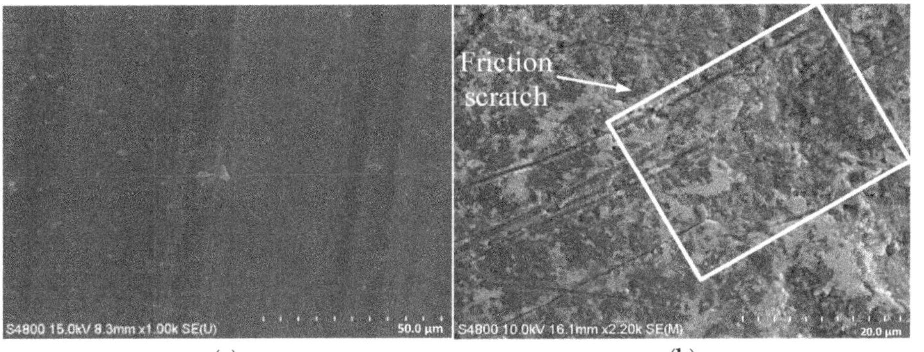

Figure 7. Surface morphology before and after the experiment. (**a**) G5 Ceramic ball surface before testing, (**b**) G5 Ceramic ball surface after testing.

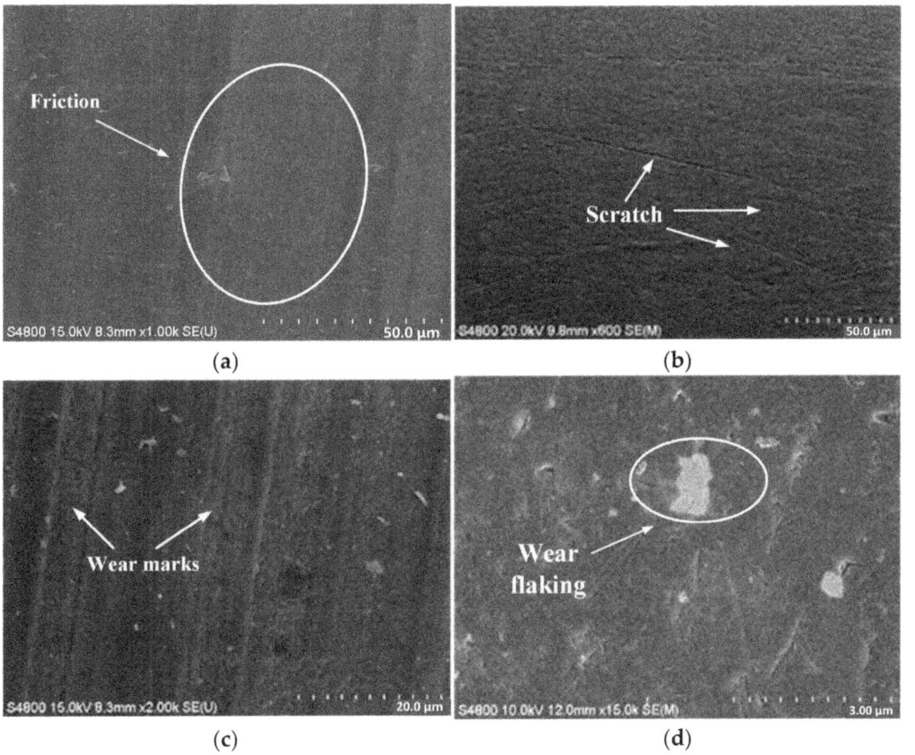

Figure 8. *Cont.*

Figure 8. Surface morphology before and after the experiment.(**a**) G3 V = 500 r/min, F = 35N, (**b**) V = 500 r/min, F = 35N, (**c**) G10 V = 500 r/min, F = 35N, (**d**) G10 V = 100 r/min, F = 75N, (**e**) G10 V = 500 r/min, F = 75N, (**f**) G10 V = 1000 r/min, F = 75N.

3.2. Dynamic Model of Ceramic Bearing under Self-Lubrication

Bearings are components that can bear loads and rotate at the same time. During the rotation, the rolling elements will revolve around the central axis of the bearing; at the same time, they will also rotate around their axis [20]. To better analyze the heat generation state during the bearing's operation, it is assumed that there is no relative displacement between the shaft and the bearing's inner ring when establishing the model; and the slight influence of contact deformation and rotational speed on inertial force and contact friction force is not considered.

Figure 9 shows the running shape of the internal components of the ball bearing during the movement. The motion state of the rolling element of the bearing has a certain influence on the frictional heat. The rolling element in the bearing has a self-rotation around its axis and a revolution around the center of the inner ring.

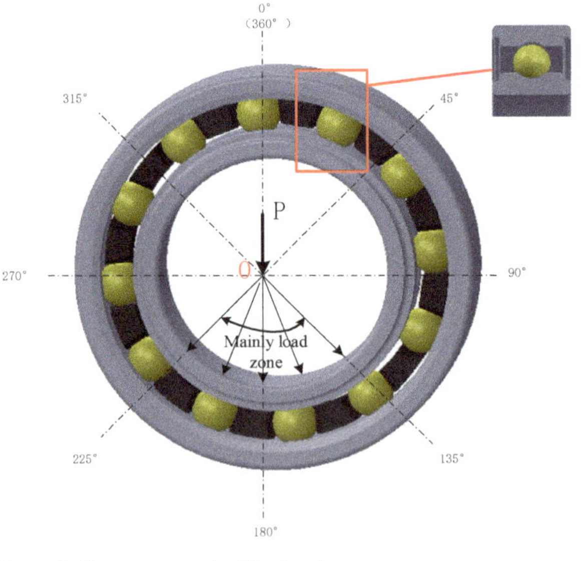

Figure 9. The structure of rolling bearing.

The orbital speed of the rolling is

$$V_m = \frac{1}{2} \cdot \frac{2\pi D_{pw} n_c}{60} \tag{1}$$

D_{pw} is the diameter of the bearing's pitch circle, and n_c is the revolution speed of the rolling body.

The revolution speed of the rolling body can be obtained from Formula (1)

$$n_c = \frac{1}{2}[n_i(1-r) + n_e(1+r)] \tag{2}$$

n_i is the speed of the inner ring of the bearing, n_e is the speed of the outer ring of the bearing, $r = D_m \cdot cos\alpha / D_{pw}$ and α is the contact angle of the bearing.

Because the rotational speed of the rolling body is equal to the relative speed of the inner ring in the contact area, the rotational speed of the bearing's rolling body can be given as

$$n_m = \frac{D_{pw}}{2D_m}(n_e - n_i)(1+r)(1-r) \tag{3}$$

where D_m is the diameter of the rolling body.

Combined with Formula (3), it is obtained that the revolution angular velocity of the rolling body is

$$W_c = \frac{\pi}{60}[n_i(1-r) + n_e(1+r)] \tag{4}$$

The following expression can determine the centrifugal force and gyroscopic moment of the rolling body

$$\begin{cases} F_{ej} = 0.5 d_m m \omega_{cj}^2 \\ M_{gj} = J\omega_{bj}\omega_{cj} sin\beta_j \end{cases} \tag{5}$$

K_e and K_i are the load deformation coefficients of the rolling body, and outer and inner ring.

$$Q_{ij} = K_{ij}\delta_{ij}^{3/2};\ Q_{ej} = K_{ej}\delta_{ej}^{3/2} \tag{6}$$

With the force balance of the bearing's rolling body under the combined load, the following force balance equation can be obtained

$$F_a - \sum_{j=1}^{z} Q_{ij} sin\alpha_{ij} = 0 \tag{7}$$

$$F_r - \sum_{j=1}^{z} Q_{ij} cos\alpha_{ij} = 0 \tag{8}$$

$$M - \sum_{j=1}^{z} \frac{d_m}{2} Q_{ij} cos\alpha_{ij} = 0 \tag{9}$$

F_a is the axial preload acting on the bearing; F_r is the radial preload; M is the overturning moment.

The bearing's rolling body is contacted with the inner and outer ring and the cage, and the contact situation is shown in Figure 10. Due to the role of bearing clearance, there is an eccentricity in the rotation center of the inner and outer ring as shown in Figure 9. The existence of the eccentricity of O_0O_1 in Figure 10 makes the force acting on the rolling body different in one rotation period [21]. It can be analyzed from Figure 8 that the rolling body is subjected to a relatively large load in the 180° area of the bearing. There is close contact between the inner and outer ring and the rolling body so that the temperature rise of the bearing here is maximum, as shown in Figure 9. On the other hand, when the bearing is located at 0° (360°), the contact between the rolling body and the groove is relatively small, and the load is the smallest so that the temperature of the bearing in this area is relatively low as shown in Figure 9.

$$\begin{cases} Q_{ij}sina_{ij} - Q_{ej}sina_{ej} + F_{ij}cosa_{ij} - F_{ej}cosa_{ej} = 0 \\ Q_{ij}cosa_{ij} - Q_{ej}cosa_{ej} + F_{ij}sina_{ij} - F_{ej}sina_{ej} + F_{ej} = 0 \\ F_{ej} + F_{ij} - 2M_{gj}/D = 0 \end{cases} \quad (10)$$

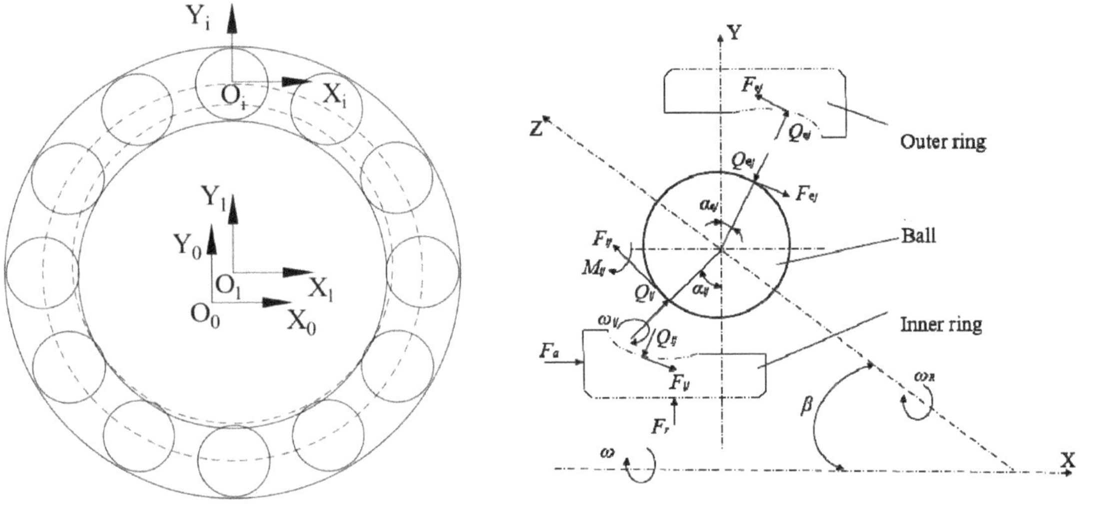

Figure 10. Bearing mechanics diagram.

Bearing heat generation is mainly the rolling body and bearing inside and outer ring friction heat generation. Palmgren summarized the calculation formula of the friction torque of the bearing under unload and load through many experiments [19].

The frictional moment generated by lubricant viscosity during the bearing's empty load is shown by:

$$n < 2000$$
$$M_0 = 160 \times 10^{-7} f_0 D_{pw}^3 \quad (11)$$
$$\nu n \geq 2000$$
$$M_0 = 10^{-7} f_0 (\nu n)^{2/3} D_{pw}^3 \quad (12)$$

f_0 is the coefficient related to bearing type and lubrication, ν is the lubricant's dynamic viscosity, and n is the bearing's speed.

For the load-dependent friction moment, M_1

$$M_1 = f_1 P_1 D_{pw} \quad (13)$$

f_1 is the coefficient related to the bearing's type and load, and P_1 is the bearing's equivalent dynamic load.

Therefore, the friction moment of the bearing formula is $M = M_0 + M_1$.

Because ceramic ball bearings are mostly used in high-speed rotation conditions, the spin sliding between the rolling body and the bearing's internal roller channel is also one of the main movements. The heat generated by spin friction is added on the basis of Formulas (11)–(13) so that the empirical formula is more accurate [22].

$$M_s = \frac{3\mu Q \alpha E_{(\eta)}}{8} \quad (14)$$

M_s is the spin friction moment of the rolling body, α is the bearing's roller contact long half shaft, μ is the friction coefficient between the bearing's roller body and rolling body, Q is the normal contact load between the bearing's rolling body and rolling body, and $E_{(\eta)}$ is the second type of elliptic integral of the roller contact zone.

Heat generated in the inner ring

$$H_i = 10^{-3} W_c \cdot M_i + 1.047 \times 10^{-4} M_{si} \cdot n_m \cdot z \tag{15}$$

Heat generated in the outer circle

$$H_e = 10^{-3} W_c \cdot M_e + 1.047 \times 10^{-4} M_{se} \cdot n_m \cdot z \tag{16}$$

z is the number of rolling bodies.

3.3. Heat Transfer Model

The bearing, rotating shaft and bearing seat are symmetrical rotation bodies. In the absence of radial torque load, the friction heat generated on the internal and outer circle remains unchanged along the circumference, and the friction heat generation and heat transfer model of the rolling body on any azimuth angle are similar [23]. Therefore, the approximate one-dimensional model is used to describe the heat transfer of the bearing.

Burton and Steph have found that the friction heat inside the bearing occurs only between the rolling body and the inner and outer ring of the bearing [24]. It is recommended that half of the heat generated go into the rolling body and half into the bearing ring. Figure 11 is the temperature node diagram of the key components of the internal bearing, and Figure 12 shows the model diagram of the bearing's heat transfer heat resistance network.

The heat transfer equations can be obtained, which include three unknown temperatures, T_{ce}, T_b and T_{ci}

$$\begin{cases} \frac{1}{2}H_i = \frac{T_{ci} - T_\infty}{R_i + R_s} + \frac{T_{ci} - T_{L\infty}}{R_{ci}} \\ \frac{1}{2}H_e = \frac{T_{ce} - T_\infty}{R_e + R_h} + \frac{T_{ce} - T_{L\infty}}{R_{ce}} \\ \frac{1}{2}H_i + \frac{1}{2}H_i = \frac{T_b - T_{L\infty}}{R_b} \end{cases} \tag{17}$$

T_{ci} and T_{ce} are the surface temperatures of the bearing's inner and outer channel, respectively. T_∞ is the external ambient temperature of the bearing. $T_{L\infty}$ is the bearing's internal air temperature. T_b is the rolling body temperature. R_i is the bearing's inner ring heat resistance. R_e is the bearing's outer ring heat resistance. R_{ci} is the convective heat resistance of the bearing. R_{ce} is the convective heat resistance of the bearing's outer surface. R_b is the rolling body surface facing the flow heat resistance. R_s is the rotating shaft heat resistance. R_h is the heat resistance for the bearing seat.

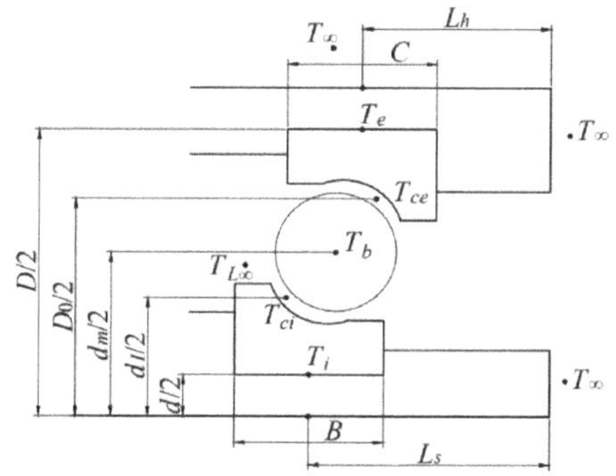

Figure 11. Temperature node diagram of key parts of bearing's internal components.

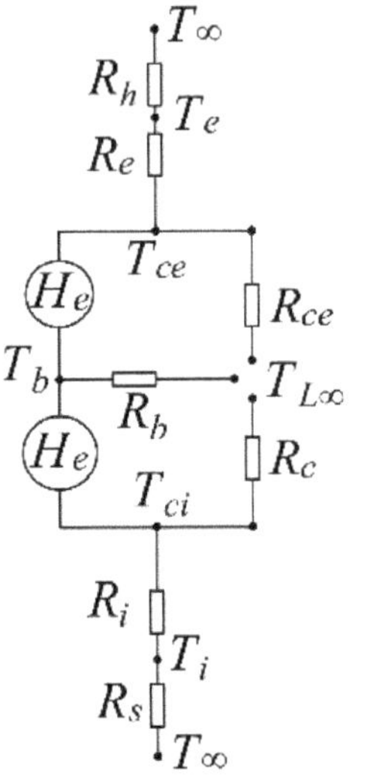

Figure 12. Thermal resistance network model of bearing's internal heat transfer.

3.4. Thermal Resistance of Full-Ceramic Bearing

For calculating the heat resistance of the bearing's inner and outer rings, the thickness of the bearing's inner ring and the outer bearing ring is much less than the overall width of the bearing, so the inner and outer bearing ring is regarded as a thin ring.

Under self-lubricating conditions, the air is the convection heat exchange medium around the bearing. The proposed average convection heat transfer coefficient between the internal bearing element and the heat exchange medium is

$$h = 0.0986\left[\frac{n}{v}\left(1 \pm \frac{d\cos\alpha}{D_{pw}}\right)\right]^{1/2} \tag{18}$$

In formula (18), "+" indicates the outer bearing ring rotation; "−" means the inner bearing ring rotation. *d* is the inner bearing diameter. *k* is the material.

3.5. Thermal Analysis and Experimental Study of the Silicon Nitride Full-Ceramic Bearing

To better analyze the silicon nitride bearing under self-lubrication, the full-ceramic silicon nitride angular contact ball bearing was used in the test. Its specifications and parameters are shown in Table 4, and the test apparatus is shown in Figure 13. The level of ceramic balls used in the ceramic bearings was G5. The characteristics of the bearing cage affect the heating of the bearing, but due to the diversity of the cage, the heat generation is calculated without considering the influence of the cage [25]. The working conditions of the theoretical simulation of bearings are shown in Table 5.

Table 4. Bearing parameters.

Parameters	Values
Inner diameter d/mm	35
Outer diameter D/mm	62
Rolling body diameter Dpw/mm	48.5
Ball diameter Dm/mm	7.938
Contact angle α/°	15
Number of rolling bodies z/psc	14

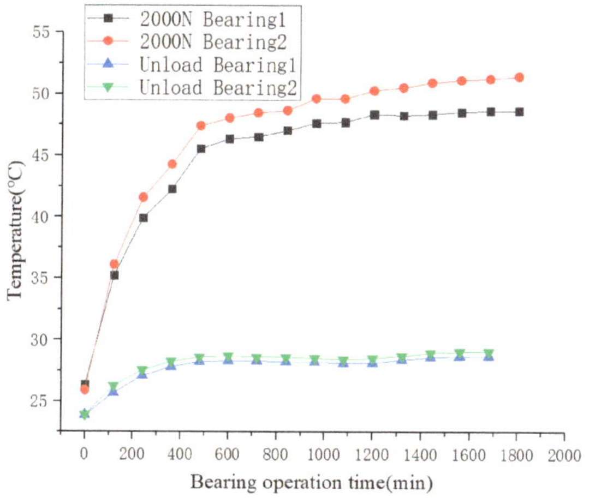

Figure 13. Relation between bearing's temperature rise and time.

Table 5. Working condition.

Working Condition	Values
Bearing's rotation speed/(r·min^{-1})	8000~24,000
Axial load/N	0~2500
Air Prandtl number	0.699
Environment temperature/°C	24

Palmgren's empirical formula is modified based on silicon nitride ceramics' properties. Thermodynamic analysis of the ceramic bearing under self-lubricating conditions is based on modified f_0 (as Formula (12)). To verify the correctness of the thermal model, Figure 13 studies the time when the temperature rise characteristics of bearings tend to a stable state under fixed working parameters. The bearing's operation must consider the steady-state temperature field [26]. Through experimental analysis, it can be seen that for the bearing to reach its steady-state operating temperature, the time is about 400 min when it runs under unloading, and the time is about 500 min when the bearing is under load. Therefore, the temperature of the bearing's operation after 500 min is selected in the test. Under the rotating speed of 8000 r/min, the load test of the bearing is carried out using the life test machine. At the same time, the temperature change is calculated by using the model. At the speed of 8000 r/min, the test temperature, model calculation temperature of the bearing and the error are shown in Table 6.

Table 6. Comparison of temperature results.

Load/N	Simulation Result (°C)	Test Result (°C)	Error Percentage (%)
0	27.697	30.2	8.29
500	30.941	32.9	4.80
1000	36.472	37.9	3.77
1500	43.041	44.9	4.14
2000	50.387	52.8	4.57

Figure 14 shows the surface's working temperature outside the bearing's outer ring. The stable measured temperature of the bearing is consistent with the trend of temperature change in theoretical analysis when the bearing's speed is 8000 r/min. The temperature test temperature in the low load area is slightly different from the theoretical analysis. This is because with the lack of application of an external load leads to roller rotation torque that cannot provide an accurate calculation; therefore, part of the thermal error is generated. With the increase in load, the error between the test and theoretical analysis temperature decreases.

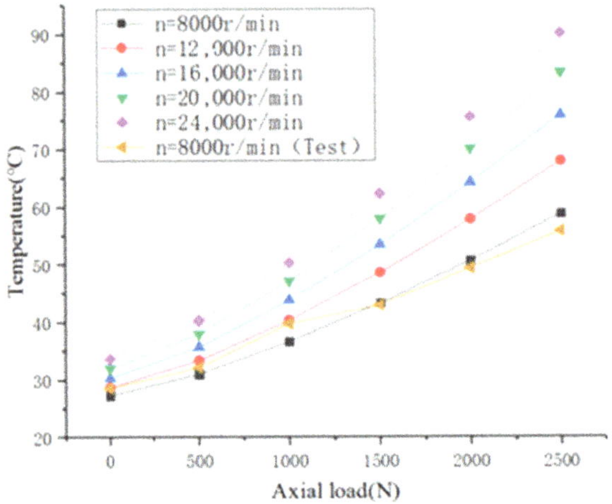

Figure 14. The surface temperature of the bearing's outer ring.

The simulation surface temperature of the bearing's outer ring increases with the load and the bearing's speed increase. When the speed is 24,000 r/min, The amplitude of the bearing's temperature rise is decreased. This is because under high-speed and heavy-load conditions, the friction coefficient and friction frequency between the bearing's rolling element and the inner and outer rings are increased. The increase in the friction coefficient and friction frequency leads to the increase in the bearing's temperature. When the load is 2500 N, the amplitude of the temperature increases is decreased with the increase in the rotation speed. This is because under a heavy load, the heat generated by the friction torque is the main heat, and the proportion of the heat generated by the rotation speed decreases. When the speed is 8000 r/min, the bearing's test and theoretical analysis tend to be consistent, but there is a slight fluctuation.

Figure 15 shows the curve of the operating simulation temperature of the bearing's inner raceway changing with speed and axial load. The variation law is the same as that of the operating temperature of the outer ring, but the temperature is higher than that of the outer ring [27]. This is because the diameter of the bearing's outer ring is larger than that of the bearing's inner ring, and the heat dissipation of the bearing's outer ring is better than that of the inner ring so that the friction heat between the rolling and the bearing's inner ring channels is easier to concentrate.

Figure 16 shows the working simulation temperature curve with the rotation speed and axial load changes. The rolling body temperature rises the highest because the rolling body is simultaneously in contact with the inner and outer channels. When the load is 2500 N and the speed is 24,000 r/min, the temperature of the bearing ball is higher than the inner ring and the higher temperature is above 40°. Under the condition of a high-speed bearing, the rolling body is heated in both directions, which is called the largest heat source in the bearing.

Through comparative analysis, it can be seen that under the same working conditions, the temperature of the bearing's rolling body is the highest, followed by the inner raceway, and the temperature of the outer ring is the lowest.

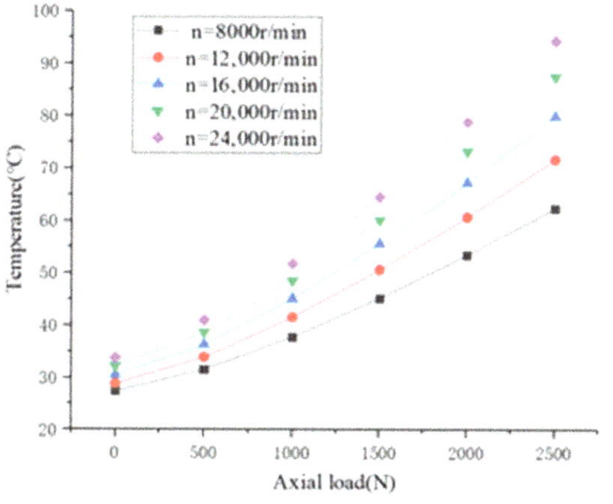

Figure 15. Raceway simulation temperature of bearing's inner ring.

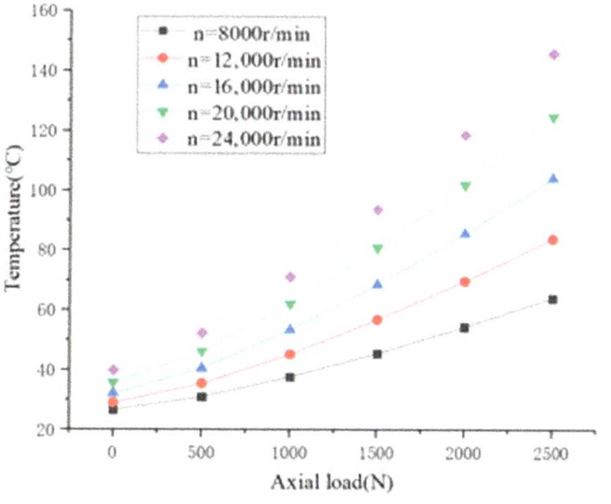

Figure 16. The temperature of rolling body simulation results.

4. Conclusions

Following are the conclusions that have been obtained after performing both the theoretical calculation and experimental comparison analysis of the self-lubricating heat generation of full-ceramic angular contact ball bearings:

(1) Through the analysis of the friction test, it can be seen that the high-grade rolling elements have a low friction coefficient and small friction torque in the process of operation, which leads to lower friction heat generation. During the friction process of G3 and G5 ceramic balls, the surface temperature fluctuation is small, between 5 °C and 10 °C. However, with G10, the fluctuation range reaches about 40 °C. Therefore, in the process of simulating bearing's heat generation, the bearing's rolling element's accuracy is considered.

(2) Based on the comparison between the calculated value of the heat generation model of the silicon nitride full-ceramic bearing and the test, when the axial load is 0 n, the error is the largest, reaching 8.29%. After loading the bearing, the minimum error is 3.77% when the axial load is 1000 N. Furthermore, the correctness of the heat generation model of full-ceramic silicon nitride bearings is explained.

(3) Under self-lubrication, the main heat source of full-ceramic silicon nitride ceramic ball bearings is the rolling element, followed by the inner ring, and finally, the outer

ring. Under heavy-load operation, the influence of speed on temperature rise tends to decrease. Under high-speed operation, the temperature rise increases with the increase in load. Therefore, the load borne by the bearing is the main factor affecting the bearing's temperature rise.

Author Contributions: Conceptualization, J.T. and Y.W.; methodology, J.S.; software, Z.X.; validation, K.R.; formal analysis, J.T.; investigation, H.W.; resources, Y.W.; data curation, J.T.; writing—original draft preparation, J.Y.; writing—review and editing, J.T.; visu-alization, Z.X.; supervision, S.L.; project administration, Y.W.; funding acquisition, Y.W. All authors have read and agreed to the published version of the manuscript.

Funding: This research was funded by the National Natural Science Foundation of China (NSFC) (grant numbers 52105196); National Defense Science and Technology Innovation Special Zone Program (grant numbers 20-163-00-TS-006-002-11); Young and Middle-aged Innovation Team in Shenyang (grant number RC210343); PhD Fund of Liaoning Province of China (grant number 2020-BS-159).

Institutional Review Board Statement: Not applicable.

Informed Consent Statement: Not applicable.

Data Availability Statement: Not applicable.

Conflicts of Interest: The authors declare no conflict of interest.

References

1. Li, S.; Wei, C.; Wang, Y. Fabrication and service of all-ceramic ball bearings for extreme conditions applications. *IOP Conf. Ser. Mater. Sci. Eng.* **2021**, *1009*, 12032. (In Romania) [CrossRef]
2. Xia, Z.; Wu, Y.; Wei, H.; Ren, K.; Gao, L.; Sun, J.; Li, S. Experimental Research on the Influence of Working Conditions on Vibration and Temperature Rise of Si3N4 Full-Ceramic Bearing Motors. *Shock Vib.* **2021**, *2021*, 1–16. (In Chinese) [CrossRef]
3. Ebert, F.J. An overview of performance characteristics, experiences and trends of aerospace engine bearings technologies. *Chin. J. Aeronaut.* **2007**, *4*, 378–384. (In Germany) [CrossRef]
4. Yan, H.P.; Wu, Y.H.; Li, S.H.; Zhang, L.X.; Zhang, K. The effect of factors on the radiation noise of high-speed full ceramic angular contact ball bearings. *J. Shock. Vib.* **2018**, *2018*, 1645878. (In Chinese) [CrossRef]
5. Nassef, M.G.A.; Soliman, M.; Nassef, B.G.; Daha, M.A.; Nassef, G.A. Impact of Graphene Nano-Additives to Lithium Grease on the Dynamic and Tribological Behavior of Rolling Bearings. *Lubricants* **2022**, *10*, 29. (In Egypt) [CrossRef]
6. Wang, Y.; Cao, J.C.; Tong, Q.B.; An, G.P.; Liu, R.F.; Zhang, Y.H.; Yan, H. Study on the thermal performance and temperature distribution of Ball bearings in the traction motor of a high-speed EMU. *Appl. Sci.* **2020**, *10*, 4373. (In Chinese) [CrossRef]
7. Shpenev, A.G. Friction and Wear of Fiber Composites with Abrasive Particles on Contact Surface. *J. Frict. Wear* **2018**, *39*, 188–194. (In Russia) [CrossRef]
8. Zhang, Y.; Chen, G.; Wang, L. Effects of thermal and elastic deformations on lubricating properties of the textured journal bearing. *Adv. Mech. Eng.* **2019**, *11*, 1687814019883790. (In Chinese) [CrossRef]
9. Sun, J.; Wu, Y.H.; Zhou, P.; Li, S.H.; Zhang, L.X.; Zhang, K. Research on the cracks extending mechanism of grinding silicon nitride ceramics. *Rev. Romana Mater. Rom. J. Mater.* **2018**, *48*, 198–203. (In Chinese)
10. Ma, S.; Zhang, X.; Yan, K.; Zhu, Y.; Hong, J. A Study on Bearing Dynamic Features under the Condition of Multiball–Cage Collision. *Lubricants* **2022**, *10*, 9. (In Chinese) [CrossRef]
11. Kannel, J.W.; Barber, S.A. Estimate of Surface Temperatures During Rolling Contac. *Tribol. Trans.* **1989**, *3*, 305–310. [CrossRef]
12. Hannon, W.M. Rolling-Element Bearing Heat Transfer—Part I: Analytic Model. *J. Tribol.* **2015**, *137*, 31102. [CrossRef]
13. Hannon, W.M. Rolling-Element Bearing Heat Transfer—Part III: Experimental Validation. *J. Tribol.* **2015**, *137*, 31104. [CrossRef]
14. Li, S.H.; Sui, Y.H.; Sun, J.; Xia, Z.X.; Wang, J.H. Analysis on Friction and Wear Characteristics and Transfer Film Formation Mechanism of Si_3N_4/PTFE under Dry Friction Condition. *Lubr. Eng.* **2022**, *47*, 53–59. (In Chinese)
15. Wu, C.; Yang, K.; Chen, Y.; Ni, J.; Yao, L.D.; Li, X.L. Investigation of friction and vibration performance of lithium complex grease containing nano-particles on rolling bearing. *Tribol. Int.* **2021**, *155*, 106761. (In Chinese) [CrossRef]
16. Jin, C.; Wu, B.; Hu, Y.M. Heat generation modeling of ball bearing based on internal load distribution. *Tribol. Int.* **2012**, *45*, 8–15. (In Chinese) [CrossRef]
17. Garrido, D.S.; Leventini, S.; Martini, A. Effect of Temperature and Surface Roughness on the Tribological Behavior of Electric Motor Greases for Hybrid Bearing Materials. *Lubricants* **2021**, *9*, 59. [CrossRef]
18. Li, X.; Lv, Y.; Yan, K.; Liu, J.; Hong, J. Study on the influence of thermal characteristics of rolling bearings and spindle resulted in condition of improper assembly. *Appl. Therm. Eng.* **2017**, *114*, 221–233. (In Chinese) [CrossRef]
19. Palmgren, A. *Ball and Roller Design Engineering*; Burkbank: Philadelphia, PA, USA, 1959.

20. Wu, Y.; Yan, H.; Li, S.; Zhang, K.; Zhang, L. Calculation on the radiation noise of ceramic ball bearings based on dynamic model considering nonlinear contact stiffness and damping. *J. Sound Vib.* **2020**, *479*, 115374. (In Chinese) [CrossRef]
21. Manzoor, S.; Wani, M.; Saleem, S. Effect of load on the friction and wear behaviour of silicon nitride and silicon nitride titanium carbide ceramic composite. *Mater. Today: Proc.* **2019**, *19*, 474–477. (In India) [CrossRef]
22. Jiang, X.Q.; Ma, J.; Zhao, L. Thermal analysis of high-speed precision angular contact ball bearings. *Bearing* **2000**, *8*, 1–4. (In Chinese)
23. Abdullah, O.I.; Schlattmann, J.; Majeed, M.H.; Sabri, L.A. The distribution of frictional heat generated between the contacting surfaces of the friction clutch system. *Int. J. Interact. Des. Manuf.* **2018**, *13*, 487–498. [CrossRef]
24. Burton, R.A.; Staph, H.E. Thermally Activated Seizure of Angular Contact Bearing, Asle Transactions. *Elsevier* **1967**, *10*, 408–417.
25. Wen, C.; Meng, X.; Fang, C.; Gu, J.; Xiao, L.; Jiang, S. Dynamic behaviors of angular contact ball bearing with a localized surface defect considering the influence of cage and oil lubrication. *Mech. Mach. Theory* **2021**, *162*, 104352. (In Chinese) [CrossRef]
26. Li, Z.F.; Shen, J.L.; Ji, J.W.; Liu, X.J. Numerical Simulation and Experimental Study on Steady-State Temperature Field of High Speed Angular Contact Ball Bearings. *Lubr. Eng.* **2021**, *46*, 45–50. (In Chinese)
27. Zhou, X.W.; Zhang, H.; Hao, X.; Liao, X.; Han, Q.K. Investigation on thermal behavior and temperature distribution of bearing inner and outer rings. *Tribol. Int.* **2018**, *130*, 289–298. (In Chinese) [CrossRef]

lubricants

MDPI

Article

Prediction of Remaining Service Life of Rolling Bearings Based on Convolutional and Bidirectional Long- and Short-Term Memory Neural Networks

Zhidan Zhong *, **Yao Zhao, Aoyu Yang, Haobo Zhang and Zhihui Zhang**

School of Mechatronics Engineering, Henan University of Science and Technology, Luoyang 471003, China; zhaoyao2020115@163.com (Y.Z.); scotthw919@163.com (A.Y.); zhb156909@163.com (H.Z.); zzhui2022@163.com (Z.Z.)

* Correspondence: zzd@haust.edu.cn

Abstract: Predicting the remaining useful life (RUL) of a bearing can prevent sudden downtime of rotating machinery, thereby improving economic efficiency and protecting human safety. Two important steps in RUL prediction are the construction of a health indicator (HI) and the prediction of life. Traditional methods simply use the time-series characteristics of the vibration signal, for example, using root mean square (RMS) as HI, but this HI does not reflect the true degradation of the bearing. Meanwhile, existing prediction models often cannot consider both the time and space characteristics of the signal, thus limiting prediction accuracy. To address the above problems, in this study, wavelet packet transform (DWPT) and kernel principal component analysis (KPCA) were combined to extract HI from the original vibration signal. Then, a CNN-BiLSTM (convolutional and bidirectional long- and short-term memory) prediction network with root mean square as input and HI as output was constructed by combining convolutional neural network (CNN) and bi-directional long- and short-term memory neural network (BiLSTM). The network improved prediction accuracy by considering the temporal and spatial characteristics of the input signal. Experimental results on the PHM2012 dataset showed that the method proposed in this paper outperformed existing methods.

Keywords: wavelet packet transform; kernel principal component analysis; remaining service life of rolling bearings; convolutional neural network; bidirectional long- and short-term memory neural network

Citation: Zhong, Z.; Zhao, Y.; Yang, A.; Zhang, H.; Zhang, Z. Prediction of Remaining Service Life of Rolling Bearings Based on Convolutional and Bidirectional Long- and Short-Term Memory Neural Networks. *Lubricants* **2022**, *10*, 170. https://doi.org/10.3390/lubricants10080170

Received: 31 May 2022
Accepted: 19 July 2022
Published: 26 July 2022

1. Introduction

Bearing is a key component in rotating machinery, known as the joint of machinery, and its failure may lead to downtime of industrial production or even cause casualties [1]. According to a survey, rolling bearing failure is one of the most important factors of rotating machinery failure, accounting for 45–55% of cases [2]. A reasonable and effective bearing remaining useful life prediction (RUL) method can help technicians develop maintenance plans for predictive maintenance [3]. Therefore, it is important to predict the remaining service life of bearings to avoid accidents and reduce economic losses [4].

Generally speaking, methods for RUL prediction of bearings fall into two main categories: model-based (physical/mathematical) methods [5,6] and data-driven methods [7]. Wang et al. [8] proposed a mechanical state prediction method based on a probabilistic model with particle filters, which was successfully used for the state prediction of wind power bearings. El-Tawil et al. [9] developed a method based on a nonlinear damage law to determine the RUL of the system. Ma et al. [5] analyzed the interaction between various parts of the bearing by modeling the angle of relative sliding velocity between the rolling element and the bearing raceway and the bearing dynamics. However, model-based methods require complex physical or mathematical models, which require researchers with extensive knowledge base and are often difficult to develop due to complex working conditions [10].

With the development of sensor technology and computer technology, data-driven methods based on data have been developed [11]. As a data-driven method, deep learning can learn the bearing degradation trend spontaneously from sensor data and establish the mapping relationship between data and bearing health status with remarkable application [12]. Deep-learning-based RUL prediction methods are mainly divided into steps such as data acquisition, health factor (HI) construction, and remaining service life prediction [13]. Liu et al. [14] proposed a rolling bearing RUL prediction method based on regularized LSTM networks and verified the advantages of the method with the dataset of PRONOSTIA platform [15]. Ning et al. [16] first performed feature screening of signals and then predicted the remaining service life of bearings using RNN models. Network models such as RNN and LSTM tend to ignore spatial features, although they can learn the degradation trend and temporal characteristics of bearings from the data [17]. Wang et al. [18] used 1d-CNN to process fused signals and learn fault features using the powerful feature extraction capability of the network. However, a single CNN network tends to ignore the temporal features of the data and is unable to learn signal features at multiple scales [19]. HI can reflect the degradation trend of bearings, and it is critical to obtain excellent HI labels for training prediction models [20]. For example, Zhang et al. [21] used the time-domain feature RMS of the vibration signal as the main performance degradation indicator. Singleton et al. [22] used the variance of the vibration signal as HI. Zhang et al. [23] used the kurtosis of the vibration signal after band-pass filtering as HI. In the literature [24], the ratio of current life to total life of the bearing is used as HI of the bearing. However, HI constructed by the above methods cannot fully describe the bearing degradation trend. Because the bearing signal is nonlinear, we can pay attention to the transient changes of the signal by analyzing the signal with different resolutions in time–frequency domain. Time–frequency analysis technology DWPT is often used in the analysis of bearing vibration signals [25].

In response to the above problems, a new method for HI and RUL prediction of rolling bearings is proposed in this paper. Firstly, discrete wavelet packet transform (DWPT) was performed on the time-domain vibration signal to extract RMS features from the obtained sub-bands, and the HI was then obtained by fusing the RMS of each sub-band through kernel principal component analysis (KPCA). Based on this, the convolutional bidirectional long- and short-term memory neural network (CNN-BiLSTM) was proposed for lifetime prediction. Finally, the feasibility of the method was verified by bearing experimental data. The main contributions are as follows.

1. Discrete wavelet packet transform (DWPT) and principal component analysis (KPCA) were combined to construct new health indicators to solve the labeling problem of RUL prediction. Compared to the life-percentage-style linear HI, this HI can better reflect the bearing degradation trend and retain the time–frequency characteristics of the signal, which is beneficial to the learning of the prediction model.
2. A convolutional bidirectional long- and short-term memory neural network (CNN-BiLSTM) was designed for RUL prediction. Convolution can extract signal features from different scales, and combined with the BiLSTM network, the model can take into account both temporal and spatial features of the signal to improve prediction accuracy.
3. Experimental data based on rolling bearing dataset were used to verify the effectiveness of the method.

The remainder of this paper is organized as follows. Section 2 provides the theoretical background. Section 3 introduces the method proposed in this paper. Section 4 describes the experimental procedure and the analysis of the results in detail. Section 5 concludes the paper.

2. Theoretical Background

2.1. Discrete Wavelet Packet Transform (DWPT)

The discrete wavelet transform can describe the local characteristics of vibration signals in the time and frequency domains and is a very effective signal analysis method, which is often used for signal preprocessing for bearing fault diagnosis and life prediction [26].

In this study, the bearing vibration signal was preprocessed based on DWPT in order to construct HI. The algorithms for wavelet packet decomposition and reconstruction are shown in Equations (1) and (2), respectively.

$$\begin{cases} d_l^{j,2n} = \sum\limits_k p_{k-2l} d_k^{j-1,n} \\ d_l^{j,2n+1} = \sum\limits_k q_{k-2l} d_k^{j-1,n} \end{cases} \tag{1}$$

$$d_l^{j-1,n} = \sum_k p_{l-2k} d_k^{j,2n} + \sum_k q_{l-2k} d_k^{j,2n+1} \tag{2}$$

where p and q are filter coefficients; d is the wavelet packet decomposition coefficient; k and l are the number of decomposition layers; and j and n are wavelet packet node numbers.

Figure 1 is a schematic diagram of the three-layer wavelet packet decomposition structure, where S0 is the original signal, S10 is the low-frequency part of the original signal, S11 is the high-frequency part of the original signal, and so on. As can be seen from the figure, the wavelet packet transform can decompose both the low- and high-frequency parts of the signal uniformly and has higher time–frequency resolution than the wavelet transform, making it more effective in analyzing nonsmooth signals (e.g., bearing vibration signals) [20]. In this study, the db4 mother wavelet was used to decompose the original vibration signal into three levels of wavelet packets to obtain eight sub-bands.

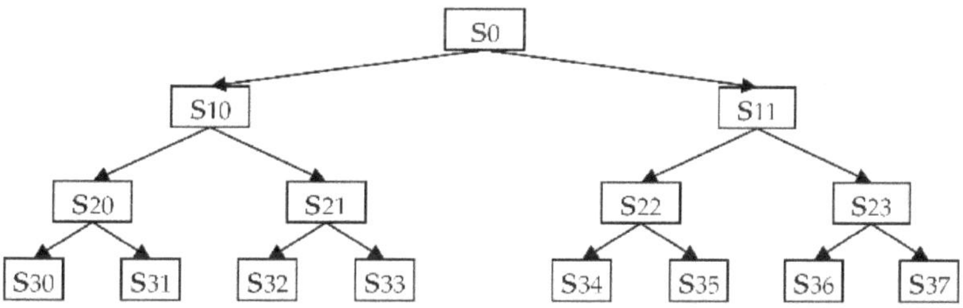

Figure 1. Structural scheme of DWPT.

2.2. Kernel Principal Component Analysis (KPCA)

Kernel principal component analysis (KPCA) [27] is a nonlinear feature extraction method that is often used for feature extraction and fusion of bearing signals [28]. The kernel function was first introduced to map the original data space to a high-dimensional feature space, and PCA was then performed to reduce the dimensionality of the analysis. The quality of the nonlinear features thus extracted was much better.

Let the data set with M samples be $\{x_1, x_2, \cdots, x_i\}(i = 1, 2, \ldots, M)$, $x_i \in R^N$ and the sample dimension be N. Normalize the high-dimensional spatial data so that it satisfies the following:

$$\frac{1}{M}\sum_{i=1}^{M} \varphi(x_i) = 0 \tag{3}$$

where φ is a nonlinear mapping function that enables the mapping of the low-dimensional spatial feature x_i to the higher dimensional space $F : \varphi(x_i)$.

The covariance matrix of F space is expressed as follows:

$$C = \frac{1}{M}\sum_{i=1}^{M} \varphi(x_i)\varphi(x_i)^T \tag{4}$$

The eigenvalues of the covariance matrix are λ, and the eigenvectors are v, both of which satisfy the following:

$$Cv = \lambda v \tag{5}$$

After transforming each sample into $\varphi(x_k)$, make inner product with Equation (5):

$$\varphi(x_k)Cv = \lambda\varphi(x_k)v \tag{6}$$

The linear representation of the feature vector is as follows:

$$v = \sum_{i=1}^{M} \alpha_i \varphi(x_i) \tag{7}$$

Simultaneous Formulas (4)–(7) can be obtained:

$$\frac{1}{M}\sum_{i=1}^{M} \alpha_i \sum_{j=1}^{M} [\varphi(x_k)\varphi(x_j)]\,[\varphi(x_j)\varphi(x_i)] = \lambda\sum_{i=1}^{M} \alpha_i[\varphi(x_k)\varphi(x_i)] \tag{8}$$

Nonlinear mapping from input space to high-dimensional feature space can be realized by kernel function inner product operation. The kernel function selected in this study is a Gaussian radial basis function, whose expression is as follows:

$$k(x_i, x_j) = \exp(-\gamma\|x_i - x_j\|^2) \tag{9}$$

where parameter γ is used to control the range of action of the kernel function.

Define the $M \times M$ dimensional matrix K, where the elements can be represented using the following kernel function:

$$K = \begin{bmatrix} k(x_1, x_1) & \cdots & k(x_1, x_m) \\ \vdots & \ddots & \vdots \\ k(x_m, x_1) & \cdots & k(x_m, x_m) \end{bmatrix} \tag{10}$$

The kernel matrix is used to represent Equation (8), which can be simplified as follows: $K\alpha = M\lambda\alpha$. The eigenvalues and eigenvectors of the kernel matrix can be derived from the simplified Equation (8), which in turn leads to the normalized eigenvector $v^k(k = 1, 2, \ldots, M)$ of the covariance matrix. Then, the k-th linear principal element of the sample x can be obtained as follows:

$$h_k = v^k\varphi(x) = \sum_{i=1}^{M} \alpha_i^k K(x_i, x) \tag{11}$$

The cumulative contribution of features is calculated and the principal element is selected as follows:

$$\sum_{k=1}^{p} \lambda_k / \sum_{i=1}^{m} \lambda_i \geq 0.90 \tag{12}$$

where $\lambda_1 \geq \lambda_2 \geq \lambda_3 \ldots \geq \lambda_m$ is the eigenvalue of the kernel matrix.

2.3. *Bidirectional Long Short-Term Memory Neural Network (BiLSTM)*

Long short-term memory neural network (LSTM) [29] is an improvement on the recurrent neural network (RNN). It solves the RNN gradient disappearance and gradient explosion problems by introducing forgetting gates and can efficiently learn the nonlinear features of time series.

As a deep learning neural network model, each neuron of LSTM is a memory cell with three gates, which are forgetting gate f_t, input gate i_t, and output gate o_t. The whole update process is shown in Equations (13)–(18).

The forgetting gate f_t determines what information is discarded and is determined by both the current input and the output of the previous sequence.

$$f_t = \sigma\left(W_f \cdot (h_{t-1}, x_t) + b_f\right) \tag{13}$$

where σ is the sigmoid activation function; W_f is the weight vector; b_f is the base vector; and C_{t-1} denotes the cell state, which is used to store the memory information of the previous moment.

Update gate i_t determines what information is stored and updates the cell state as follows:

$$i_t = \sigma(W_i \cdot (h_{t-1}, x_t) + b_c \tag{14}$$

$$\widetilde{C}_t = tanh(W_c \cdot (h_{t-1}, x_t) + b_c \tag{15}$$

$$C_t = f_t \cdot C_{t-1} + i_t \cdot \widetilde{C}_t \tag{16}$$

where $tanh$ is the activation function, $\widetilde{C}_t$ is the candidate vector for the current new state information; $f_t \cdot C_{t-1}$ denotes the information to be forgotten; $i_t \cdot \widetilde{C}_t$ is the information to be retained; and C_t is the current cell state.

$$o_t = \sigma(W_o \cdot (h_{t-1}, x_t) + b_o) \tag{17}$$

$$h_t = o_t \cdot tanh(C_t) \tag{18}$$

where o_t represents the output of information from the output gate, and h_t is the output of the memory cell, which will also be input in the next LSTM cell.

The BiLSTM [30,31] consists of two LSTMs that pass information from the forward and reverse directions, respectively, compared to the LSTM and can associate both past and future states. The Bi-LSTM structure is shown in Figure 2, and its output is as follows [32].

$$h_t = [\overrightarrow{h_t}, \overleftarrow{h_t}] \tag{19}$$

where $\overrightarrow{h_t}$ is the result of forward propagation, and $\overleftarrow{h_t}$ is the result of backward propagation.

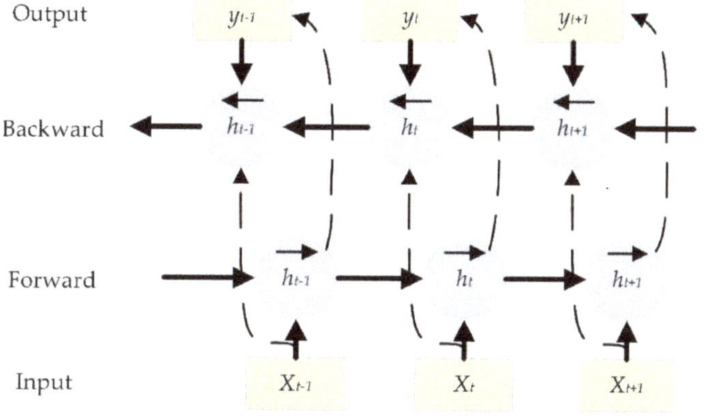

Figure 2. BiLSTM network architecture.

2.4. Convolutional Neural Network (CNN)

Convolutional neural networks [33] have the characteristics of local connectivity and weight sharing. One-dimensional convolutional neural networks can perform feature extraction on time-domain signals and are commonly used in the field of bearing fault diagnosis [34].

Convolutional neural networks usually consist of three types of network layers: convolutional layer, pooling layer, and fully connected layer. The convolutional layer can implement convolutional operations for feature extraction, the pooling layer can reduce the feature dimensionality and prevent overfitting, and the fully connected layer can perform nonlinear combination of the extracted features.

The formula for one-dimensional convolution is as follows:

$$Z^{l+1} = \left[Z^l * w^{l+1}\right] + b = \sum_{x=1}^{f}\left[Z_k^l(s_0 + x)w_k^{l+1}(x)\right] + b \tag{20}$$

The maximum pooling equation is as follows:

$$A_i^{l+1}(j) = \max_{(j-1)W+1 \leq jW}\left\{F_i^l(t)\right\} \tag{21}$$

where Z^l is the convolutional input of layer l + 1, and Z^{l+1} is the output of layer l + 1; b is the amount of variance; w_k^{l+1} is the weight of layer l + 1; f is the convolutional kernel size; s_0 is the convolutional step size; $F_i^l(t)$ is the value of the t-th neuron in the i-th feature of layer l; W is the pooling region; and A_i^{l+1} is the output of the neuron of layer l + 1.

3. The Proposed Framework

The overall block diagram of the proposed method is shown in Figure 3. Firstly, the original vibration signal was subjected to discrete wavelet packet transform (DWPT) to obtain eight sub-bands and extract the RMS values of different sub-bands. Then, KPCA was used to downscale the multidimensional RMS to obtain HI. Finally, the remaining lifetime prediction was performed by CNN-BiLSTM network.

(1) Data acquisition: The accelerometers were placed on the horizontal and vertical axes with sampling frequency of 25.6 KHZ, sampling interval of 10 s, and sampling time of 0.1 s. Sampling was performed under three working conditions.

(2) Building health indicators: The original vibration signal was decomposed into eight sub-bands using DWPT. The sub-bands were reconstructed according to the coefficients, and RMS values were extracted. The multidimensional RMS values were dimensionalized using the KPCA algorithm, and low-dimensional sensitive features were selected as HI and used as training labels for the prediction network.

(3) Proposed neural network: The spatial features of the signal were extracted using a convolutional network followed by a BiLSTM layer to extract temporal features from the forward and reverse directions. The global average pooling layer in the model can pay attention to the overall information, which is conducive to model prediction [35]. The mean square error was used as the loss function, and the optimizer was Adam. The input to the network was RMS at the current moment, and the output was the HI at the future moment.

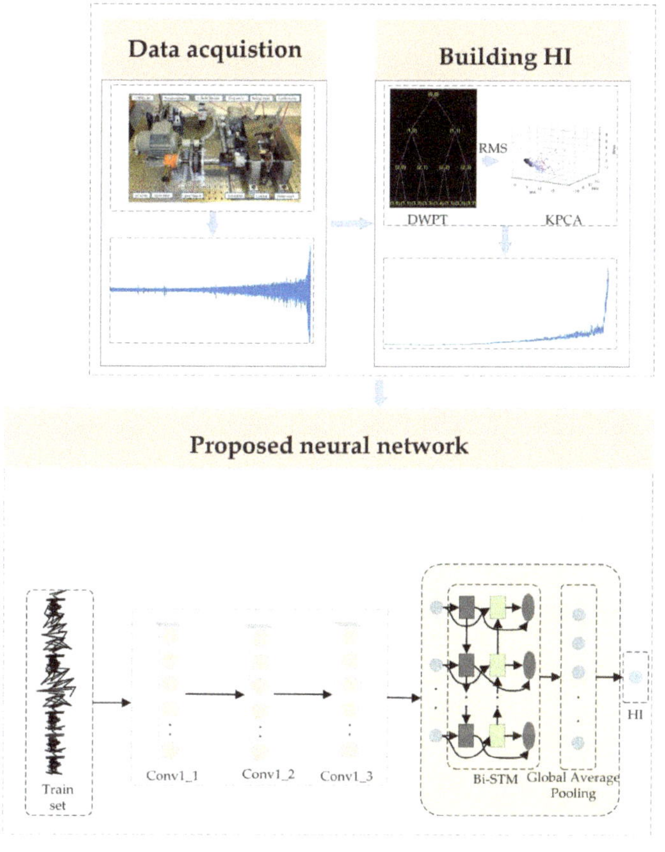

Figure 3. A flowchart of the proposed method.

4. Experiments and Results

4.1. Data Description

To verify the effectiveness of the proposed method, the PHM 2012 Challenge dataset was used in this study. The data was collected and obtained from the PRONOSTIA testbed, as shown in Figure 4.

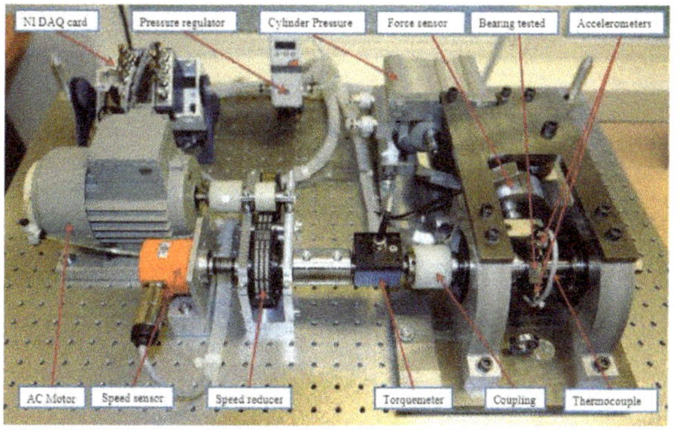

Figure 4. Pronostia bearing testbed.

The acquisition device was used for 17 full-life cycle experiments of the bearing under three operating conditions, and a total of 17 sets of data were collected in the horizontal and vertical directions of the bearing using an accelerometer. The sampling frequency of the accelerometer sensor was 25.6 kHZ, and the experimental setup recorded the vibration signals at 10 s intervals with a sampling time of 0.1 s [36]. Under working condition I, bearings 1-1 to 1-7 were tested with motor speed of 1800 rpm and load of 4000 N. Under working condition II, bearings 2-1 to 2-7 were tested with motor speed of 1650 rpm and load of 4200 N. Under working condition III, bearings 3-1 to 3-3 were tested with motor speed of 1500 rpm and load of 5000 N. Details of the data are shown in Table 1.

Table 1. Operating condition of the PHM2012 dataset.

Working Condition	Load (N)	Rotation Speed	Dataset
1	4000	1800 rpm	Bearing 1-1 (train) Bearing 1-2 (train) Bearing 1-3 (test) Bearing 1-4 (test) Bearing 1-5 (test) Bearing 1-6 (test) Bearing 1-7 (test)
2	4200	1650 rpm	Bearing 2-1 (train) Bearing 2-2 (train) Bearing 2-3 (test) Bearing 2-4 (test) Bearing 2-5 (test) Bearing 2-6 (test) Bearing 2-7 (test)
3	5000	1500 rpm	Bearing 3-1 (train) Bearing 3-2 (train) Bearing 3-3 (test)

4.2. Construction of Health Indicators

In this section, we describe the process of HI construction in detail. Figure 5 shows the original vibration signal of bearing 1-1. As can be seen, the vibration signal amplitude of bearing 1-1 initially fluctuated roughly smoothly. The vibration signal showed a gradual upward trend with increase in time and increased sharply at the later stage.

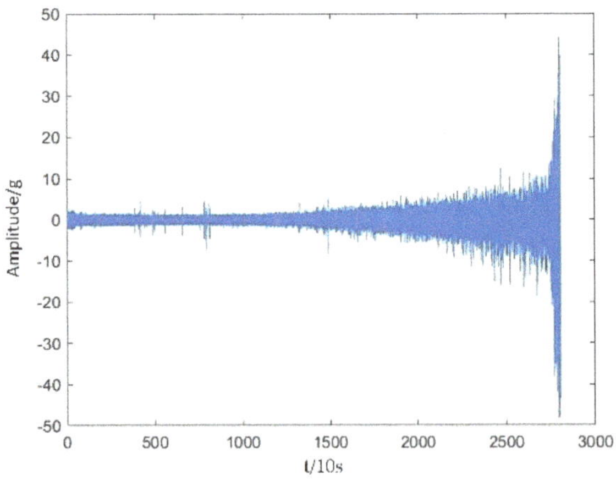

Figure 5. Original signal of bearing 1-1.

In the original time-domain signal shown in Figure 5, every 2560 consecutive points constitute a sample. These samples were processed by fast Fourier transform to obtain the corresponding frequency-domain samples. In Figure 6, 10 frequency-domain samples are shown. In the figure, the time corresponding to these samples increases with the sample number. For example, sample 1 corresponds to the beginning of the bearing life cycle, and sample 10 corresponds to the end of the bearing life cycle.

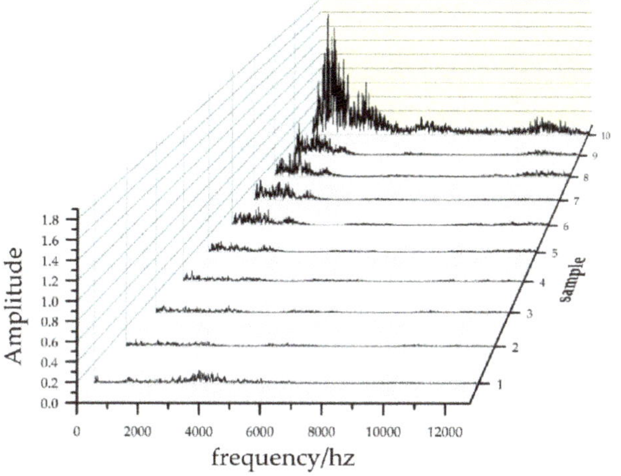

Figure 6. Bearing 1-1 partial sample frequency-domain signal.

As can be seen from Figure 6, the bearings had different degrees of amplitude increase in different frequency sections. The low-frequency vibration was due to rotation frequency, rolling body, and internal and external fault frequency of rolling. The high-frequency vibration was caused by the inherent frequency of each component of the bearing. When the bearing failed, the shape and quality of the component changed, affecting the high-frequency vibration.

To construct the HI, the original vibration signal was first decomposed into wavelet packets using db4 wavelets for three-level decomposition [20]. Reconstruction was performed according to the reconstruction coefficients to obtain eight sub-bands. The reconstructed sub-bands are shown in Figure 7. It can be seen that the eight sub-bands exhibited different trends, and each contained degradation characteristics of different frequency bands.

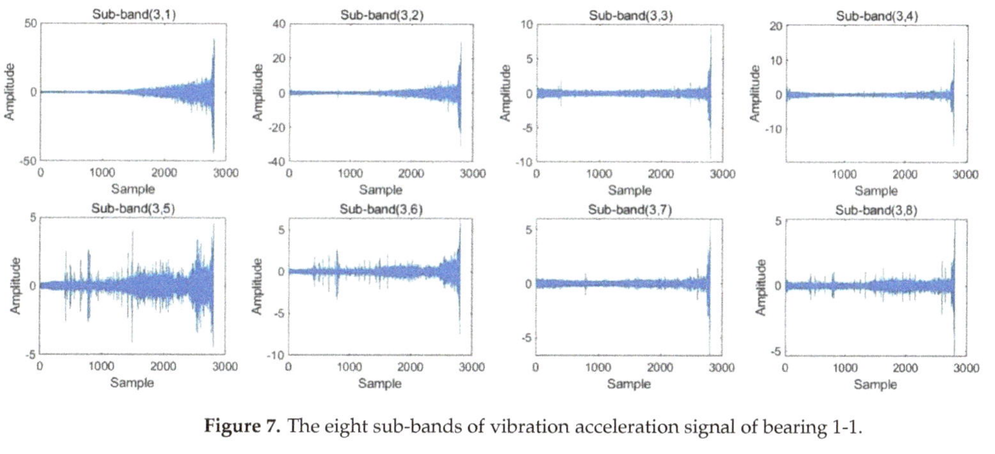

Figure 7. The eight sub-bands of vibration acceleration signal of bearing 1-1.

The RMS was extracted for each of the eight sub-bands, and the RMS values for each sub-band were obtained, as shown in Figure 8.

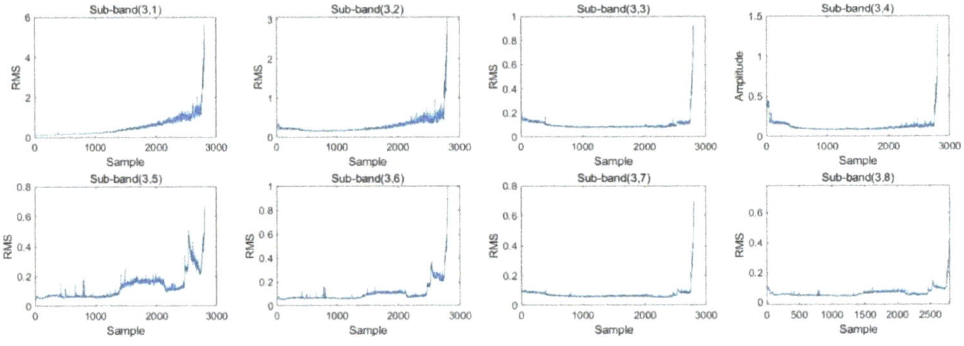

Figure 8. RMS trends extracted from eight sub-bands of the vibration acceleration signals for bearing 1-1.

As can be seen from Figure 8, the RMS extracted from the eight sub-bands had different trends. During the whole life cycle of the bearing, the RMS of some sub-bands showed an increasing trend, while the RMS of other sub-bands showed significant fluctuations. The RMS of all sub-bands showed a steep upward trend in the last part of the life cycle, while the RMS of some sub-bands showed sensitivity at the beginning of the wear. This indicates that different sub-bands carry different degradation information.

In order to fuse the most important degradation information exhibited by all sub-bands, the RMS of the eight sub-bands were feature fused using the KPCA algorithm. First, the eight sub-band RMS sequences shown in Figure 8 were selected to construct an eight-dimensional high-dimensional feature set. Then, the KPCA algorithm was used to reduce the dimensionality of the feature set. Finally, the first principal element (contribution rate >90%) was selected as the final HI. Table 2 shows the contribution rates of the principal elements. The final construction results are shown in Figure 9.

Table 2. Contribution rate of partial principal components.

Principal Component Serial Number	Contribution Rate	Cumulative Contribution Rate
1	0.919	0.919
2	0.061	0.980
3	0.019	0.999

Figure 9 shows a comparison of the values of RMS extracted from the sub-bands and RMS extracted using the method proposed in this study. The sub-band RMS fluctuated a lot, and the curves were messy. The method proposed in this paper could fuse the important characteristics of each sub-band, such as the sudden increase in RMS of sub-band (3,5) near sample 2500, which reflected the sensitivity at the early stage of wear; the gentle fluctuation of RMS of sub-band (3,7) at sample 2600 and the decrease in HI, which reflected the sensitivity at the recovery period of wear; and the sharp increase in RMS of sub-band (3,1) at the last part of the life cycle, which reflected the sensitivity of the late wear period. The proposed HI thus contained more comprehensive information on bearing degradation and the curve was smoother, reflecting its superiority as an HI. To further illustrate the superiority of the proposed HI, the HI of bearings 2-1, 2-6, 3-2, and 3-3 under different operating conditions were extracted, and the results are shown in Figure 10.

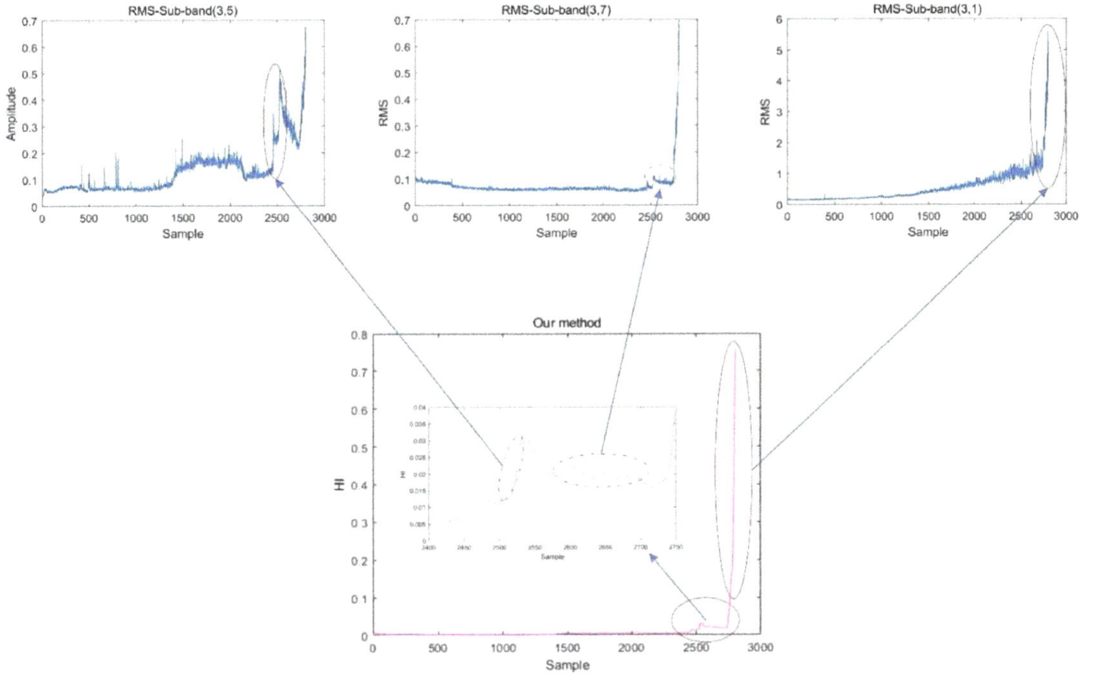

Figure 9. Prominent feature accumulation process of RMS values of different sub-bands of bearing 1-1.

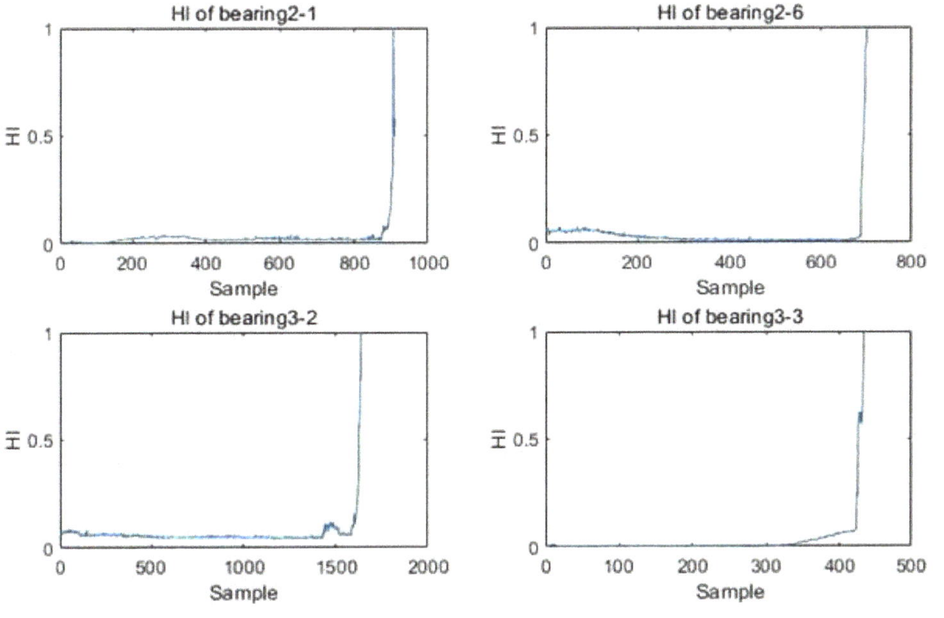

Figure 10. HI of different bearings.

The health factor, which reflects the degradation trend of the bearing, needs to have a strong correlation with the degradation trend of the bearing. The physical degradation process of the bearing is irreversible, so the health factor should also have a similar monotonic change trend [37]. Monotonicity indexes are widely used in the evaluation of health factor performance. Yang et al. [38] obtained the optimal HI by optimizing the monotonicity index of HI. Lin et al. [39] used ensemble stacked autoencoders to construct health factors for bearings and evaluated their performance using monotonicity metrics.

We compared the performance of the proposed HI with the original RMS using the metric of monotonicity according to the following equation:

$$Mon(X) = \frac{1}{K-1}|No.ofd/d_x > 0 - No.ofd/d_x < 0| \tag{22}$$

where X denotes the feature sequence; K denotes the total number of features; and $No.ofd/d_x > 0$ and $No.ofd/d_x < 0$ denote the number of positive and negative variances, respectively. The higher the *Mon* score, the better the monotonicity and the better the index performance. The results are shown in Table 3.

Table 3. HI performance analysis using monotonicity.

Bearing	Monotonicity of RMS	Monotonicity of Proposed HI
1-1	0.161	0.961
1-2	0.102	0.962
1-3	0.047	0.936
1-4	0.051	0.917
1-5	0.101	0.959
1-6	0.099	0.908
1-7	0.149	0.933
2-1	0.087	0.915
2-2-	0.132	0.941
2-3	0.141	0.907
2-4	0.097	0.913
2-5	0.089	0.968
2-6	0.127	0.947
2-7	0.152	0.903
3-1	0.074	0.931
3-2	0.046	0.903
3-3	0.047	0.933

From the table, we can see that the proposed HI had better monotonicity, which proves the superiority of the proposed method.

4.3. RUL Prediction

In this section, we outline the development of a CNN-BiLSTM prediction model with RMS input and HI labels and discuss the effect of model parameters on prediction accuracy. Bearing 1-1 was used as an example to develop a detailed description.

4.3.1. Input Selection

It is convenient to process raw vibration signals of bearings in the time domain, and features such as Rms, Peak2Peak, Kurtosis, Impulse Factor, Var, and Clearance Factor are commonly used in the analysis of the remaining service life of bearings and as inputs to prediction networks [4,40].

We can extract many features, such as time domain and frequency domain, from vibration signals. These features have different representation abilities for vibration signals. Some features are not helpful or even cause interference for characterizing signals. Therefore, it is very important to select appropriate features [41]. The degradation process is an accumulation of random fatigue failure processes, so it should have a certain overall increasing or decreasing trend on the time axis, i.e., the characteristic quantity should have certain monotonicity [37]. Zhang et al. [42] extracted signal time-domain, frequency-domain, and time–frequency-domain correlation features and defined metrics such as monotonicity for feature selection based on the trend and residuals of the features. Tian et al. extracted 10 features of bearing vibration signals and used the monotonicity index to screen good features as input to the neural network [43].

Different time-domain features have different characterization capabilities for the original signal, and monotonicity continues to be used to assess the characterization capabilities of time-domain features.

The monotonicity score was calculated for the time-domain features, and the results are shown in Figure 11.

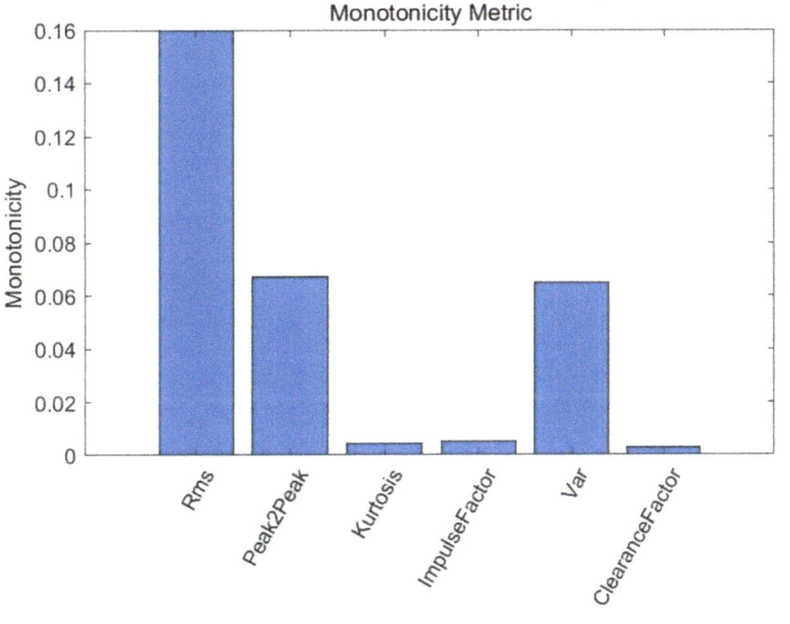

Figure 11. Monotonicity index of six vibration signal characteristics.

As can be seen from Figure 11, the Rms monotonicity of the original vibration signal was the best, and Rms was chosen here as input to the prediction network.

4.3.2. Training and Test of CNN-BiLSTM Model

Before training and testing a prediction model, it is necessary to construct the dataset and determine the correspondence between the input Rms and output HI labels. Suppose the Rms sequence is $[X_1, X_2, X_3, X_4]$ and the HI sequence is $[Y_1, Y_2, Y_3, Y_4]$, then the prediction relationship of the network is $F([X_1, X_2]) = Y_3$, $F([X_2, X_3]) = Y_4$. A sliding window was used to take the value of the Rms sequence with a window width of 64 and a sliding step of 1. The specific correspondence is shown in Figure 12. The final sample format (number of samples, time step, and dimension) obtained was (2739, 64, and 1).

For the CNN-BiLSTM model, Adam optimizer with an initial learning rate of 0.001 was used, and in order to maximize the optimization of the network parameters, a decreasing learning strategy was used to reduce the learning rate by 10^{-6} per round. The training results of the model are shown in Figure 13, and the prediction results of the training set were almost identical to the real HI labels.

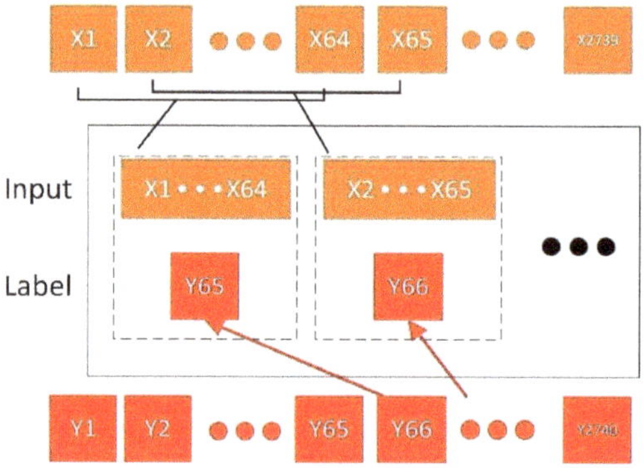

Figure 12. Sliding window sampling method.

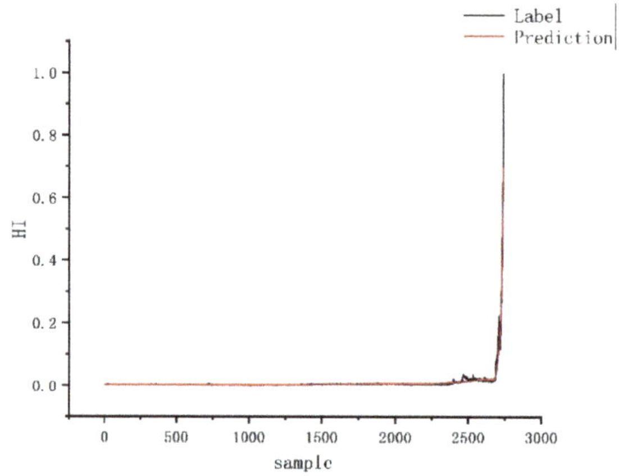

Figure 13. Training of the model.

Signal processing was performed as described in the previous section for HI construction, and finally the trained CNN-BiLSTM model was used for prediction of bearing 1-5. The prediction results are shown in Figure 14. As can be seen, the prediction results are basically consistent with the real HI labels, which verifies the effectiveness of the method.

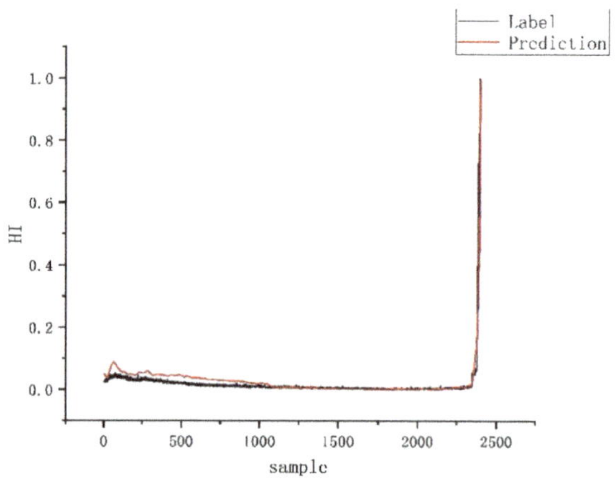

Figure 14. Testing of the model.

4.3.3. Selection of Hyper Parameters

In prediction models, the width of the sliding window and the size of the batch size are key hyperparameters that affect the performance of the model. This section discusses the effects of both parameters on the model performance.

The 2^n facilitated the computer processor for optimization, and window widths of 8, 16, 32, 64, and 128 were used to shape the samples. The prediction model was then trained and tested. The models were evaluated using mean square error (MSE), root mean square error (RMSE), and mean absolute error (MAE). The errors on the training and test sets are shown in Figure 15.

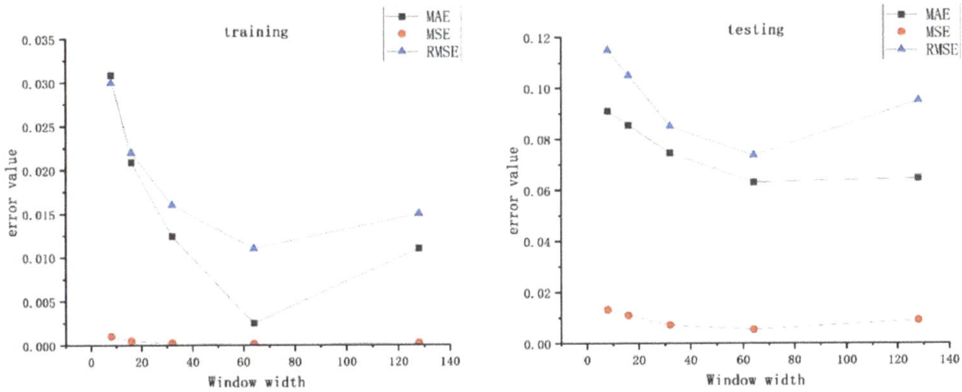

Figure 15. The prediction error for window width.

The above metrics can be described as follows.

$$\text{RMSE} = \sqrt{\frac{1}{N}\sum_{i=1}^{N}(y_i - \hat{y}_i)^2} \tag{23}$$

$$\text{MSE} = \frac{1}{N}\sum_{i=1}^{N}(y_i - \hat{y}_i)^2 \tag{24}$$

$$\text{MAE} = \frac{1}{N}\sum_{i=1}^{N}|y_i - \hat{y}_i| \tag{25}$$

where y_i is the true label; $\hat{y}_i$ is the predicted label; $\overline{y}_i$ is the mean of the actual labels; and N denotes the total number of samples.

As can be seen from Figure 15, on the training set, with the increase in the window width, each error showed a decreasing–increasing trend and the smallest error was at the window width of 64. On the test set, with the increase in the window width, the error showed irregular fluctuations and the smallest error was at the window width of 64. Therefore, the window width of 64 was chosen.

The batch size represents the number of data samples crawled in one training session. The batch size affects the training speed and model optimization. Batch sizes of 8, 16, 32, 64, 128, and 256 were selected, and the relationship between the three types of errors and batch size was observed. The results are shown in Figure 16.

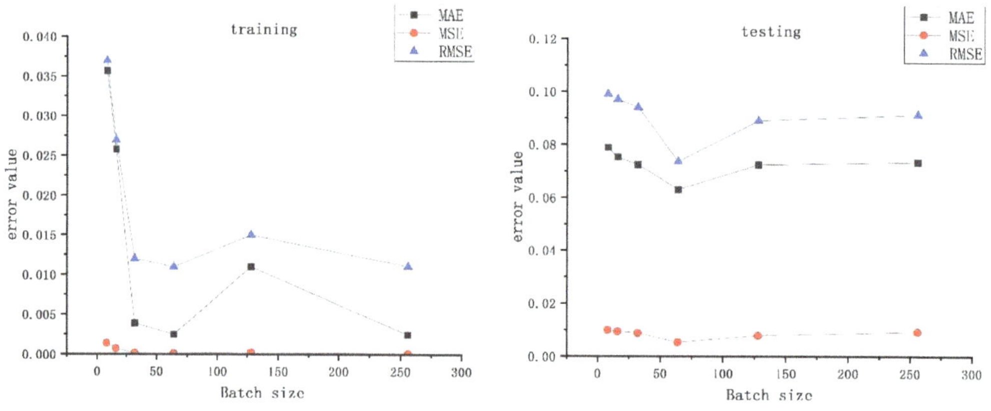

Figure 16. The prediction error for batch size.

From Figure 16, it can be seen that the three errors on the training set basically showed a decreasing trend with the increase of Batch size. However, in the test set, the three errors first decreased and then increased as the batch size increased. The errors increased significantly when the batch size exceeded 64, so 64 was chosen as the batch size.

4.3.4. Results of Different Models

This section discusses the prediction effects of CNN, LSTM, and BiLSTM models on the PHM2012 dataset and compares them with the method proposed in this paper.

Prediction experiments were conducted on the dataset using each of the three models mentioned above, and the prediction results for bearings 1-1, 1-3, and 1-5 were visualized. The results are shown in Figure 17.

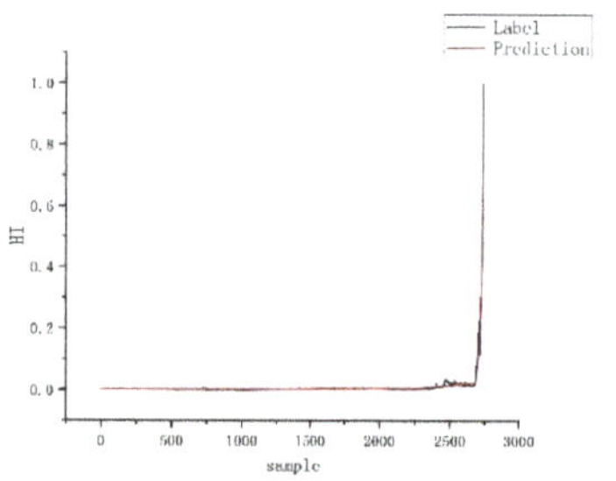

(a) our method prediction result of bearing1-1

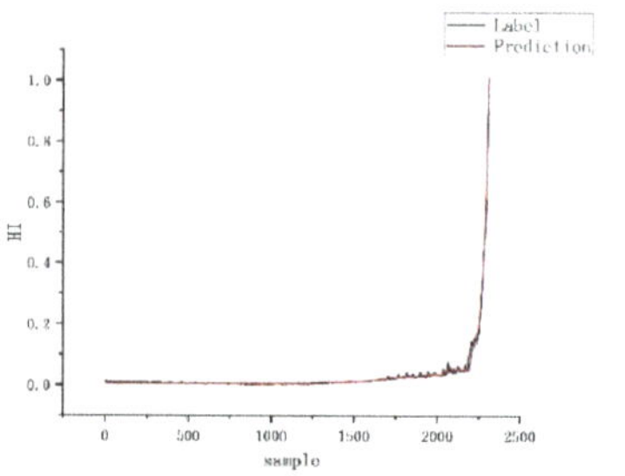

(b) our method prediction result of bearing1-3

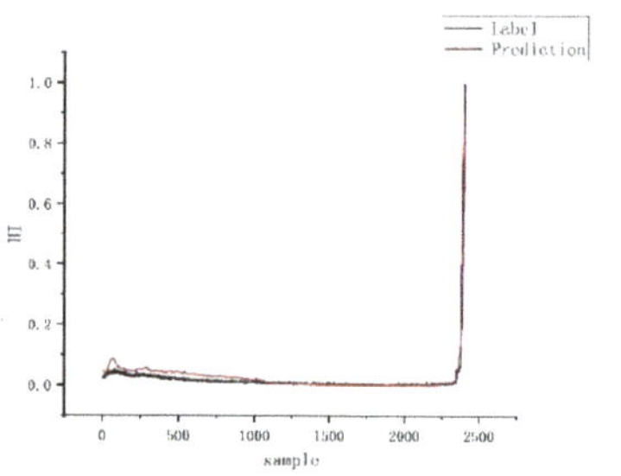

(c) our method prediction result of bearing1-5

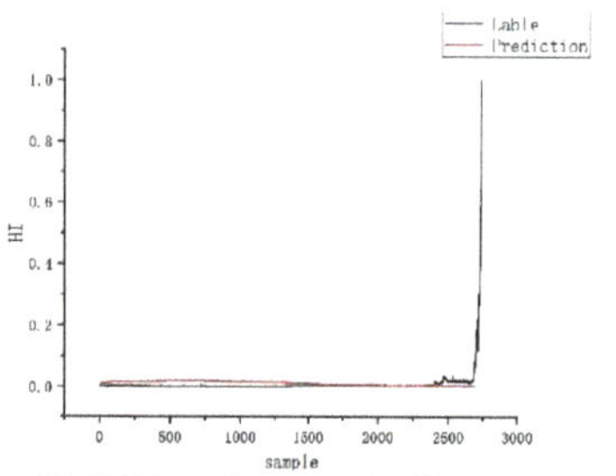

(d) LSTM prediction result of bearing1-1

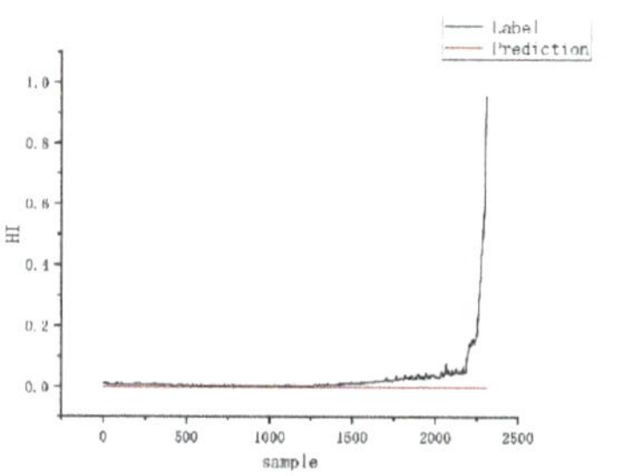

(e) LSTM prediction result of bearing1-3

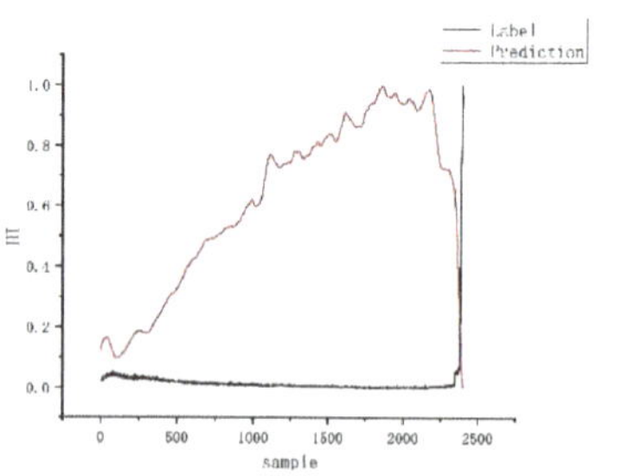

(f) LSTM prediction result of bearing1-5

Figure 17. *Cont.*

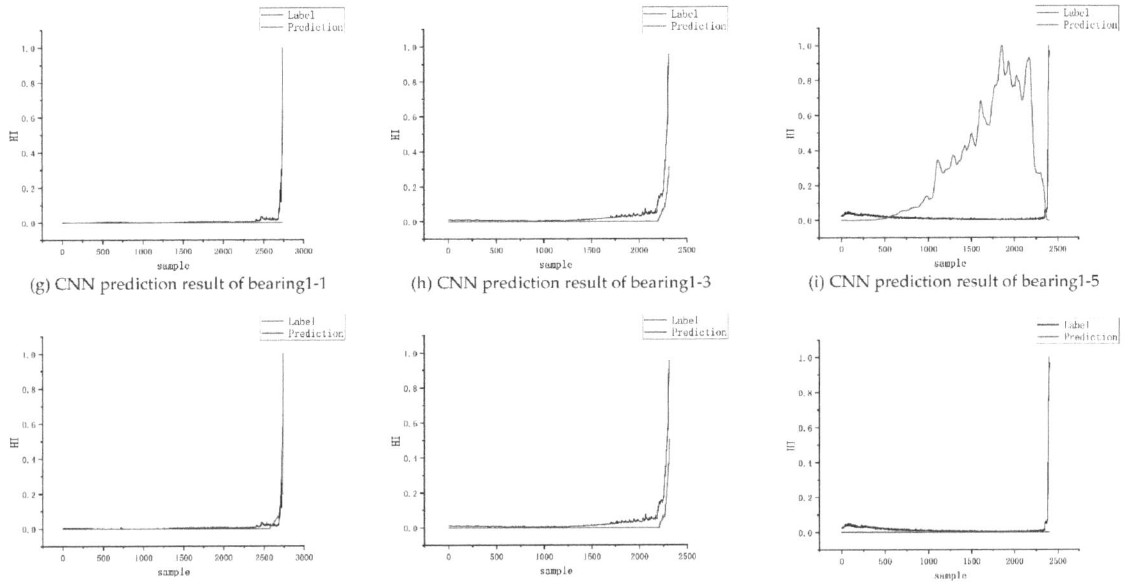

(g) CNN prediction result of bearing1-1

(h) CNN prediction result of bearing1-3

(i) CNN prediction result of bearing1-5

(j) BiLSTM prediction result of bearing1-1

(k) BiLSTM prediction result of bearing1-3

(l) BiLSTM prediction result of bearing1-5

Figure 17. Prediction effects of different models. (**a–c**) is the prediction result of our method. (**d–f**) is the prediction result of CNN model. (**g–i**) is the prediction result of LSTM model. (**j–l**) is the predicted result of BiLSTM.

From Figure 17, it can be seen that the proposed method could accurately predict the degradation trend of the bearing, and the rest of the models had different degrees of problems. For example, both the LSTM and CNN models could not well predict the rapid degradation stage of bearing 1-1. The prediction effect of BiLSTM was better, but the fluctuation trend at the beginning of degradation could not be well predicted. Both the LSTM and CNN models could not well predict the rapid degradation stage of bearing 1-1. The prediction effect of BiLSTM was better, but the fluctuation trend at the beginning of degradation could not be well predicted. The BiLSTM model could not predict the rapid degradation trend of the bearing at the end of cycle 1-5. The CNN could roughly predict the degradation trend of the bearing from 1-3, but the accuracy was insufficient.

As shown in Table 4, the proposed model achieved the smallest prediction error on almost all bearings using MSE as an indicator, thereby showing the superiority of the proposed method.

Table 4. MSE for different prediction models.

Model (MSE)	Bearing 1-3	Bearing 1-4	Bearing 1-5	Bearing 1-6	Bearing 1-7	Bearing 2-3	Bearing 2-4	Bearing 2-5	Bearing 2-6	Bearing 2-7	Bearing 3-3
Proposed	0.0054	0.0031	0.007 6	0.0044	0.0048	0.0621	0.0091	0.0143	0.0087	0.1026	0.0078
CNN	0.1431	0.1101	0.1923	0.0942	0.2721	0.0623	0.0089	0.2641	0.3101	0.2176	0.1739
LSTM	0.2145	0.0671	0.2167	0.1743	0.0074	0.1473	0.1012	0.1149	0.2031	0.0098	0.2147
BiLSTM	0.0097	0.1431	0.1016	0.2497	0.1364	0.1824	0.2107	0.1087	0.1047	0.4102	0.3006

5. Conclusions

The remaining service life prediction of bearings is a research focus. In this study, we established a suitable health indicator (HI) and proposed a prediction network combining CNN and BiLSTM. First, wavelet packet transform was performed on the original vibration signal of the bearing to obtain eight sub-bands, and the RMS of each of the eight sub-bands were extracted. The KPCA algorithm was then used to fuse the features of the extracted RMS of the eight sub-bands to obtain the HI of the bearing life cycle. The CNN-BiLSTM prediction network was then developed, which can extract the spatiotemporal features of the signal at the same time to improve prediction accuracy. The prediction network uses the RMS of the original signal at the current moment as the network input and the HI of the future moment as the network output. Experiments were conducted on the PHM2012 dataset to verify the effectiveness of the proposed prediction model.

Author Contributions: Data curation, A.Y., H.Z. and Z.Z. (Zhihui Zhang); funding acquisition, Z.Z. (Zhidan Zhong); investigation, Y.Z.; project administration, Z.Z. (Zhidan Zhong); software, Y.Z.; supervision, Z.Z. (Zhidan Zhong); writing—original draft, Y.Z. All authors have read and agreed to the published version of the manuscript.

Funding: This research was supported by the National Natural Science Foundation of China (No.: 52105182) and the Major Science and Technology Project of Henan Province (project name: Research and industrialization of key technologies of high-end bearings for major equipment).

Data Availability Statement: Public datasets used in our paper: https://github.com/wkzs111/phm-ieee-2012-data-challenge-dataset (accessed on 7 January 2022).

Conflicts of Interest: The authors declare no conflict of interest.

References

1. Pech, M.; Vrchota, J.; Bednář, J. Predictive Maintenance and Intelligent Sensors in Smart Factory: Review. *Sensors* **2021**, *21*, 1470. [CrossRef] [PubMed]
2. Mao, W.; Liu, Y.; Ding, L.; Safian, A.; Liang, X. A New Structured Domain Adversarial Neural Network for Transfer Fault Diagnosis of Rolling Bearings Under Different Working Conditions. *IEEE Trans. Instrum. Meas.* **2021**, *70*, 3509013. [CrossRef]
3. Hamadache, M.; Jung, J.H.; Park, J.; Youn, B.D. A comprehensive review of artificial intelligence-based approaches for rolling element bearing PHM: Shallow and deep learning. *JMST Adv.* **2019**, *1*, 125–151. [CrossRef]
4. Cheng, H.; Kong, X.; Chen, G.; Wang, Q.; Wang, R. Transferable convolutional neural network based remaining useful life prediction of bearing under multiple failure behaviors. *Measurement* **2021**, *168*, 108286. [CrossRef]
5. Ma, S.; Zhang, X.; Yan, K.; Zhu, Y.; Hong, J. A Study on Bearing Dynamic Features under the Condition of Multiball–Cage Collision. *Lubricants* **2022**, *10*, 9. [CrossRef]
6. Qian, Y.; Yan, R.; Gao, R.X. A multi-time scale approach to remaining useful life prediction in rolling bearing. *Mech. Syst. Signal Process.* **2017**, *83*, 549–567. [CrossRef]
7. Xu, J.; Duan, S.; Chen, W.; Wang, D.; Fan, Y. SACGNet: A Remaining Useful Life Prediction of Bearing with Self-Attention Augmented Convolution GRU Network. *Lubricants* **2022**, *10*, 21. [CrossRef]
8. Bousdekis, A.; Magoutas, B.; Apostolou, D.; Mentzas, G. Review, analysis and synthesis of prognostic-based decision support methods for condition based maintenance. *J. Intell. Manuf.* **2018**, *29*, 1303–1316. [CrossRef]
9. El-Tawil, K.; Jaoude, A.A. Stochastic and nonlinear-based prognostic model. *Syst. Sci. Control Eng.* **2013**, *1*, 66–81. [CrossRef]
10. Rathore, M.S.; Harsha, S.P. Prognostics Analysis of Rolling Bearing Based on Bi-Directional LSTM and Attention Mechanism. *J. Fail. Anal. Prev.* **2022**, *22*, 704–723. [CrossRef]
11. Cerrada, M.; Sánchez, R.-V.; Li, C.; Pacheco, F.; Cabrera, D.; de Oliveira, J.V.; Vásquez, R.E. A review on data-driven fault severity assessment in rolling bearings. *Mech. Syst. Signal Process.* **2018**, *99*, 169–196. [CrossRef]
12. Mushtaq, S.; Islam, M.M.M.; Sohaib, M. Deep Learning Aided Data-Driven Fault Diagnosis of Rotatory Machine: A Comprehensive Review. *Energies* **2021**, *14*, 5150. [CrossRef]
13. Chen, D.; Qin, Y.; Wang, Y.; Zhou, J. Health indicator construction by quadratic function-based deep convolutional auto-encoder and its application into bearing RUL prediction. *ISA Trans.* **2021**, *114*, 44–56. [CrossRef] [PubMed]
14. Liu, Z.-H.; Meng, X.-D.; Wei, H.-L.; Chen, L.; Lu, B.-L.; Wang, Z.-H.; Chen, L. A Regularized LSTM Method for Predicting Remaining Useful Life of Rolling Bearings. *Int. J. Autom. Comput.* **2021**, *18*, 581–593. [CrossRef]
15. Nectoux, P.; Gouriveau, R.; Medjaher, K.; Ramasso, E.; Chebel-Morello, B.; Zerhouni, N.; Varnier, C. PRONOSTIA: An experimental platform for bearings accelerated degradation tests. In Proceedings of the IEEE International Conference on Prognostics and Health Management (PHM'12), Beijing, China, 23–25 May 2012.
16. Ning, Y.; Wang, G.; Yu, J.; Jiang, H. A Feature Selection Algorithm Based on Variable Correlation and Time Correlation for Predicting Remaining Useful Life of Equipment Using RNN. In Proceedings of the 2018 IEEE Condition Monitoring and Diagnosis (CMD), Perth, Australia, 23–26 September 2018.
17. Huang, G.; Li, H.; Ou, J.; Zhang, Y.; Zhang, M. A Reliable Prognosis Approach for Degradation Evaluation of Rolling Bearing Using MCLSTM. *Sensors* **2020**, *20*, 1864. [CrossRef]

18. Wang, X.; Mao, D.; Li, X. Bearing fault diagnosis based on vibro-acoustic data fusion and 1D-CNN network. *Measurement* **2021**, *173*, 108518. [CrossRef]
19. Zhang, X.; Cong, Y.; Yuan, Z.; Zhang, T.; Bai, X. Early Fault Detection Method of Rolling Bearing Based on MCNN and GRU Network with an Attention Mechanism. *Shock Vib.* **2021**, *2021*, 6660243. [CrossRef]
20. Duong, B.P.; Khan, S.A.; Shon, D.; Im, K.; Park, J.; Lim, D.-S.; Jang, B.; Kim, J.-M. A Reliable Health Indicator for Fault Prognosis of Bearings. *Sensors* **2018**, *18*, 3740. [CrossRef]
21. Zhang, N.; Wu, L.; Wang, Z.; Guan, Y. Bearing Remaining Useful Life Prediction Based on Naive Bayes and Weibull Distributions. *Entropy* **2018**, *20*, 944. [CrossRef]
22. Singleton, R.K.; Strangas, E.G.; Aviyente, S. Extended Kalman filtering for remaining-useful-life estimation of bearings. *IEEE Trans. Ind. Electron.* **2015**, *62*, 1781–1790. [CrossRef]
23. Zhang, Z.-X.; Si, X.-S.; Hu, C.-H. An Age- and State-Dependent Nonlinear Prognostic Model for Degrading Systems. *IEEE Trans. Reliab.* **2015**, *64*, 1214–1228. [CrossRef]
24. Hu, L.; Hu, N.-Q.; Fan, B.; Gu, F.-S.; Zhang, X.-Y. Modeling the Relationship between Vibration Features and Condition Parameters Using Relevance Vector Machines for Health Monitoring of Rolling Element Bearings under Varying Operation Conditions. *Math. Probl. Eng.* **2015**, *2015*, 123730. [CrossRef]
25. Song, L.; Wang, H.; Chen, P. Vibration-based intelligent fault diagnosis for roller bearings in low-speed rotating machinery. *IEEE Trans. Instrum. Meas.* **2018**, *67*, 1887–1899. [CrossRef]
26. Nguyen, H.N.; Kim, J.; Kim, J.M. Optimal sub-band analysis based on the envelope power Spectrum for effective fault detection in bearing under variable, low speeds. *Sensors* **2018**, *18*, 1389. [CrossRef]
27. Schölkopf, B.; Smola, A.; Mueller, K.-R. Nonlinear Component Analysis as a Kernel Eigenvalue Problem. *Neural Comput.* **1998**, *10*, 1299–1319. [CrossRef]
28. Shen, J.; Xu, F. Method of fault feature selection and fusion based on poll mode and optimized weighted KPCA for bearings. *Measurement* **2022**, *194*, 110950. [CrossRef]
29. Hochreiter, S.; Schmidhuber, J. Long Short-Term Memory. *Neural Comput.* **1997**, *9*, 1735–1780. [CrossRef]
30. Schuster, M.; Paliwal, K.K. Bidirectional recurrent neural networks. *IEEE Trans. Signal Proces.* **1997**, *45*, 2673–2681. [CrossRef]
31. Graves, A.; Schmidhuber, J. Framewise phoneme classification with bidirectional LSTM and other neural network architectures. *Neural Netw.* **2005**, *18*, 602–610. [CrossRef]
32. Zhan, Y.; Sun, S.; Li, X.; Wang, F. Combined Remaining Life Prediction of Multiple Bearings Based on EEMD-BILSTM. *Symmetry* **2022**, *14*, 251. [CrossRef]
33. Lecun, Y.; Bottou, L.; Bengio, Y.; Haffner, P. Gradient-based learning applied to document recognition. *Proc. IEEE* **1998**, *86*, 2278–2324. [CrossRef]
34. Xie, W.; Li, Z.; Xu, Y.; Gardoni, P.; Li, W. Evaluation of Different Bearing Fault Classifiers in Utilizing CNN Feature Extraction Ability. *Sensors* **2022**, *22*, 3314. [CrossRef]
35. Zhang, X.; He, C.; Lu, Y.; Chen, B.; Zhu, L.; Zhang, L. Fault diagnosis for small samples based on attention mechanism. *Measurement* **2022**, *187*, 110242. [CrossRef]
36. Ren, L.; Sun, Y.; Wang, H.; Zhang, L. Prediction of Bearing Remaining Useful Life with Deep Convolution Neural Network. *IEEE Access* **2018**, *6*, 13041–13049. [CrossRef]
37. Lei, Y.; Li, N.; Guo, L.; Li, N.; Yan, T.; Lin, J. Machinery health prognostics: A systematic review from data acquisition to RUL prediction. *Mech. Syst. Signal Process.* **2017**, *104*, 799–834. [CrossRef]
38. Yang, F.; Habibullah, M.S.; Zhang, T.; Xu, Z.; Lim, P.; Nadarajan, S. Health Index-Based Prognostics for Remaining Useful Life Predictions in Electrical Machines. *IEEE Trans. Ind. Electron.* **2016**, *63*, 2633–2644. [CrossRef]
39. Lin, P.; Tao, J. A novel bearing health indicator construction method based on ensemble stacked autoencoder. In Proceedings of the 2019 IEEE International Conference on Prognostics and Health Management, San Francisco, CA, USA, 17–20 June 2019.
40. Guo, L.; Li, N.; Jia, F.; Lei, Y.; Lin, J. A recurrent neural network based health indicator for remaining useful life prediction of bearings. *Neurocomputing* **2017**, *240*, 98–109. [CrossRef]
41. Huang, H.-Z.; Wang, H.-K.; Li, Y.-F.; Zhang, L.; Liu, Z. Support vector machine based estimation of remaining useful life: Current research status and future trends. *J. Mech. Sci. Technol.* **2015**, *29*, 151–163. [CrossRef]
42. Zhang, B.; Zhang, L.; Xu, J. Degradation Feature Selection for Remaining Useful Life Prediction of Rolling Element Bearings. *Qual. Reliab. Eng. Int.* **2016**, *32*, 547–554. [CrossRef]
43. Han, T.; Pang, J.; Tan, A.C. Remaining useful life prediction of bearing based on stacked autoencoder and recurrent neural network. *J. Manuf. Syst.* **2021**, *61*, 576–591. [CrossRef]

lubricants

MDPI

Article

Flow Mechanisms and Lubrication Performance of Water-Lubricated Thrust Bearings with Herringbone Grooves

Haifeng Ran [1], Peng Dai [2], Shuping Yan [1], Fengtao Wang [1], Xingjia Yao [3], Jianping Wang [1,*] and Guizhong Zuo [4]

1 School of Mechanical Engineering, Anhui Polytechnic University, Wuhu 241000, China
2 State Key Laboratory of Engines, Tianjin University, Tianjin 300072, China
3 Anhui Huadong Polytechnic Institute, Wuhu 241000, China
4 Institute of Plasma Physics, Chinese Academy of Sciences, Hefei 230031, China
* Correspondence: tjqhd@163.com

Abstract: Due to their excellent stability and zero leakage capability, thrust bearings with herringbone spiral grooves are frequently used in transmission mechanisms. However, the lubrication mechanism of thrust bearings has not been clearly understood and explained, preventing the optimization of the bearing performance. Thus, this paper is devoted to solving this problem by building a three-dimensional finite element flow model. In this model, the change in viscosity temperature is considered using Roelands equation, and the turbulence and cavitation are taken into consideration. Using the established model, the influence of parameters such as spiral angle, groove width ratio, and rotational speed on the cavitation area of the thrust bearing are analyzed. The pressure contour and speed distribution are obtained inside the clearance, as well as the volume fraction of the gas phase at the end face. Finally, according to the analysis results, the optimum structural parameter for the herringbone spiral groove structure is proposed, which enables higher bearing stability and provides a reference for engineering practice.

Keywords: cavitation effect; herringbone spiral groove; optimum structural parameters; turbulence; water lubrication

Citation: Ran, H.; Dai, P.; Yan, S.; Wang, F.; Yao, X.; Wang, J.; Zuo, G. Flow Mechanisms and Lubrication Performance of Water-Lubricated Thrust Bearings with Herringbone Grooves. *Lubricants* **2022**, *10*, 182. https://doi.org/10.3390/lubricants10080182

Received: 27 June 2022
Accepted: 5 August 2022
Published: 10 August 2022

Publisher's Note: MDPI stays neutral with regard to jurisdictional claims in published maps and institutional affiliations.

1. Introduction

Herringbone groove thrust bearings (HGTBs) have broad application prospects in hydraulic machinery, machine tool spindles, and other industrial fields [1–3], due to the advantages of good stability, high stiffness, and zero leakage. In the bearing, oil and air are commonly used as lubricative mediums. However, the gas-lubricated HGTB has low stiffness, and the oil-lubricated HGTB has a rapid temperature rise. On the opposite, the water-lubricated [4,5] HGTB has excellent dynamic stability and low frictional power consumption. Therefore, water-lubricated HGTBs are more suitable as spindle supports for ultra-precision machine tools under high speeds [6–8]. Cavitation, however, commonly occurs in water lubrication, reducing bearing performance and service life. It is valuable for establishing a three-dimensional water-lubricated flow field model considering the cavitation effect, which is used to explore the influence of structural changes on the bearing performance.

In recent years, more and more scholars [9,10] have achieved bearing resistance reduction, and the bearing stability was improved. In some research, it was shown that changes in the lubricating properties of bearings were closely related to the design of the surface texture [11–13] (distribution, shape, and size). Guenat [14,15] used the experiment to study the effect of the herringbone groove geometry on lubrication performance deeply. Wang [16] developed a thermo-hydrodynamic model of a combined bearing, and the effect of the groove structure on the bearing characteristics was investigated. If the bearing worked at high speed, the cavitation inception number of water is much higher. Thus, cavitation effects cannot be ignored. Some researchers treat cavitation lubricants as gas–liquid mixed fluids. According to the different calculation methods of the gas volume fraction, the cavitation models in the mixed fluid model can be divided into three types: (1) Model based on the R-P equation [17,18]; (2) Equation model based on gas solubility and surface tension of bubbles [19,20]; (3) Model of the Transport Equation based on Gas Volume Fraction [21,22]. Moreover, some researchers used CFD to analyze the cavitation. In this theory, the journal bearings were investigated numerically in three dimensions by using FLUENT software [23], and the viscosity changes have a significant impact on the pressure distribution, cavity volume, and load-bearing capacity. Furthermore, based on the multi-phase flow cavitation model and the dynamic mesh technology, the cavitation characteristics of the liquid film in the dynamic pressure mechanical seal gap were investigated [24]. Meanwhile, Xu et al. [25] determined the optimal mass transfer coefficient in the Lee phase transition mass transfer equation, and the influence of groove structure parameters on the vaporization characteristics was studied.

In the above literature summary, the herringbone groove structure has been widely used and studied in thrust bearing design. However, the effects of temperature change and cavitation on lubrication states have rarely been studied. In this paper, CFD is used for modeling and analysis. The cavitation effects of the herringbone groove under changes in viscosity and temperature are studied, and its sealing performance and load-bearing performance are analyzed. The lubrication mechanism of the herringbone groove seal is explored, and the influence of the groove structure on the cavitation effect is summarized.

2. Governing Equations

2.1. Conservation Equations

Fluid domains are analyzed using CFD methods, and the relative equations can be found as following.

The mass conservation equation is expressed by Equation (1):

$$\frac{\partial \rho}{\partial t} + \nabla \cdot (\rho \vec{v}) = 0 \tag{1}$$

where, ρ represents the fluid density, $\rho \vec{v}$ is the gravitational body force, and t is the time.

The momentum equation is presented as Equation (2):

$$\rho \frac{d\vec{v}}{dt} = -\nabla p + \nabla \cdot \tau \tag{2}$$

where, $\vec{v}$ is the fluid velocity vector, p represents the static pressure, and τ is the stress. The energy equation is given in Equation (3):

$$\frac{\partial(\rho E)}{\partial t} + \nabla \bullet \left[\vec{u}(\rho E + p)\right] = \nabla \bullet \left[k_{\text{eff}} \nabla T - \sum_j h_j J_j + \left(\tau_{\text{eff}} \bullet \vec{u}\right)\right] + S_h \tag{3}$$

where, E represents the total energy of the fluid, h_j is the enthalpy of component j, k_{eff} is the effective heat transfer coefficient, $k_{\text{eff}} = k + k_t$, k_t denotes the turbulent heat transfer coefficient, J is the diffusion flux of component j, and S_h represents the volumetric heat source term.

2.2. Cavitation Equations

The cavitation phenomenon easily appears in the high-speed rotation of the water film. When the temperature changes drastically, the water rises to the boiling point, the originally small bubbles rapidly expand and escape from the surface of the liquid. In this study, the phase change process is mainly controlled by temperature. The volume of fluid (VOF) model is chosen for the tracing of the gas–liquid interface.

The interphase interface is tracked by solving the continuity equation. This equation is solved by using the volume fraction of one (or more) phases. For the qth phase, this equation has the following form:

$$\frac{1}{\rho_q}\left[\frac{\partial}{\partial t}(\alpha_q\rho_q) + \nabla\cdot\left(\alpha_q\rho_q\vec{v}_q\right) = S_{\alpha_q} + \sum_{p=1}^{n}(\dot{m}_{pq} - \dot{m}_{qp})\right] \tag{4}$$

In Equation (4), α_v is the phase q volume fraction, ρ_q is the phase q density, $\dot{m}_{qp}$ is the mass transfer from phase q to phase p, and $\dot{m}_{qp}$ is the mass transfer from phase p to phase q. The source term on the right-hand side of Equation (4) ($S\alpha_p$) is zero.

Among various cavitation models, the Lee model is relatively mature and matches the VOF model well. Therefore, this study adopted the Lee model as the phase transition model. By simplifying Equation (4), the transport equation of the bubble phase is obtained.

$$\frac{\partial}{\partial t}(\alpha_v\rho_v) + \nabla\cdot(\alpha_v\rho_v\vec{v}) = R_e - R_c \tag{5}$$

In Equation (5), R_e and R_c represent the evaporation and condensation terms in the phase transition process, respectively. They are respectively determined by using the gas and liquid phase medium temperatures, the saturation vaporization temperature is considered, and the relevant equations are as follows:

$$\begin{gathered} T_l > T_{\text{sat}},\ R_e = \text{coeff} \times \alpha\rho_l\frac{(T_l - T_{\text{sat}})}{T_{\text{sat}}} \\ T_v < T_{\text{sat}},\ R_c = \text{coeff} \times (1-\alpha)\rho_v\frac{(T_{\text{sat}} - T_v)}{T_{\text{sat}}} \end{gathered} \tag{6}$$

where, T represents the local saturation vaporization temperature; coeff is the evaporative condensation coefficient derived from the interfacial concentration under vesicular flow.

$$\text{coeff} = \frac{6}{d_b}\eta\sqrt{\frac{M}{2\pi RT_{\text{sat}}}}L\left(\frac{\alpha_v\rho_v}{\rho_l - \rho_v}\right) \tag{7}$$

where, d is the vapor bubble diameter, and η represents called the coefficient of adaptation and characterizes the adsorption of vapor molecules on the surface of the liquid, approximated by 1 under equilibrium conditions. L is the latent heat of vaporization, M is the molar mass, and R is the gas constant.

3. Model Analysis

3.1. Geometric Model

The simplified structure of the herringbone thrust bearing is shown in Figures 1 and 2. Compared with the general single-sided spiral groove structure, the lubricating medium at the inner and outer diameters is pumped into the groove when the bearing is in operation. This structure has excellent lubrication and sealing performance. Its structural expressions in the r–θ coordinate system are given by Equations (8) and (9):

$$R_{\text{in}} = R_1 e^{\theta\cot\beta} \tag{8}$$

$$R_{\text{out}} = R_2 e^{-\theta\cot\beta} \tag{9}$$

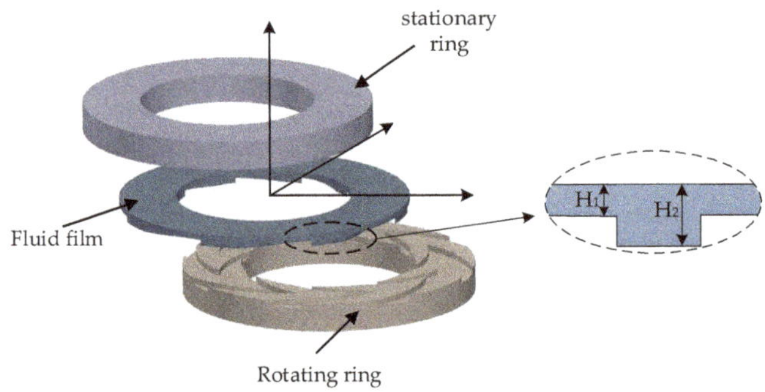

Figure 1. Bearing structures.

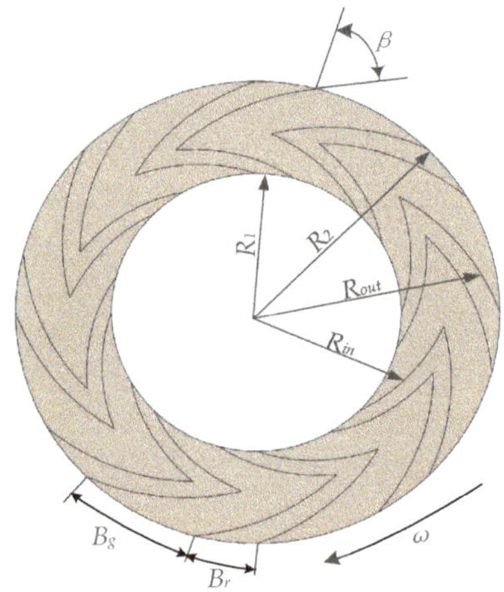

Figure 2. Surface texture model.

In Equations (8) and (9), R_1 is the inner diameter, R_2 represents the outer diameter, β is the spiral angle, θ denotes the angle, R_{in} is the inner spiral, R_{out} is the outer spiral, B_g is the groove width, B_r is the ridge width, B is the groove width ratio, $B = B_g/B_r$, H_2 is the water film depth, and H_1 is the bearing clearance.

3.2. CFD Model

The calculation model is simplified on the premise of assuring the accuracy of the calculation results and given the following assumptions: the effects of body force and inertial force are ignored, the speed slip is neglected, the effect of surface roughness is ignored.

A CFD model is built by considering the above assumptions. Solidworks is used for geometric modeling, and ANSYS is used for the meshing. Finally, the case is solved in Fluent. The film thickness direction is enlarged 100 times for facilitating observation. The grid and boundary conditions for the clearance water film are presented in Figure 3. The three-dimensional geometry of the clearance was discretized with Hexahedral elements, the quality of the grid was assessed by checking the elements' quality (0.98), aspect ratio (1.1), and skewness (0.04). Then, it is indicating that the grid was of high quality. Due to the existence of fluid viscosity, there will be a boundary layer when the fluid moves in the near-wall region. As shown in Figure 3b, grid refinements are made on both sides of the wall. The inlet and outlet of the three-dimensional model are set as pressure boundaries, the walls in contact with the rotating ring are set as rotating walls, and the others are set as static walls.

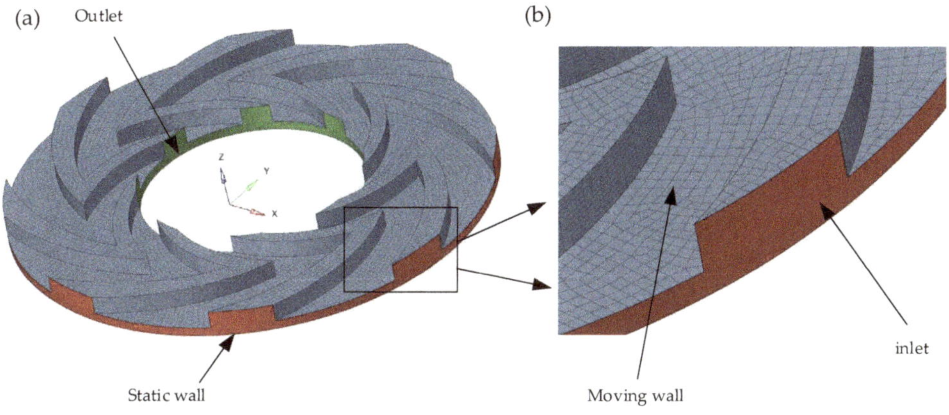

Figure 3. Grid and boundary conditions for the water film. (**a**) grid of the water film; (**b**) detailed information on the grid.

3.3. Viscosity–Temperature Relationship

For the high-temperature sealing medium, the change in medium viscosity with temperature must be considered. Roelands equation [26] is as follows:

$$\mu = \mu_0 \exp[\varepsilon(\frac{1}{T} - \frac{1}{T_0})] \tag{10}$$

where μ is the dynamic viscosity of water, T is the water temperature, and μ_0 is the viscosity at the reference temperature T_0.

3.4. Solver Settings

A mixture model was selected and analyzed. Then, the implicit method was adopted. In this study, liquid water was set as the primary phase, and water vapor was set as the secondary phase. The two phases were embedded in the mass transfer mechanism. "Dispersed" was selected as the interface type. The Zwart–Gerber–Belamri model was used in the simulation.

A pressure-based solver was chosen for the numerical simulation. The SIMPLE algorithm was used, which is an algorithm based on the pressure-velocity coupling principle. Then, the default state was used in other settings. A tolerance of 10^{-3} was set for the residual terms.

3.5. Verification of the Model

The work in Ref. [27] is similar to that of this paper, and the research method has been proved by experimental tests. Therefore, the results in Ref. [27] can be used to confirm the model established in this paper, and the result is given in Figure 4. It is found that the cavitation position obtained from the model established in this paper is consistent with the result in the literature [27].

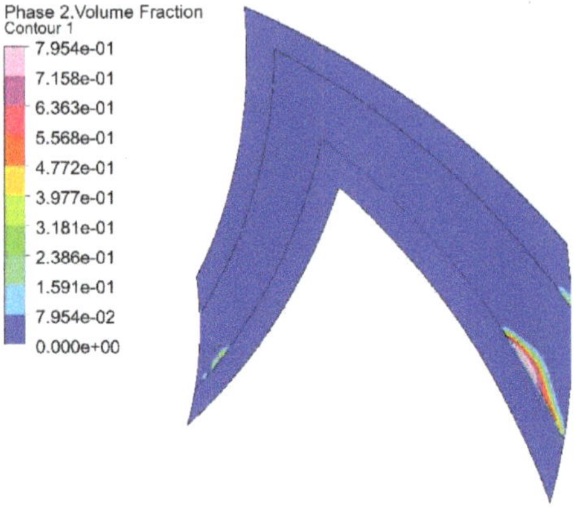

Figure 4. Vapor phase volume fraction distribution.

The leakage was obtained experimentally and validated as a quantitative indicator. As shown in Figure 5, the results of current work calculation are analogous with theoretical values and tests in the literature [27]. The relative error between this study and Ref. [27] is less than 5%, which proves the reliability of the proposed model.

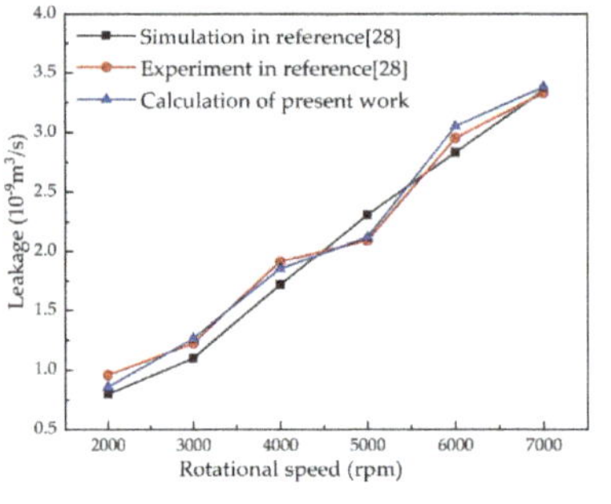

Figure 5. Verification of the model by using reference [27,28].

4. Results and Discussion

The subsequent simulation solution takes a single cycle model caused by the flow's periodicity. Then, computer resources are saved, and the accuracy of solution is improved. The geometric and working parameters of the face seal are shown in Table 1.

Table 1. Geometric and operating parameters.

Parameters	Value
Temperature T (K)	300
Groove number N	18
rotational speed n (rpm)	6000
Spiral groove angle β (°)	65
Inner radius R_1 (mm)	15
Outer radius R_2 (mm)	20
Pressure inlet P_{in} (MPa)	0.1
Pressure outlet P_{out} (MPa)	0.2
Water film depth H_2 (μm)	30
Bearing clearance H_1 (μm)	10
Density ρ (kg/m^3)	998.2
Viscosity μ (Pa·s)	8.49×10^{-4}

4.1. Liquid Film Flow Mechanism

The flow, which is inside a micro gap, is difficult to capture. Thus, its flow characteristics cannot be accurately understood. In this paper, a cross-sectional approach was chosen to demonstrate micro gap flow. The method is as follows: the actual film thickness is used in the simulations, but it is scaled up in the direction of the film thickness when the detailed flow is charactered.

When HGTB is working, the lubricating medium for groove structure is pumped into the top. Under the obstruction of the ridge, a part of the fluid slows down and increases the pressure, and the other part of the fluid climbs over the ridge. Thus, the most complex area of flow is at the top of the groove. Figures 6 and 7 show the pressure distribution and velocity field distribution on the radius section at the top of the groove, respectively. As shown in Figure 6, the pressure in this section is concentrated at the top of the groove. Because of the extremely small gap size, there is no significant pressure change in the thickness direction of the film. From the results in Figure 7, eddy currents exist in the gap. The fluid is divided into two parts: one part creates eddy currents in the groove, and the other part goes to lubricate the bearing clearance.

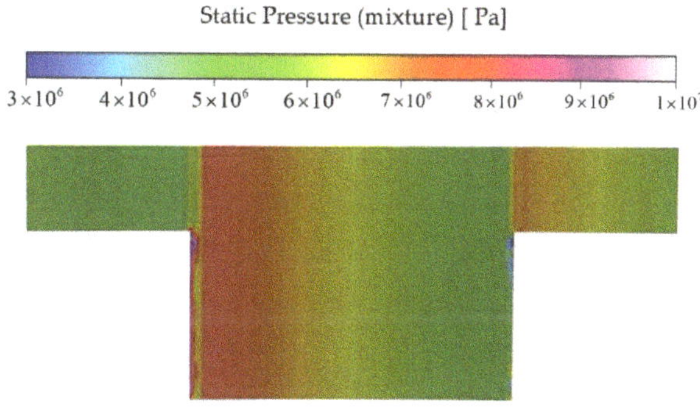

Figure 6. Pressure distribution of cross-section in the gap.

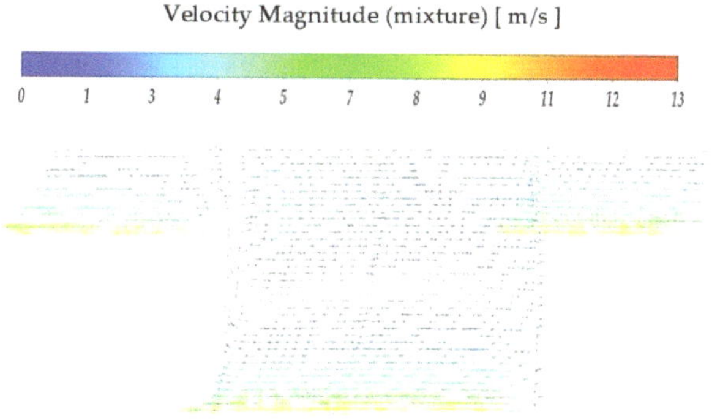

Figure 7. Velocity distribution of cross-section in the gap.

In Figure 8, when the groove surface moves, the fluid in the groove is pushed from one side to the other. Fluid flow characteristics in the micro gap are abundant. The fluid is squeezed at the front face, and a pressure peak is generated. The fluid after the collision is divided into two parts: one forms a vortex in the groove, and the other flows to the bearing gap. The fluid in the clearance acts as a lubricating medium, and the bearing surfaces are pushed apart. Thus, the load-carrying capacity is increasing, and the friction coefficient is reduced [28].

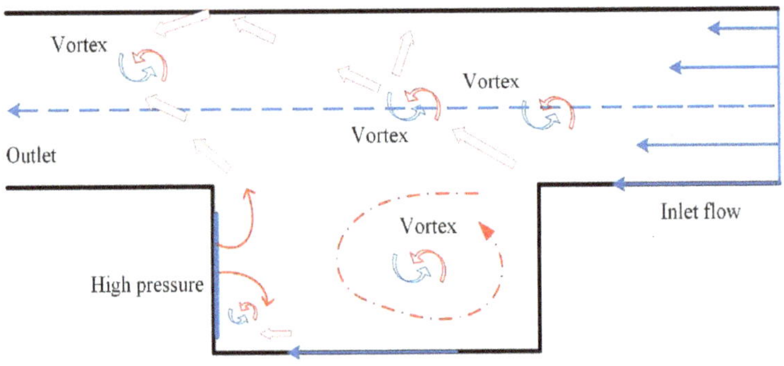

Figure 8. Schematic representation of the fluid motion within bearing clearance.

4.2. Influence of Rotational Speed on the Flow Field

When the speed is changing, and other parameters remain constant. The fluid film pressure distribution can be seen in Figure 9. A high-pressure area appears at the tip of the herringbone groove. Low-pressure areas appear at the groove root of the outer diameter side of the herringbone groove, the groove root, and side wall of the inner diameter side groove. As shown in Figure 10, cavitation takes place at the root of the outer diameter side groove and the side wall of the inner diameter.

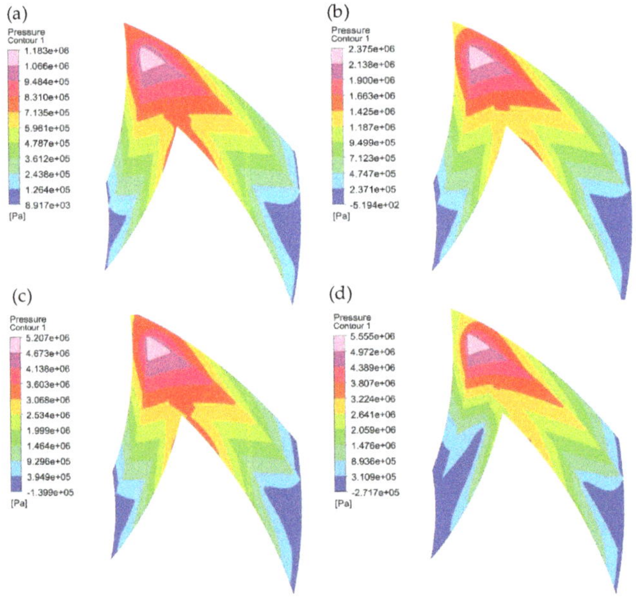

Figure 9. Liquid film pressure distribution. (**a**) 2000 rpm; (**b**) 4000 rpm; (**c**) 6000 rpm; (**d**) 8000 rpm.

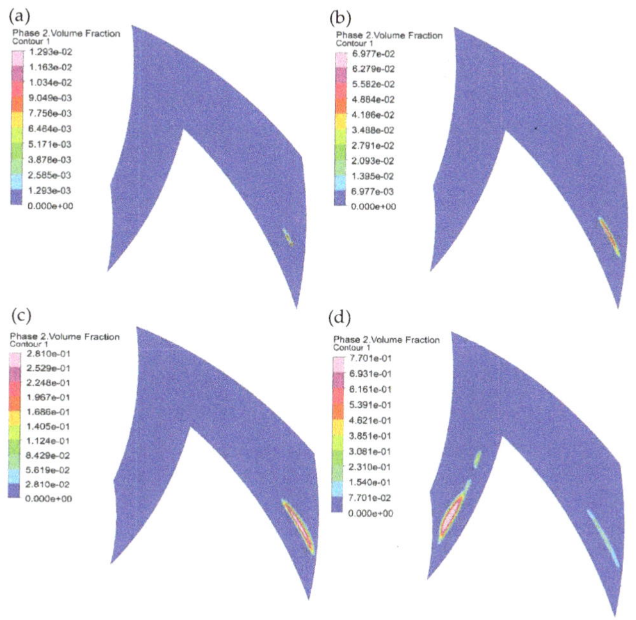

Figure 10. Liquid film gas phase volume fraction distribution. (**a**) 2000 rpm; (**b**) 4000 rpm; (**c**) 6000 rpm; (**d**) 8000 rpm.

As the rotational speed increases, the maximum pressure of the fluid film magnifies, and the pressure distribution law of the fluid film is unchanged. The cavitation area gradually expands with the increasement of the rotational speed, which indicates that the cavitation area is extensible.

4.3. Influence of Structural Parameters on the Flow Field

4.3.1. Spiral Angle

The spiral angle β directly affects the boosting effect of the thrust bearing. In the Figures 11 and 12, the effect of the spiral angle on the bearing's load-carrying capacity and leakage rate was investigated. The load capacity appears to increase with spiral angle from 50° to 70°. Then, it decreases as the spiral angle increases. When the spiral angle is about 68°, the bearing capacity reaches the maximum value. When the spiral angle is 0° or 90°, the dynamic pressure effect and step effect are greatly reduced, and the bearing capacity is almost lost. Thus, the leakage exhibits the downtrend.

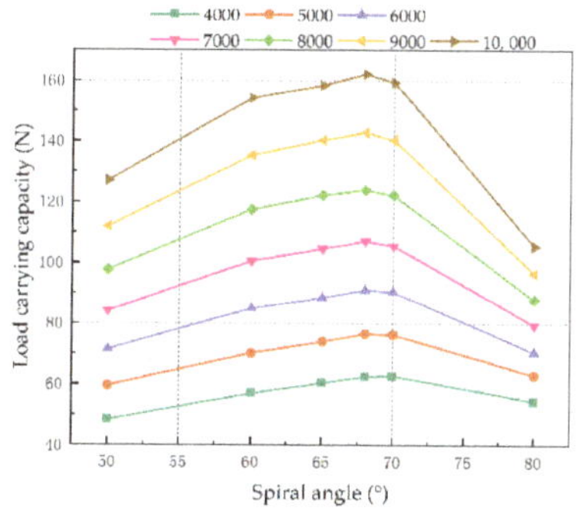

Figure 11. Effect of the spiral angle on the load-carrying capacity.

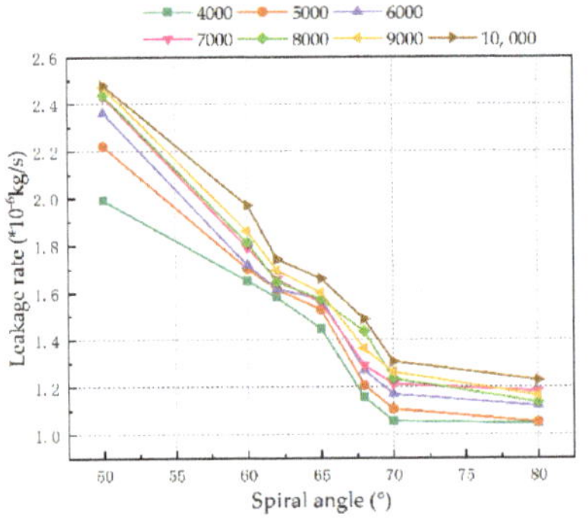

Figure 12. Effect of the spiral angle on the leakage rate.

4.3.2. Groove Width Ratio

From Figure 13, as the groove width ratio increases, the axial load capacity of the thrust bearing gradually increases. When the groove width ratio is about 1, the bearing capacity reaches the maximum value, and then gradually decreases. If the groove width ratio is equal to 0 or infinite, the bearing end face becomes flat, and the dynamic pressure effect and step effect cannot be found. Therefore, there is a suitable groove width ratio to maximize the bearing capacity. As seen in Figure 14, when the groove width ratio increases, the leakage also increases and tends toward a stable value. The magnified groove width ratio corresponds indirectly increases the clearance of the bearing, and the leakage is also increasing.

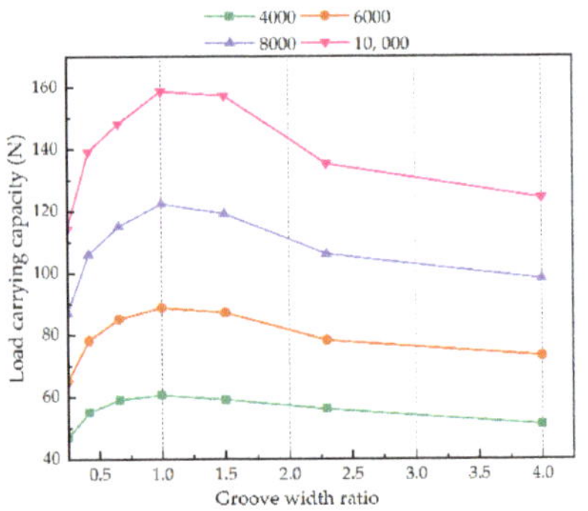

Figure 13. Effect of the groove width ratio on the load-carrying capacity.

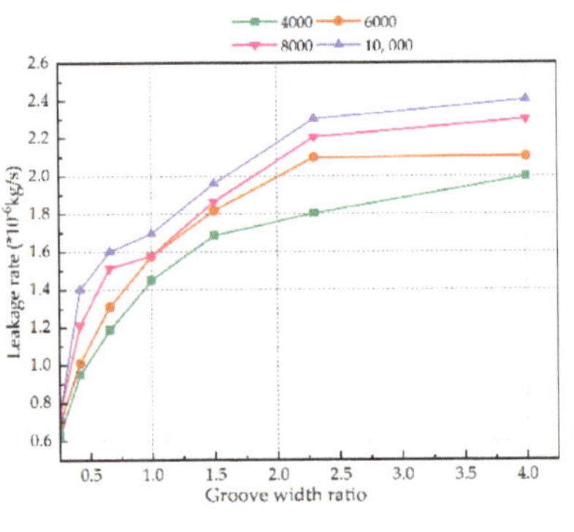

Figure 14. Effect of the groove width ratio on the leakage rate.

5. Conclusions

In this study, a finite element flow field model is established to understand and explain the flow mechanisms of HGTBs. The influence of structural parameters on the bearing performance was investigated. In this model, viscous temperature changes, turbulence, and cavitation are considered. The main conclusions are summarized as follows:

(1) Water is pumped into the spiral groove from both ends by vicious and differential pressure forces. The lubricating medium is converged at the top of the herringbone groove. Then, one part of the fluid goes over the ridge, and the other part forms vortices in the groove. The lubrication performance of the thrust bearing is affected by the dynamic pressure and the step.

(2) The cavitation is concentrated at the root of the outer diameter side of the herringbone groove and the inner diameter wall. The cavitated area is extended by the increasing rotational speed. There is an optimum value for the groove width ratio B and the helix angle β, which can improve the bearing performance.

(3) Optimum values for the geometrical parameters of the thrust bearing have been determined using the load-carrying capacity and leakage. When the bearing clearance is 10 μm, the groove width ratio is taken as 1, and the spiral angle β can be selected from 65° to 70°.

Author Contributions: Conceptualization, H.R.; methodology, H.R.; software, H.R.; validation, J.W. and S.Y.; formal analysis, F.W.; investigation, F.W.; resources, P.D. and J.W.; data curation, P.D. and S.Y.; writing—original draft preparation, H.R.; writing—review and editing, F.W. and J.W.; visualization, P.D. and G.Z.; supervision, X.Y. and S.Y.; project administration, X.Y. All authors have read and agreed to the published version of the manuscript.

Funding: The work described in this paper was supported by the National Natural Science Foundation of China (No. 51905001), the Science and Technology of Wuhu (No. 2021yf09), and the Anhui Province Science and Technology Major Project (202103a05020033), Research Project of Anhui Polytechnic University (No. Xjky2022012).

Institutional Review Board Statement: Not applicable.

Informed Consent Statement: Not applicable.

Data Availability Statement: The data that support the findings of this study are available from the author, H.R., upon reasonable request.

Conflicts of Interest: The authors declare no conflict of interest.

References

1. Wang, C.-C.; Lin, J.-T. Numerical study of hydrodynamic herringbone-grooved journal bearings combined with thrust bearings considering thermal effects. *J. Mech.* **2022**, *38*, 13–21. [CrossRef]
2. Yu, Y.; Pu, G.; Jiang, T.; Jiang, K. Discontinuous grooves in thrust air bearings designed with CAPSO algorithm. *Int. J. Mech. Sci.* **2020**, *165*, 105197. [CrossRef]
3. Wagner, P.H.; Van Herle, J.; Schiffmann, J. Theoretical and Experimental Investigation of a Small-Scale, High-Speed, and Oil-Free Radial Anode Off-Gas Recirculation Fan for Solid Oxide Fuel Cell Systems. *J. Eng. Gas Turbines Power* **2020**, *142*, 041023. [CrossRef]
4. Qiao, J.; Zhou, G.; Pu, W.; Li, R.; He, M. Coupling analysis of turbulent and mixed lubrication of water-lubricated rubber bearings. *Tribol. Int.* **2022**, *172*, 107644. [CrossRef]
5. Liu, Q.; Ouyang, W.; Cheng, Q.; Li, J.; Cheng, Q.; Li, R. Influences of bidirectional shaft inclination on lubrication and dynamic characteristics of the water-lubricated stern bearing. *Mech. Syst. Signal Process.* **2022**, *169*, 108623. [CrossRef]
6. Lin, X.; Wang, S.; Jiang, S.; Zhang, S. Thermohydrodynamic Model of Cavitating Flow and Dynamic Characteristic Calculation for High-Speed Water-Lubricated Pump-Out Spiral Groove Bearing. *Tribol. Trans.* **2020**, *63*, 736–755. [CrossRef]
7. Zhang, S.; Jiang, S.; Lin, X. Static and dynamic characteristics of high-speed water-lubricated spiral-groove thrust bearing considering cavitating and centrifugal effects. *Tribol. Int.* **2020**, *145*, 106159. [CrossRef]
8. Wang, D.; Mo, J.; Wang, Z.; Chen, G.; Ouyang, H.; Zhou, Z. Numerical study of friction-induced vibration and noise on groove-textured surface. *Tribol. Int.* **2013**, *64*, 1–7. [CrossRef]
9. Feng, H.; Jiang, S.; Shang-Guan, Y. Three-dimensional computational fluid dynamic analysis of high-speed water-lubricated hydrodynamic journal bearing with groove texture considering turbulence. *Proc. Inst. Mech. Eng. Part J. J. Eng. Tribol.* **2021**, *235*, 2272–2286. [CrossRef]

10. Guo, Z.; Xie, X.; Yuan, C.; Bai, X. Study on influence of micro convex textures on tribological performances of UHMWPE material under the water-lubricated conditions. *Wear* **2019**, *426–427*, 1327–1335. [CrossRef]
11. Chang, T.; Guo, Z.; Yuan, C. Study on influence of Koch snowflake surface texture on tribological performance for marine water-lubricated bearings. *Tribol. Int.* **2019**, *129*, 29–37. [CrossRef]
12. Feng, H.; Peng, L. Numerical analysis of water-lubricated thrust bearing with groove texture considering turbulence and cavitation. *Ind. Lubr. Tribol.* **2018**, *70*, 1127–1136. [CrossRef]
13. Xie, Z.; Shen, N.; Ge, J.; Zhu, W.; Song, P.; Liu, H.; Hao, L.; Tian, W. Analysis of the flow noises of the nuclear main pump caused by the high temperature liquid Sodium in the two-circuit main loop liquid Sodium pump system. *Ann. Nucl. Energy* **2020**, *145*, 107550. [CrossRef]
14. Guenat, E.; Schiffmann, J. Dynamic force coefficients identification on air-lubricated herringbone grooved journal bearing. *Mech. Syst. Signal Process.* **2020**, *136*, 106498. [CrossRef]
15. Guenat, E.; Schiffmann, J. Performance potential of gas foil thrust bearings enhanced with spiral grooves. *Tribol. Int.* **2019**, *131*, 438–445. [CrossRef]
16. Wang, L.; Han, Y.; Tang, D.; Cai, J. Numerical and Experimental Investigations on the Wear Behavior of Water-Lubricated Bearings with Different Materials. *Ind. Lubr. Tribol.* **2022**, *74*, 134–143. [CrossRef]
17. Song, Y.; Gu, C.-W. Development and Validation of a Three-Dimensional Computational Fluid Dynamics Analysis for Journal Bearings Considering Cavitation and Conjugate Heat Transfer. *J. Eng. Gas Turbines Power* **2015**, *137*, 122502. [CrossRef]
18. Wang, Y.; Yin, Z.; Jiang, D.; Gao, G.; Zhang, X. Study of the lubrication performance of water-lubricated journal bearings with CFD and FSI method. *Ind. Lubr. Tribol.* **2016**, *68*, 341–348. [CrossRef]
19. Grando, F.P.; Priest, M.; Prata, A.T. A two-phase flow approach to cavitation modelling in journal bearings. *Tribol. Lett.* **2006**, *21*, 233–244. [CrossRef]
20. Wang, L.; Zeng, Q.; Lu, C.; Liang, P. A numerical analysis and experimental investigation of three oil grooves sleeve bearing performance. *Ind. Lubr. Tribol.* **2019**, *71*, 181–187. [CrossRef]
21. Zhang, X.; Yin, Z.; Gao, G.; Li, Z. Determination of stiffness coefficients of hydrodynamic water-lubricated plain journal bearings. *Tribol. Int.* **2015**, *85*, 37–47. [CrossRef]
22. Chen, Y.; Sun, Y.; He, Q.; Feng, J. Elastohydrodynamic Behavior Analysis of Journal Bearing Using Fluid–Structure Interaction Considering Cavitation. *Arab. J. Sci. Eng.* **2019**, *44*, 1305–1320. [CrossRef]
23. Taghipour, Y.; Akbarzadeh, P.; Moradgholi, F.; Yazdi, M.E. Numerical study of the cavitation effect on plain bearings in constant and variable viscosity states. *Meccanica* **2021**, *56*, 2507–2516. [CrossRef]
24. Chen, H.; Wu, Q.; Xu, C.; Zuo, M. Research on Cavitation Regions of Upstream Pumping Mechanical Seal Based on Dynamic Mesh Technique. *Adv. Mech. Eng.* **2014**, *6*, 821058. [CrossRef]
25. Xu, X.D.; Ma, C.B.; Sun, J.J.; Zhang, Y.Y.; Yu, Q.P. Influence and optimization of groove structure parameters on vaporization characteristics of liquid film mechanical seals based on optimal mass transfer coefficient. *Chem. J.* **2022**, *73*, 1147–1156.
26. Bair, S. A Routine High-Pressure Viscometer for Accurate Measurements to 1 GPa. *Tribol. Trans.* **2004**, *47*, 356–360. [CrossRef]
27. Yang, D.D. Steady Character Research and Software Development of Liquid Film Seal with Herringbone Grooves. Master's Thesis, China University of Petroleum, Beijing, China, 2016. Available online: https://kns.cnki.net/KCMS/detail/detail.aspx?dbname=CMFD201801&filename=1018814320.nh (accessed on 8 July 2022).
28. Xie, Z.; Zhang, Y.; Zhou, J.; Zhu, W. Theoretical and experimental research on the micro interface lubrication regime of water lubricated bearing. *Mech. Syst. Signal Process.* **2021**, *151*, 107422. [CrossRef]

lubricants

Article

Rheological and Frictional Properties of Lithium Complex Grease with Graphene Additives

Yanshuang Wang [1,2], Xudong Gao [1,2], Jianghai Lin [1,2,*] and Pu Zhang [1,2]

1 School of Mechanical and Automotive Engineering, Qilu University of Technology (Shandong Academy of Sciences), Jinan 250353, China; wys2021@qlu.edu.cn (Y.W.); 201601011161@stu.qlu.edu.cn (X.G.); 10431200115@stu.qlu.edu.cn (P.Z.)
2 Shandong Institute of Mechanical Design and Research, Jinan 250031, China
* Correspondence: ljh2021@qlu.edu.cn

Abstract: Few-layer graphene (FLG) was added as a nano-additive to lithium complex grease (LCG) to explore the influence of FLG on the microstructure, viscoelasticity, friction and wear properties of LCG. Studies have found that the addition of FLG makes the microstructure of the thickener more compact, which in turn leads to an increase in the viscoelasticity of LCG. FLG additives can improve the viscosity-temperature properties of the grease and change the elastic deformation response to temperature changes. Among the temperatures selected in this article, the effect of graphene is more obvious at 70 °C. During the friction process, a proper amount of FLG can quickly form a boundary film and is not easily damaged, thereby optimizing the friction and wear performance of LCG.

Keywords: few-layer graphene; lithium complex grease; viscoelasticity; friction and wear properties

Citation: Wang, Y.; Gao, X.; Lin, J.; Zhang, P. Rheological and Frictional Properties of Lithium Complex Grease with Graphene Additives. *Lubricants* **2022**, *10*, 57. https://doi.org/10.3390/lubricants10040057

Received: 10 March 2022
Accepted: 29 March 2022
Published: 1 April 2022

1. Introduction

Grease is an indispensable part of bearings [1]. Recently, lubricating grease has received extensive attention among engineers and scientists in academia as a result of its complex and unique rheological properties. Generally, a series of rheological parameters are selected to ascertain the rheological properties of grease in both qualitative and quantitative terms. Lithium-based grease is one of the most common greases, and the factors affecting its rheology have attracted the attention of many scholars. Enhui Zhang et al. [2] prepared four lithium-based greases with paraffin oil, naphthenic oil, poly-α-olefin and polyol ester, respectively, and studied the influence of base oil on the rheological and frictional properties of lithium-based greases. It was found that naphthenic oil-based grease exhibited the best colloidal stability, and the thickener fibers in the polyethylene-based grease show a relatively flat network. Compared with the planar network structure, the lithium soap fiber with a three-dimensional entangled structure has a stronger friction reducing ability. Xiaoqiang Fan et al. [3] studied the influence of different thickeners on the friction characteristics of greases. They found that the excellent lubricating performance of the grease is mainly attributed to the properties and structure of the thickener and the synergistic effect of the grease film composed of sulfate and complex oxides and the tribochemical reaction film. Although additives account for a relatively low proportion of the grease by mass or volume, they have a greater influence on the rheological properties thereof.

Many nano-materials can be used as solid lubricants and can also be used as additives in lubricating oils and greases. Nikhil Kumar et al. added nanoparticles of Talc [4] and PTFE particles [5] into lithium-based grease to study its rheology and friction properties. It is found that the smaller the size of the additive material, the better the friction performance of the lithium-based grease. Compared with other shapes, the spherical particles can improve the friction performance of the lithium-based grease. Few-layer graphene (FLG) is a kind of nano material with many excellent properties. It has good thermal conductivity and can significantly improve the cooling performance of lubricants [6]. As a coating, it can improve the corrosion resistance of metals [7]. After investigation, it is found that as a lubricating additive, the ease of the peeling of the FLG facilitates the formation of a boundary film under a frictional load, which not only prevents direct contact between friction pairs, and improves the flexibility and adaptive ability of the friction interface, but, as the FLG spacing in the boundary film is relatively large, and the van der Waals force between the layers is low, this can promote lubrication by reducing the energy barrier between layers [8–10].

Some studies have found that adding FLG to lubricating oil can significantly improve the lubricating performance and friction performance of the lubricating oil; the friction coefficient (COF), wear scar diameter (WSD), and wear volume have been significantly reduced [11–16]. Tiancheng Ouyang et al. [17] added FLG to PAO6 base oil. Under low speed and heavy load conditions (4.2 mm/s and 1.0 GPa contact pressure), COF and wear volume loss can be reduced by as much as 29.1% and 55%, respectively.

For non-Newtonian greases, FLG also has a significant effect on its rheology. Mohamed et al. [18] added FLG and multi-walled carbon nanotubes to a composite calcium-based grease to make a mixed calcium nano-grease, and conducted rheological experiments using the Brookfield DV-III programmable rheometer, and found that the shear stress and viscosity increased with the addition of nano-additives. In recent years, researchers have added FLG to lithium grease to study the friction characteristics, and found that the addition of FLG has improved their anti-wear and anti-friction properties. Zhanjun Li et al. [19] added FLG to lithium-based grease and found that when the FLG content is 0.1%, the COF reduction rate is the largest, and the wear scar diameter (WSD) is the smallest, and FLG has the most obvious effect under a larger load and lower speed. The main component of the chemical reaction film on the friction surface of pure lithium grease is Fe_3O_4. After adding FLG, the main components of the boundary film become Fe_2O_3 and adsorbed FLG. Jin Zhang et al. [20] found through research that, compared with pure lithium grease with an FLG content of 2 wt%, the average friction coefficient is reduced by 27%, and the solder joint and load wear index are 1.6 and 1.4 times that of pure lithium grease, respectively. Tiancheng Ouyang et al. [21] found that the addition of FLG can significantly improve the anti-friction and anti-wear ability of the grease, thereby making the wear surface smoother. In particular, compared with lithium grease, the wear amount and COF of lithium grease containing 0.3 wt% FLG can be reduced by 52.0% and 20.3%, respectively. Ruslan Aziev et al. [22] found that after FLG was added to LCG, the tribochemical reaction between 12-hydroxystearate molecules and the surface of FLG occurred, resulting in an increase in the welding load and a decrease in the WSD. Bo Lin et al. [23] added graphene to lithium-based grease to study its tribological behavior under three operating conditions: slow moving heavy load, medium load speed and high speed light load. The study found that under different working conditions, the optimum content of graphene that makes the grease with the best anti-wear and anti-friction ability is different.

FLG composite materials can also improve the lubricating ability and anti-wear and anti-wear ability of lithium-based greases. Jin, Bao et al. [24] developed a Mn_3O_4/FLG (Mn_3O_4#G) nanocomposite and used it as a nano-additive for lithium-based grease. The study found that under the conditions of 100 °C and 150N load, when the concentration of Mn_3O_4#G is 0.03 wt%, COF and wear scar depth can be reduced by 35% and 76%, respectively, and the addition of Mn_3O_4#G enhances the ability of lithium-based grease to withstand high temperatures. Liu Xiaolong et al. [25] obtained FLG oxide by Hummers oxidation method and added it to lithium grease. The study found that FLG oxide as a grease additive can effectively reduce COF, reduce wear, extend lubrication time, and improve lubricating performance. FLG can also enhance its lubricating ability, anti-wear and anti-wear ability for other types of grease [26,27].

Most of the aforementioned projects involved the investigation of the influence of FLG additives on the viscosity of lubricating oils and greases, however, there is a lack of research on the influence of FLG additives on other rheological parameters such as yield strength, cross-stress, and elastic modulus, and there is no literature comprehensively covering the rheological properties parameters of greases with different FLG contents across a range of temperatures. There is no relevant report on research into the influence mechanism of FLG additives on the rheological properties and friction properties of LCG. In the present study, FLG was used as an additive to prepare LCG. Amplitude scanning and shear scanning experiments were conducted using a rheometer to measure the rheological properties of LCG containing different amounts of FLG at different temperatures. The SRV-4 friction tester was used to conduct friction experiments on LCG to study the influence of the content of FLG on the tribological behavior of LCG. A scanning electron microscope (SEM) was employed to characterize the micro-morphology of LCG, analyze the influence of FLG on the micro-morphology of LCG, and reveal the mechanism of the influence of FLG additives on the rheological and friction properties of LCG.

2. Materials and Methods

2.1. Material Processing and Sample Preparation

The thickener material of the LCG is 12-hydroxystearic acid (China Petrochemical Corporation Tianjin Branch, Tianjin, China) with the contents of 15%, and the base oil is mineral oil and PAO oil (China Petrochemical Corporation Tianjin Branch, Tianjin, China) with a viscosity of 220 cst. Lithium hydroxide aqueous solution (China Petrochemical Corporation Tianjin Branch, Tianjin, China) is added for saponification reaction to generate LCG. The additive is Changzhou sixth element graphene SE1231 (Changzhou Sixth Element Materials Technology Co., Ltd., Changzhou, China). The SE1231 graphene is a black powder, and the median particle size D_{50} is less than 10.0 μm (equivalent spherical diameter). Taking 50 g of mineral oil and PAO oil, we added a certain amount of FLG, mixed them and placed it in an ultrasonic cleaning machine for 2 h. We added 350 g of the LCG into the aforementioned mineral oil and PAO oil containing FLG to prepare an FLG mixed LCG with mass fractions of 0.5, 1, and 2 wt%. We used an SDF1100-type dispersing sand mill (Suzhou Qile Electronic Technology Co., Ltd., Suzhou, China) to mix and grind the specimens at 900 rpm for 2 h, so that the FLG is uniformly mixed within the LCG.

2.2. Experimental Procedure

The MCR302 rheometer (Anton Paar Instruments, Graz, Steiermark, Austria) was used to measure the rheological properties of the four LCG specimens. The experiment is divided into two parts: amplitude scan and shear scan. Lubricating grease is a viscoelastic fluid, and the shear scan uses the rheometer-rotation mode with the cp25-1 cone-plate test system with a diameter of 25 mm and an angle of 1.0°; the amplitude sweep uses the rheometer-oscillation mode with the pp25 parallel plate test system with a diameter of 25 mm. The experiment was conducted at three temperatures, respectively 30 °C, 70 °C and 130 °C (±0.1 °C). The rotation mode used in shear scanning is to apply strain or stress with continuous rotation in the same direction to obtain a constant shear-strain rate (i.e., a steady-state test). The controlled shear-strain rate (CSR) rotation mode is selected and the shear-strain rate $\dot{\gamma}$ (s-1) is set to obtain a certain shear strength τ (Pa) of the grease. The oscillation mode used in amplitude sweeps is a reciprocating mode of oscillation designed to apply constant strain or stress (i.e., a dynamic test). The controlled strain (CS) mode of oscillation is selected, and the shear strain is set to obtain a certain shear strength τ (Pa) of the grease in an oscillating state of strain. Before each test, pre-shearing is applied for between five and fifteen minutes to eliminate the residual stress in the grease.

The SRV-4 friction and wear tester (Optimol Instruments, Munich, Germany) was used to measure the friction properties of the four LCG specimens. The test steel ball has an accuracy of G10, a diameter of 10.318 mm, and a set load of 90 N. According to the Hertz formula, the working stress is 2.04 GPa. The test adopts a reciprocating mode with a frequency of 10 Hz and an amplitude of four mm. The test temperature was set to 30 °C and test time to 0.5 h. COF curves of the four LCG were obtained. Three friction tests were performed for each grease. The MIT300 metallurgical microscope (Chongqing Auto Optical Instrument Co., Ltd., Chongqing, China) was used to measure the WSD on the steel ball after each test.

3. Results

3.1. Effect of FLG on the Microstructure of LCG

Figure 1 shows the FT-IR spectra of LCG with different content of FLG. The IR bands appearing at 2890–2960 cm^{-1} and 1580 cm^{-1} were ascribed to C=O groups, respectively. The bands appearing at 1460 and 1375 cm^{-1} were ascribed to the stretching vibrations of C–H and O–H groups, respectively. For the bands appearing at 3500–4000 cm^{-1}, as the FLG content increases, the fluctuation becomes more obvious, indicating that they were ascribed to the stretching vibration of the C–C group in FLG.

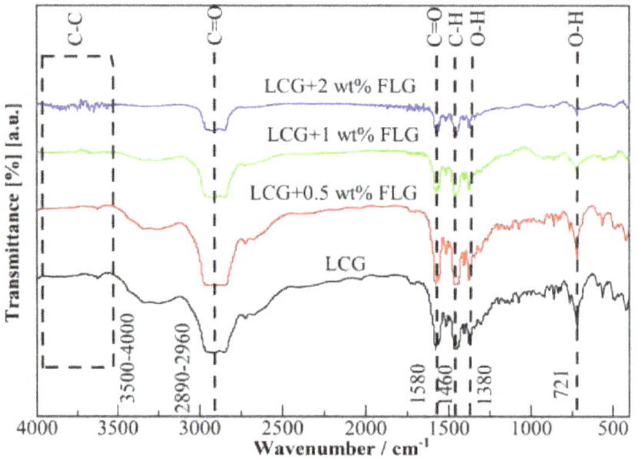

Figure 1. FT-IR spectra of LCG specimens.

Figure 2 shows SEM micrographs of the LCG with the base oil washed away and the LCG with 2 wt% FLG added. Figure 2a demonstrates that the microstructure of the LCG thickener is a fiber structure; in Figure 2b, 2 wt% FLG is added to the LCG, and it is found that the thickener fiber structure is wound on the surface of FLG; since FLG is polar, it will be preferentially adsorbed on the thickener of the same polarity when added, which increases the polarity of the thickener. Under the condition that the polarity of the base oil remains unchanged, this leads to a gap between the base oil and the thickener. The difference in polarity is greater, and the thickening agent is not easily dispersed in the base oil, resulting in a denser thickening agent.

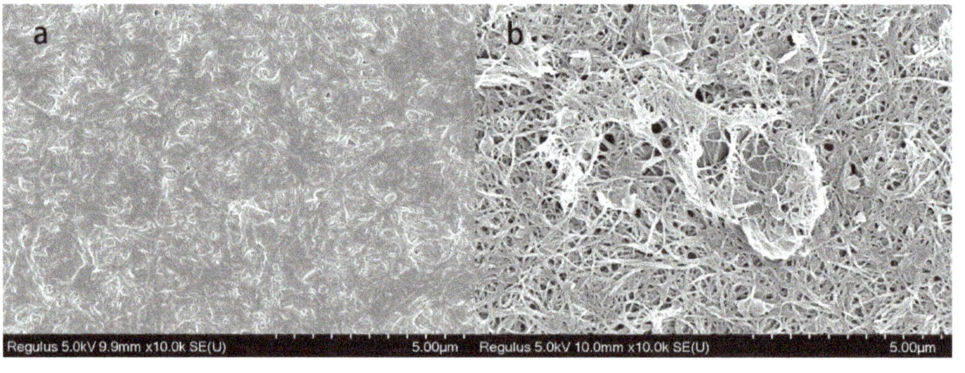

Figure 2. SEM micrograph of LCG specimens. (**a**) LCG; (**b**) LCG with 2 wt% FLG.

3.2. *Effect of FLG on the Viscoelasticity of LCG*

Grease is a semi-solid material that is both viscous and elastic. To study the viscoelastic properties of grease, the complex modulus is introduced [28]. The complex modulus of grease (G*, Pa) can be described by example (1):

$$G^* = G' + i\,G'', \tag{1}$$

where, G′ is the storage modulus, which represents the energy stored by the elastic deformation of the grease, Pa; G″ denotes the loss modulus, which represents the energy dissipated during viscous deformation of the grease, Pa.

A rheometer was used to perform amplitude sweep experiments on four LCG samples in oscillation mode, and the relationship between storage modulus G′ and loss modulus G″ with shear strain at three test temperatures (30 °C, 70 °C and 130 °C) was obtained (Figure 3). With the increase in shear stress, the storage modulus G′ continuously decreases, while the loss modulus G″ increases, before finally converging at the flow point. The stress corresponding to the flow point is called the cross-stress. Beforehand, the elasticity of the LCG plays a dominant role, and thereafter, the LCG exhibits flow characteristics, and the viscosity of the LCG plays a dominant role.

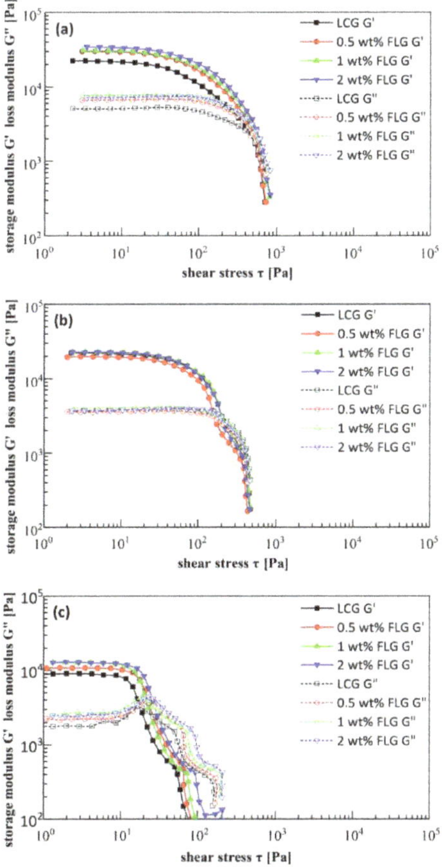

Figure 3. Changes in storage modulus G′ and loss modulus G″ of LCG with shear stress. (**a**) 30 °C; (**b**) 70 °C; (**c**) 130 °C.

It can be seen from Figures 3 and 4 that the higher the FLG content, the greater the storage modulus and loss modulus of LCG, and the dynamic yield stress and the cross stress also increased to varying degrees, which shows that FLG additives can enhance the elasticity and viscosity of LCG. Combined with the microscopic morphological examination, FLG increases the density of the LCG fiber structure and increases the ability of the grease to resist elastic deformation and viscous flow. The higher the temperature, the smaller the storage modulus and loss modulus of LCG, and the greater the reduction in loss modulus, which shows that temperature will reduce the elasticity and viscosity of LCG, and has a greater impact on viscosity. The dynamic yield stress and cross stress of the LCG also decrease with the increase in temperature, which is related to the Brownian motion of the molecules. The higher the temperature, the more intense the Brownian motion of the molecules and the smaller the intermolecular force, which leads to easier yielding of the structure of the LCG and stronger fluidity. According to Figure 4, as a whole, the more FLG content, the greater the dynamic yield stress. However, the cross stress only shows this trend at 30 °C. Under the influence of a temperature rise, the effect of FLG on the cross stress is weakened. Comparing the effects of FLG addition and temperature changes on the dynamic yield stress and cross stress, it is found that FLG has a greater impact on the dynamic yield stress, and temperature changes have a greater impact on the cross stress. It can be proved that FLG mainly affects the elasticity of LCG, and temperature mainly affects the viscosity of LCG. Furthermore, adding an appropriate amount of FLG can offset the effect of temperature on the elasticity and viscosity of the LCG, which means that FLG additives can improve the viscosity-temperature performance of the grease and the elastic deformation with temperature.

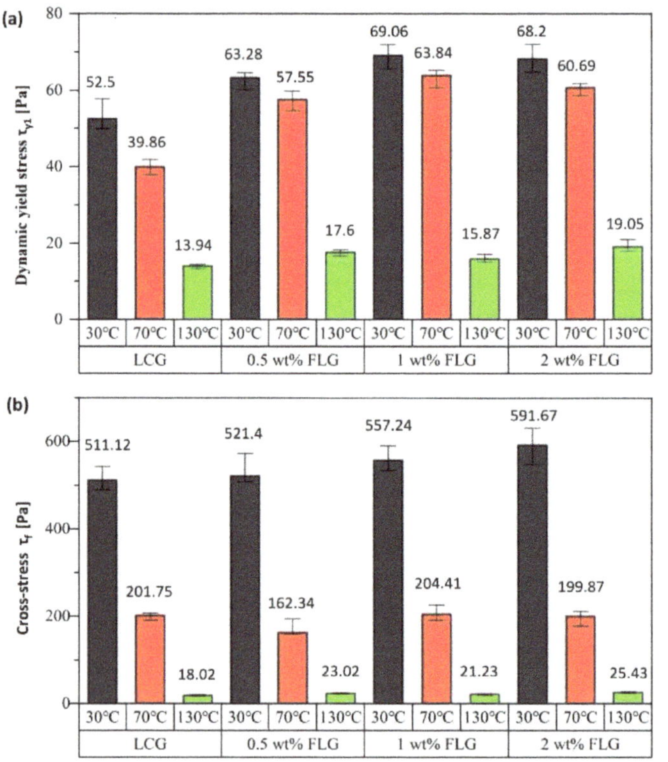

Figure 4. Dynamic yield stress and cross-stress of LCG. (**a**) Dynamic yield stress; (**b**) Cross-stress.

Figure 5 shows the variation in the shear stress versus the shear strain rate of the four greases at three temperatures, respectively: the shear strength (τ_y) of the four greases increases with the increase in the shear strain rate ($\dot{\gamma}$), indicating a non-linear trend. The regions of increasing shear stress are labeled a, c and e in Figure 5a, and a and c in Figure 5b,c. However, in area d (Figure 5a) and area b (Figure 5b,c), the increase in shear stress is much smaller than the overall growth trend, and this area is the shear yield area. The point where the shear stress begins to plateau is the yield point. Since the rotation mode is a steady state test, the corresponding shear stress is the steady state yield stress τ_{y2}. The steady-state yield stresses of the four LCGs at the three temperatures are given in Figure 6. In area b (Figure 5a), the shear stress tends to decrease (this area is the wall sliding area). The intermolecular resistance fails under the action of the external force and relative movement occurs between the molecular layers, and the ability of the LCG to resist deformation is weakened. Therefore, the measured shear stress begins to decrease, and the shear stress continues to increase until the wall-slip effect ends.

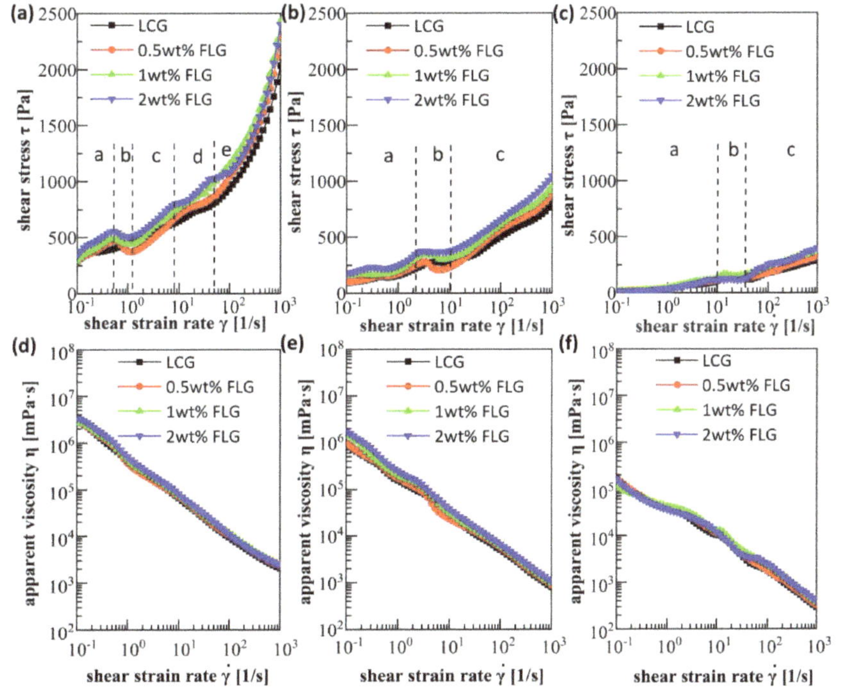

Figure 5. LCG shear stress and apparent viscosity curve: (**a–c**) shear stress; (**d–f**) apparent viscosity; (**a**,**d**) 30 °C; (**b**,**e**) 70 °C; (**c**,**f**) 130 °C.

According to Figure 5, it is found that at different temperatures, the higher the FLG content, the larger the shear stress borne by the LCG. It can be seen from Figure 2 that an increase in the amount of FLG can result in a denser microstructure of the LCG thickener and an increase in structural strength, resulting in an increase in the shear stress of the LCG as the amount of FLG increases. Through comparison of the data pertaining to each of two temperatures in Figure 5, it is found that with the increase in the shear strain rate, the shear stress of the LCG has a very obvious rising trend at 30 °C, the rising trend at 70 °C is relatively gentle, and the rising trend at 130 °C is more gentle. This means that the higher the temperature, the lower the rate at which the shear stress increases with the shear strain rate, and the lower the shear stress of the LCG. From the perspective of the influence of the viscoelasticity on shear stress, the addition of FLG is shown to increase the viscoelasticity of LCG, and an increase in the temperature will reduce its viscoelasticity.

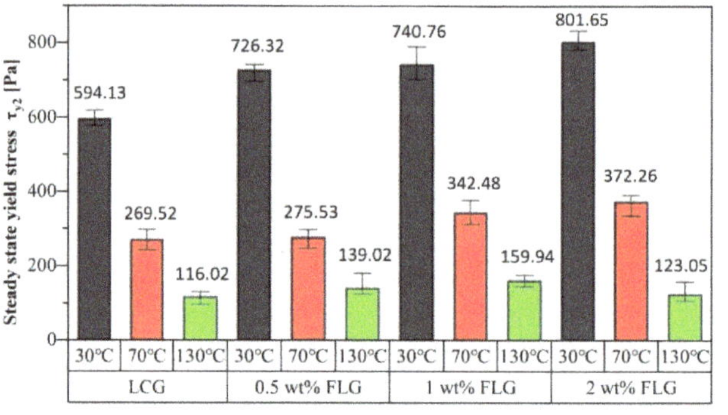

Figure 6. Steady state yield stress of LCG.

Comparing the yield zone of the LCG at different temperatures in Figure 5, combined with the steady-state yield stress data of Figure 6, it can be found that the steady-state yield stress of the same LCG is smaller at higher temperatures. Table 1 shows the effect of the FLG content on the steady-state yield stress of LCG. The steady-state yield stress of LCG with FLG contents of 0.5, one, and two is found to be increased by 22.25%, 24.68%, and 34.93%, respectively, at 30 °C compared with the LCG without FLG; the steady-state yield stress at 70 °C is increased by 2.23%, 27.07%, and 38.12%, respectively; the steady-state yield stress at 130 °C is increased by 19.82%, 38.86%, and 6.06%, respectively. On the whole, the addition of FLG at 70 °C increases the steady-state yield stress of LCG more significantly, indicating that among the three temperatures selected in the experiment, 70 °C is more suitable for FLG to play a role. And it shows that FLG additives can inhibit the temperature decrease in the yield strength of LCG.

Table 1. Steady state yield stress(τ_{y2}) of LCG at different FLG contents and temperatures.

	LCG	0.5 wt% FLG		1 wt% FLG		2 wt% FLG	
T [°C]	**τ_{y2} [Pa]**	**τ_{y2} [Pa]**	**growth rate [%]**	**τ_{y2} [Pa]**	**growth rate [%]**	**τ_{y2} [Pa]**	**growth rate [%]**
30	594.13	726.32	22.25	740.76	24.68	801.65	34.93
70	269.52	275.53	2.23	342.48	27.07	372.26	38.12
130	116.02	139.02	19.82	159.94	37.86	123.05	6.06

Viscosity is a characteristic of a substance that produces flow and irreversible deformation under stress. The apparent viscosity (η) is the ratio of the shear stress to the shear strain rate under steady flow. Figure 5a–c illustrates the changes in the apparent viscosity of the four greases with shear strain rate at different temperatures. The apparent viscosity of the four LCGs decreases with the increase in the shear strain rate, showing the phenomenon of shear thinning. This arises as some of the thickener fibers constituting the skeletal structure are oriented and unwound along the shear direction under the action of mechanical external forces, the skeleton is gradually destroyed and the shear resistance is reduced. According to Figure 5d–f, it is found that the higher the FLG content, the greater the apparent viscosity of the LCG; the higher the temperature, the lower the apparent viscosity of the LCG. This is due to the FLG additives densifying the structure of the LCG, resulting in an increase in flow resistance, which results in an increase in the apparent viscosity of the LCG. The higher the temperature, the more intense the Brownian motion of the molecules and the smaller the intermolecular forces, resulting in lower flow resistance, resulting in a decrease in the apparent viscosity of the LCG.

3.3. Effect of FLG on the Friction and Wear of LCG

The friction test used the SRV-4 friction test machine, which is lubricated with LCG containing 0 wt%, 0.5 wt%, 1 wt%, and 2 wt% FLG. The upper and lower friction pairs are made of GCr15 bearing steel. The test temperature was set to 30 °C, the load was 90N, the frequency was 10 Hz, and the amplitude was four mm. The test duration is 1800s.

Figure 7 shows the influence of FLG contents (0 wt%, 0.5 wt%, 1 wt%, and 2 wt%) on the friction curve of LCG. It can be seen that the friction curves of the LCG with FLG added are generally smaller than those of the LCG without FLG, and the friction curves of the LCG with a FLG content of 0.5 wt% are the smallest overall. The friction coefficients of the four LCG reached about 0.18 at 40 s, and then different contents of FLG began to work independently. Among them, LCG without FLG gradually formed the boundary film at 40 s, and the friction coefficient curve was relatively stable. At 220s, the boundary film gradually failed, the friction coefficient began to rise, and the friction coefficient reached about 0.2 at 320s. Then, the friction coefficient decreased slightly, indicating that the residual boundary film was still in effect, and a new boundary film was gradually formed. However, at 480s, the friction coefficient increased sharply to 0.23, and continued until the end of the friction experiment. At this moment, boundary lubrication is gradually formed. For LCG with a FLG content of 0.5 wt%, the friction curve fluctuates greatly from 40s to 160s. During this process, the disordered FLG in the LCG gradually becomes ordered under the action of the tangential force, and the boundary film is gradually formed. Then the friction curve fluctuated regularly and had a slight downward trend until the end of the friction experiment. This shows that under the action of 0.5 wt% FLG, the boundary film is more stable and can withstand longer friction without failure. The friction curve of LCG with a FLG content of 1 wt% did not fluctuate much from 40 s to the end of the test, and only fluctuated significantly between 900 s and 1000 s, indicating that 1 wt% of FLG also made the boundary film more stable. However, due to the increase in FLG content, the frictional resistance increases, resulting in the overall friction coefficient greater than that of LCG with FLG content of 0.5 wt%. LCG with a FLG content of 2 wt% began to fluctuate greatly at 40 s, and until 330 s, the friction coefficient gradually increased to about 0.21, and the effect of excessive FLG began to appear. From 330 s to the end of the test, the friction curve has been in a state of irregular fluctuation. This is due to the FLG content being too large and it is easy to agglomerate, however, under the action of the tangential force, the FLG has a tendency to change to an orderly state. FLG changes continuously in two states of agglomeration and order (as shown in Figure 8), so the friction curve fluctuates more obviously.

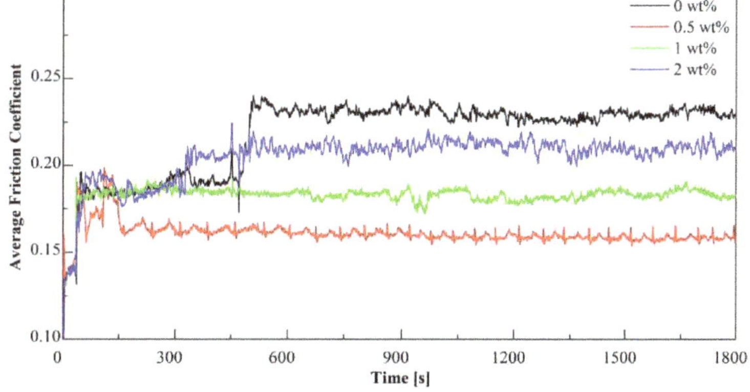

Figure 7. Average COF curve of LCG.

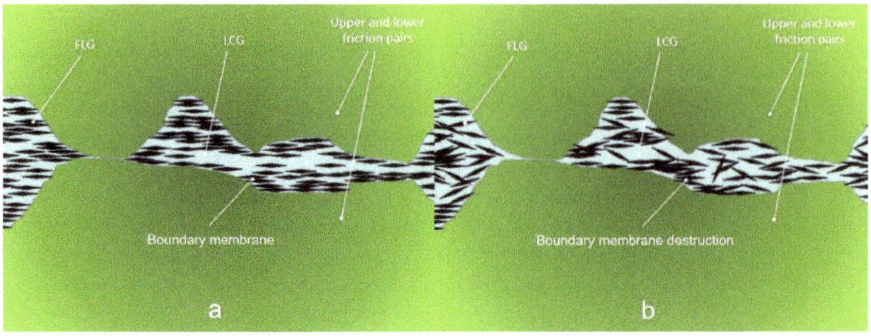

Figure 8. Two states of FLG in LCG: (**a**) Ordered arrangement; (**b**) Unordered aggregation.

Figure 9 shows the average wear scar diameter (WSD) and average friction coefficient (COF) for the four greases. It can be seen that the average WSD and average COF of LCG without FLG addition are the largest, and the average WSD and average COF of LCG with 0.5 wt% FLG content are the smallest. Figure 10 presents all the wear spot pictures of the four LCG, where Figure 10a–c are LCG, Figure 10d–f are LCG with 0.5 wt% FLG content, and Figure 10g–i are LCG with a FLG content of 1 wt%, Figure 10j–l are LCG with a FLG content of 2 wt%. According to Tables 2 and 3, LCG with a FLG content of 0.5 wt% has an increase of 52.05% in anti-friction ability and 19.60% in anti-friction ability compared with LCG without FLG. The LCG with a FLG content of 1 wt% increased the wear reduction ability by 16.06% and the anti-friction ability by 14.56%. The LCG with a FLG content of 2 wt% increased the wear reduction ability by 5.96% and the anti-friction ability by 7.09%. This shows that among the four selected lubricating greases, LCG with a FLG content of 0.5 wt% has the best anti-wear and anti-friction effect.

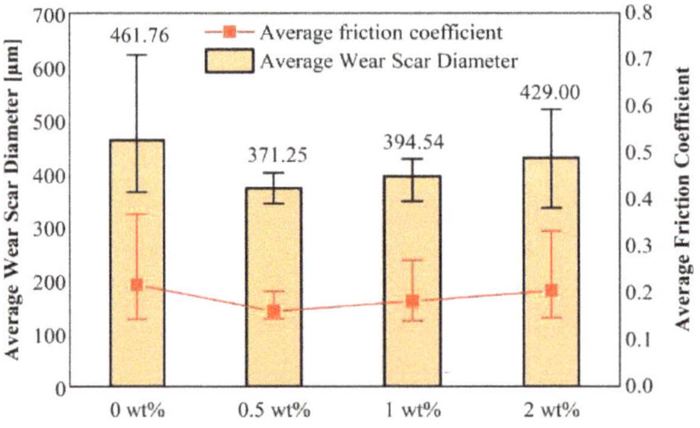

Figure 9. Average WSD and Average COF of LCG.

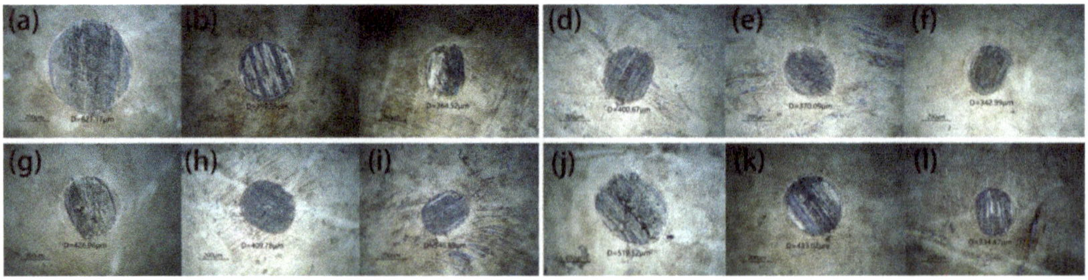

Figure 10. WSD of LCG: (**a–c**) LCG; (**d–f**) 0.5 wt% FLG; (**g–i**) 1 wt% FLG; (**j–l**) 2 wt% FLG.

Table 2. Average COF of LCG with different FLG contents.

LCG	0.5 wt% FLG		1 wt% FLG		2 wt% FLG	
COF [1]	COF [1]	Reduction Rate [%]	COF [1]	Reduction Rate [%]	COF [1]	Reduction Rate [%]
0.218	0.161	52.05	0.183	16.06	0.205	5.96

Table 3. Average WSD of LCG with different FLG contents.

LCG	0.5 wt% FLG		1 wt% FLG		2 wt% FLG	
WSD [μm]	WSD [μm]	Reduction Rate [%]	WSD [μm]	Reduction Rate [%]	WSD [μm]	Reduction Rate [%]
461.76	371.25	19.60	394.54	14.56	429.00	7.09

The wear volume of the steel ball was estimated once the height of the worn cap, h, was obtained by the following formula [23]:

$$h = r - \sqrt{r^2 - \frac{d^2}{4}} \tag{2}$$

$$V = \left(\frac{\pi h}{6}\right)\left(\frac{3d^2}{4} + h^2\right) \tag{3}$$

where r is the initial radius of the ball and d is WSD. The specific wear rate can be defined as the volume of material removed per unit load and sliding distance:

$$W = \frac{V}{P \cdot S} \tag{4}$$

where V is the calculated wear volume, P is the applied normal load, and S is the total sliding distance.

Figure 11 shows a histogram of the average wear rates for the four greases. It can be seen that the wear rate trends of the four greases are the same as those of their WSD and COF.

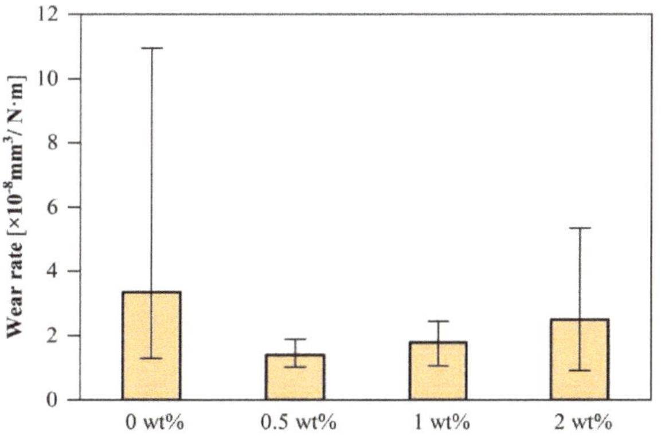

Figure 11. Wear rate of steel balls.

The above phenomenon shows that, for LCG without FLG, the boundary film that gradually formed during the friction process can reduce friction, yet is easily destroyed, resulting in relatively large wear. When the FLG content is too high, FLG can fully cover the friction surface, however, it is easy to polymerize during the friction process, so that the FLG changes continuously in the two states of disordered polymerization and ordered arrangement, which affects the anti-wear and anti-friction effect of the grease. Under the working conditions of this paper, the grease with FLG content of 0.5 wt% is the grease with the best friction and wear performance among the four greases.

4. Conclusions

1. The addition of FLG makes the fiber microstructure of the LCG more compact, resulting in greater structural strength, yield strength, apparent viscosity, storage modulus, and loss modulus of the LCG.
2. FLG can enhance the viscoelasticity of the LCG, and the increase in temperature will reduce the viscoelasticity of the LCG, and has a greater impact on the viscosity. Adding an appropriate amount of FLG can offset the effect of temperature on the elasticity and viscosity of the LCG, which means that FLG additives can improve the viscosity–temperature performance of the grease and the performance of the elastic deformation with temperature changes. Among the temperatures selected in this article, 70 °C is more suitable for FLG to play a role.
3. An appropriate amount of FLG can make the boundary film more stable, can withstand long-term friction without failure, and reduce wear. When the FLG content is too high, FLG can fully cover the friction surface, however, it is easy to polymerize during the friction process, which makes the FLG change continuously in the two states of disordered polymerization and ordered arrangement, resulting in the instability of the boundary film and affecting the friction effect. Under the working conditions of this paper, the grease with FLG content of 0.5 wt% is the grease with the best friction and wear performance among the four greases.

Author Contributions: Conceptualization, Y.W. and X.G.; methodology, Y.W.; software, X.G. and P.Z.; validation, Y.W., X.G. and P.Z.; formal analysis, Y.W., X.G. and P.Z.; investigation, X.G. and P.Z.; resources, Y.W. and J.L.; data curation, Y.W., X.G. and P.Z.; writing—original draft preparation, X.G. and P.Z.; writing—review and editing, Y.W. and J.L.; visualization, X.G.; supervision, Y.W. and J.L.; project administration, Y.W.; funding acquisition, Y.W. All authors have read and agreed to the published version of the manuscript.

Funding: This research was funded by National Natural Science Foundation of China (Grant No. 52075274 and 51475143) and Shandong Provincial Key Research and Development Program (Major Science and Technology Innovation Project) (Grant No. 2020CXGC011003).

Institutional Review Board Statement: Not applicable.

Informed Consent Statement: Not applicable.

Conflicts of Interest: The authors declare no conflict of interest.

Abbreviations

FLG	few-layer graphene
COF	friction coefficient
LCG	lithium complex grease
WSD	wear scar diameter
SEM	scanning electron microscope

References

1. Ma, S.; Zhang, X.; Yan, K.; Zhu, Y.; Hong, J. A Study on Bearing Dynamic Features under the Condition of Multiball–Cage Collision. *Lubricants.* **2022**, *10*, 9. [CrossRef]
2. Zhang, E.; Li, W.; Zhao, G.; Wang, Z.; Wang, X. A Study on Microstructure, Friction and Rheology of Four Lithium Greases Formulated with Four Different Base Oils. *Tribol. Lett.* **2021**, *69*, 98. [CrossRef]
3. Fan, X.; Li, W.; Li, H.; Zhu, M.; Xia, Y.; Wang, J. Probing the effect of thickener on tribological properties of lubricating greases. *Tribol. Int.* **2018**, *118*, 128–139. [CrossRef]

4. Kumar, N.; Saini, V.; Bijwe, J. Exploration of Talc nanoparticles to enhance the performance of Lithium grease. *Tribol. Int.* **2021**, *162*, 107107. [CrossRef]
5. Kumar, N.; Saini, V.; Bijwe, J. Performance properties of lithium greases with PTFE particles as additive: Controlling parameter-size or shape? *Tribol. Int.* **2020**, *148*, 106302. [CrossRef]
6. Huang, X.; Zhi, C.; Lin, Y.; Bao, H.; Wu, G.; Jiang, P.; Mai, Y.W. Thermal conductivity of graphene-based polymer nanocomposites. *Mater. Sci. Eng. R-Rep.* **2020**, *142*, 100577. [CrossRef]
7. Wu, W.; Chen, R.; Yang, Z.; He, Z.; Zhou, Y.; Lv, F. Corrosion resistance of 45 carbon steel enhanced by laser graphene-based coating. *Diam. Relat. Mat.* **2021**, *116*, 108370. [CrossRef]
8. Mao, J.; Chen, G.; Zhao, J.; He, Y.; Luo, J. An investigation on the tribological behaviors of steel/copper and steel/steel friction pairs via lubrication with a graphene additive. *Friction* **2021**, *9*, 228–238. [CrossRef]
9. Niu, M.; Qu, J.; Gu, L. Synthesis of titanium complex grease and effects of graphene on its tribological properties. *Tribol. Int.* **2019**, *140*, 105815. [CrossRef]
10. Wang, J.; Guo, X.; He, Y.; Jiang, M.; Gu, K. Tribological characteristics of graphene as grease additive under different contact forms. *Tribol. Int.* **2018**, *127*, 457–469. [CrossRef]
11. Zhao, J.; Huang, Y.; Li, Y.; Gao, T.; Dou, Z.; Mao, J.; Wang, H.; He, Y.; Li, S.; Luo, J. Superhigh-exfoliation graphene with a unique two-dimensional (2D) microstructure for lubrication application. *Appl. Surf. Sci.* **2020**, *513*, 145608. [CrossRef]
12. Curà, F.; Mura, A.; Adamo, F. Experimental investigation about tribological performance of grapheme-nanoplatelets as additive for lubricants. *Procedia Struct. Integr.* **2018**, *12*, 44–51. [CrossRef]
13. Zheng, D.; Cai, Z.; Shen, M.; Li, Z.; Zhu, M. Investigation of the tribology behaviour of the graphene nanosheets as oil additives on textured alloy cast iron surface. *Appl. Surf. Sci.* **2016**, *387*, 66–75. [CrossRef]
14. Wu, L.; Gu, L.; Jian, R. Lubrication mechanism of graphene nanoplates as oil additives for ceramics/steel sliding components. *Ceram. Int.* **2021**, *47*, 16935–16942. [CrossRef]
15. Radhika, P.; Sobhan, C.B.; Chakravorti, S. Improved tribological behavior of lubricating oil dispersed with hybrid nanoparticles of functionalized carbon spheres and graphene nano platelets. *Appl. Surf. Sci.* **2021**, *540*, 148402. [CrossRef]
16. Wang, W.; Zhang, G.; Xie, G. Ultralow concentration of graphene oxide nanosheets as oil-based lubricant additives. *Appl. Surf. Sci.* **2019**, *498*, 143683. [CrossRef]
17. Ouyang, T.; Shen, Y.; Lei, W.; Xu, X.; Liang, L.; Waqar, H.S.; Lin, B.; Tian, Z.; Shen, P. Reduced friction and wear enabled by arc-discharge method-prepared 3D graphene as oil additive under variable loads and speeds. *Wear* **2020**, *462*, 203495. [CrossRef]
18. Mohamed, A.; Tirth, V.; Kamel, B.M. Tribological characterization and rheology of hybrid calcium grease with graphene nanosheets and multi-walled carbon nanotubes as additives. *J. Mater. Res. Technol.* **2020**, *9*, 6178–6185. [CrossRef]
19. Li, Z.; He, Q.; Du, S.; Zhang, Y. Effect of few layer graphene additive on the tribological properties of lithium grease. *Lubr. Sci.* **2020**, *32*, 333–343. [CrossRef]
20. Zhang, J.; Wang, A.; Yin, H. Preparation of graphite nanosheets in different solvents by sand milling and their enhancement on tribological properties of lithium-based grease. *Chin. J. Chem. Eng.* **2020**, *28*, 1177–1186. [CrossRef]
21. Ouyang, T.; Shen, Y.; Yang, R.; Liang, L.; Liang, H.; Lin, B.; Tian, Z.; Shen, P. 3D hierarchical porous graphene nanosheets as an efficient grease additive to reduce wear and friction under heavy-load conditions. *Tribol. Int.* **2020**, *144*, 106118. [CrossRef]
22. Aziev, R.; Savilov, S.; Kupreenko, S.; Ivanov, A.; Stolbov, D.; Usol'tseva, N.; Lunin, V. Graphene nanoflakes as effective dopant to Li-based greases. *Funct. Mater. Lett.* **2020**, *9*, 2040006. [CrossRef]
23. Lin, B.; Rustamov, I.; Zhang, L.; Luo, J.Q.; Wan, X. Graphene-Reinforced Lithium Grease for Antifriction and Antiwear. *ACS Appl. Nano Mater.* **2020**, *3*, 10508–10521. [CrossRef]
24. Jin, B.; Zhao, J.; Chen, G.; He, Y.; Huang, Y.; Luo, J. In situ synthesis of Mn_3O_4/graphene nanocomposite and its application as a lubrication additive at high temperatures. *Appl. Surf. Sci.* **2021**, *546*, 149019. [CrossRef]
25. Liu, X.; Chen, H.; Qiao, D.; Feng, D.; Wang, H.; Meng, W. The Influence of Graphene Oxide on the Tribological Properties of Lithium-based Grease. *Sur. Tech.* **2021**, *50*, 70–78. [CrossRef]
26. Sun, Z.; Xu, C.; Peng, Y.; Shi, Y.; Zhang, Y. Fretting tribological behaviors of steel wires under lubricating grease with compound additives of graphene and graphite. *Wear* **2020**, *454*, 203333. [CrossRef]
27. Cheng, Z.; Kong, Y.; Fan, L.; Liu, Z. Ultrasound-assisted Li^+/Na^+ co-intercalated exfoliation of graphite into few-layer graphene. *Ultrason. Sonochem.* **2020**, *66*, 105108. [CrossRef]
28. Pan, J.; Cheng, Y.; Qian, M.; Zhou, B. Thermal-rheological properties and variation mechanisms of lithium lubricating grease. *Chem. Ind. Eng. Progress* **2018**, *37*, 1509–1515. [CrossRef]

lubricants

MDPI

Article

Research on the Bearing Sliding Loss Based on Time-Varying Contact Angle between Ball and Raceway

Shuaijun Ma, Ke Yan *, Mengnan Li, Yongsheng Zhu and Jun Hong

Key Laboratory of Education Ministry for Modern Design and Rotor-Bearing System, Xi'an Jiaotong University, Xi'an 710049, China
* Correspondence: yanke@mail.xjtu.edu.cn

Abstract: Based on the mechanical model, the friction loss between the ball and the raceway along the major axis of the contact ellipse is analyzed. The result shows that this part of the loss accounts for about 13.67% of the overall loss, which is mainly determined by the ball sliding length and cannot be ignored. The effects of the radial force, torque, rotational speed and groove curvature ratio on the sliding are all studied. Compared with other factors, radial force has the greatest influence on the sliding loss. As bearing speed gradually grows, the sliding on the inner raceway gradually increases while it gradually decreases on the outer raceway. Compared to the outer raceway curvature ratio, the sliding length is less sensitive to changes in the curvature ratio of the inner raceway. The paper provides theoretical guidance for the design and application of low-friction bearings.

Keywords: angular contact ball bearing; sliding length; contact ellipse major axis; friction loss; contact angle

1. Introduction

Owing to the characteristics of low friction and high precision, ball bearings are widely used in rotational systems, such as high-speed motors, precision machine tools and aero engines [1,2]. Regarding the complex internal structure of the bearing, relative motion and force conditions, etc., there are complex tribological behaviors between balls and inner/outer rings, which may cause bearing friction heat and power loss, and even accelerate bearing wear and reduce bearing operation accuracy, etc. [1]. Therefore, the bearing friction phenomenon hinders the further improvement of high-speed bearing performance [3]. Especially with the continuous improvement of energy consumption requirements of rotating equipment, the design of low-friction bearings has become a current research hot topic. It is preferred to analyze the various complex sources of bearing friction, such as the structural constraints, and to reduce their sizes in a targeted manner by improving the structural and external operating conditions. Accurate computation of bearing friction loss, operating in complex motion and force conditions, has become a prerequisite for this research.

Since the middle of the 20th century, the frictional properties of bearing are investigated by few scholars. Under different working conditions, the friction torque of different types of bearings was experimentally tested by Palmgren et al. [4,5]. The empirical formulas of bearing frictional torque were proposed based on the experimental data. However, the friction model by these empirical formulas is mainly suitable for light load and low speed conditions. Besides, the famous bearing company SKF proposed a set of models for calculating bearing frictional torque also based on experimental dates [6]. Compared to the former, the model accuracy has been improved, while it is mainly used for bearings in standard installation and standard load condition. With an experimental method, the influence of the raceway curvature radius on the frictional torque of angular contact ball bearings was experimentally analyzed by Todd et al. [7]. Similarly, Rodionov [8] studied

Citation: Ma, S.; Yan, K.; Li, M.; Zhu, Y.; Hong, J. Research on the Bearing Sliding Loss Based on Time-Varying Contact Angle between Ball and Raceway. *Lubricants* **2022**, *10*, 185. https://doi.org/10.3390/lubricants10080185

Received: 30 June 2022
Accepted: 10 August 2022
Published: 15 August 2022

the effects of surface finish quality on bearing friction torque and it showed that, compared with the surface finish quality of rings, the rolling element significantly affected the friction torque. With the development of data processing methods, some new methods were applied to the study of bearing friction torque. Based on statistical correlation theory, the parametric-nonparametric fusion method was used to analyze the bearing friction torque by Xu [9]. By measuring the friction torque, Zhang et al. [10] modeled the data with gray theory to predict bearing friction torque. By summarizing the above work, it can be found that the present friction torque models were all based on experimental methods and dates, and then the overall friction performance of the bearing was investigated. However, the bearing friction torque is actually affected by many factors inside the bearing, such as bearing structural parameters, motion characteristics, load distribution etc. Especially for ball bearing, the ball's revolution, spin, gyro and other motion forms will affect its friction performance.

Snare, Li and Wang et al. [11–13] pointed out that the friction inside the bearing mainly comes from the elastic hysteresis, differential sliding, spinning, and friction agitation. Ye et al. [14] established a simple friction model for bearing in steady state. The differential sliding and spin friction were considered while the complex motion in the contact area was ignored. In order to explore the influence of the movement of the contact area on the friction performance of the bearing, by energy conservation law, the solution results of the mechanical model were used to calculate the friction torque by Deng [15]. The results show that the complex motion of the contact zone, such as balls sliding in the contact zone, has an important impact on the friction loss. On the basis of the above studies, Todd et al. [16] used the micro-slip theory to determine the pure rolling line in the contact zone. Then the frictional resistance was solved by integrating the complex motion of the contact zone. Unfortunately, the friction torque model and mechanical model are unidirectionally coupled, thus the effect of friction torque on bearing internal motion is missing. Houpert et al. [17] analyzed the motion of the contact zone and obtained the pure rolling line between ball and raceway. Then, the integral of the contact zone was calculated to obtain the contact friction. Finally, the bidirectionally coupled friction torque was achieved based on Cao's quasi-static model [18]. Because of the implementation of bidirectional coupling and the refinement of the contact zone, an obvious change of the friction results was observed. It indicates that the complex contact behavior between bearing components has a significant impact on its total friction loss. Therefore, it is necessary to study the internal friction characteristics of the bearing parts.

For bearing internal friction analysis, the current research mainly focuses on the steady-state friction. That is, the total friction is generated by the complex motion in the contact zone between the bearing parts, such as differential sliding and spin friction, etc. As mentioned earlier, due to the motion of the ball in the ball bearing being particularly special, the friction induced by the revolution is disregarded when the rolling elements move in the circumferential direction. In other words, the current friction calculation mainly considered the motions between the balls and the rings along the contact ellipse minor axis, while the sliding friction along the major axis is ignored [19]. In fact, in conditions of high speed and combined load, the internal forces of the bearing become very complicated. In addition to the friction caused by the gyroscopic torque, the friction generated by the contact angle cannot be neglected. Due to the inconsistent force of each rolling element, the contact angle between the balls and the ring is unequal, which is the main cause of the ball slides along the major axis of the contact ellipse during bearing operation. For instance, the technical report of SKF pointed out that a large variation in contact angle has an impact on the bearing friction [20]. As the ball slides along the major axis of the contact ellipse, the collision force between the cage pocket and the ball increased, which eventually causes the increases of bearing friction. Thus, it is of great significance to investigate the sliding of the rolling element of the bearing along the major axis of the contact zone. In this paper, taking the angular contact ball bearing as the research object, a quasi-static model is established. In condition of the combined load, it is found that the friction caused by the sliding length of the bearing is discussed, such as operating parameters, structural parameters and contact angles.

2. Quasi-Static Model of Ball Bearing

2.1. Establishment of Model

Due to the complex internal structure of angular contact ball bearings, the position relationship among the bearing parts is described. As shown in Figure 1, Figure 1b is the overall structure of an angular contact ball bearing, Figure 1a is a radial sectional view of the bearing, and Figure 1c is an axial sectional view. In Figure 1, the x-axis is defined as the bearing axial direction, and the radial plane is the y-z plane. As a whole, the bearing has five degrees of freedom (δ_x, δ_y, δ_z—movement along the x, y, and z axes, θ_y, θ_z—rotation around the y, z axes). In other words, under the action of load F (F_x, F_y, F_z, M_y, M_z), the bearing may produce a generalized displacement in five directions. When a bearing is used in engineering, the inner ring always rotates and the outer ring is fixed. Therefore, the external load and rotational speed are applied to the inner ring, while the outer ring is limited to six degrees of freedom.

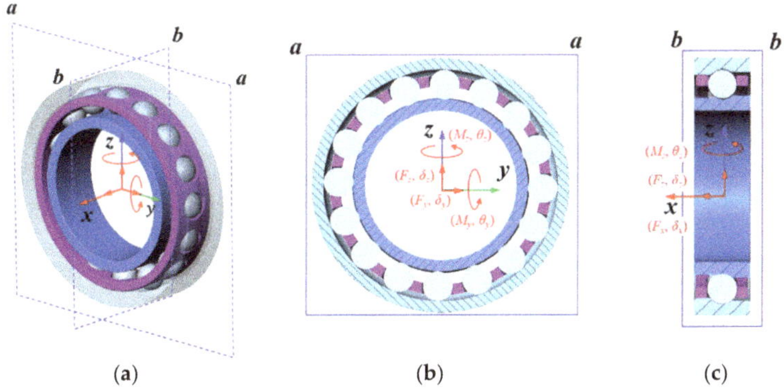

Figure 1. Internal structure of angular contact ball bearing. (**a**) Overall structure; (**b**) Radial sectional view; (**c**) Axial sectional view.

2.1.1. Relative Position Analysis of Bearing Parts

After the ball at the azimuth position φ_j is loaded, its position changes are revealed in Figure 2. Before being loaded, the three points of the curvature center of the inner raceway, the ball center and the curvature center of the outer raceway were collinear and there was no contact deformation. The distance between the curvature center, the inner raceway and the outer can be expressed as follows:

$$l = B \cdot D_w, \tag{1}$$

where $B = f_i + f_o - 1$, f_i and f_o are the curvature ratios of the inner raceway and the outer, respectively. D_w represents the ball diameter.

When the load F (F_x, F_y, F_z, M_y, M_z) and rotational speed n_i are applied to the inner ring, the relative positions change among the curvature center of the inner raceway, the ball center and the curvature center of the outer raceway and the three are no longer collinear. As the rotational speed increases, the ball-inner raceway contact angle gradually enlarges, while the contact angle of the ball-outer raceway gradually decreases.

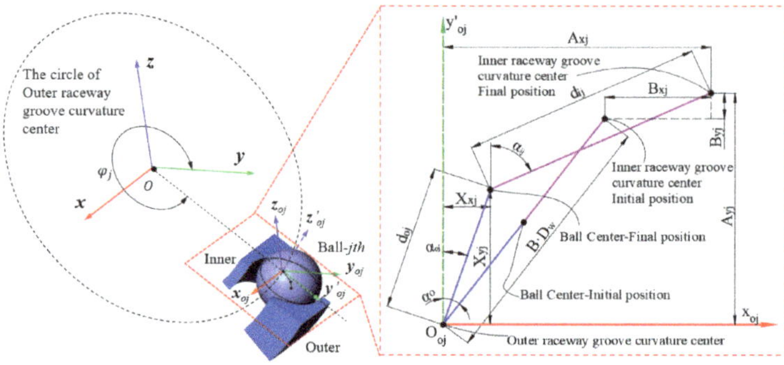

Figure 2. Position relationships among the ball center and the raceway curvature centers.

At the azimuth position φ_j, the distance between the curvature center of the inner raceway and the final position of the ball center is written as:

$$d_{ij} = (f_i - 0.5) \cdot D_w + \delta_{ij}. \tag{2}$$

Similarly, the distance of the outer ring is obtained by:

$$d_{oj} = (f_o - 0.5) \cdot D_w + \delta_{oj}, \tag{3}$$

where δ_{ij} is the contact deformation between the ball and the inner raceway while the deformation of the outer is δ_{oj}.

From Figure 2, the distances between the curvature center of the inner raceway and the outer in the horizontal and vertical directions are calculated as:

$$A_{xj} = B \cdot D_w \cdot \sin\alpha^0 + B_{xj} \tag{4}$$

$$A_{yj} = B \cdot D_w \cdot \cos\alpha^0 + B_{yj}, \tag{5}$$

where α^0 means the initial contact angle. B_{xj} and B_{yj} are the variations of the curvature center of the inner raceway before and after loading on the x-axis and y-axis, respectively, which can be expressed as:

$$B_{xj} = \delta_x + R_i \cdot \theta_y \cdot \sin\varphi_j + R_i \cdot \theta_z \cdot \cos\varphi_j \tag{6}$$

$$B_{yj} = \delta_y \cdot \sin\varphi_j + \delta_z \cdot \cos\varphi_j, \tag{7}$$

where δ_x, δ_y, δ_z are the displacements of the inner center relative to the outer center on the x, y, and z axes, respectively. θ_y, θ_z denote the angular displacements around the y and z axes. R_i represents the radius of the inner curvature center, which is determined by:

$$R_i = d_m/2 + (f_i - 0.5) \cdot D_w \cdot \cos\alpha^0, \tag{8}$$

where d_m is the pitch diameter of the bearing.

By observing Figure 2, it can be seen that these equilibrium equations, established in the horizontal and vertical directions, contain a larger number of trigonometric functions. Trigonometric functions are known to be unfavorable for numerical iterative solutions because they are periodic functions. However, refer to Jones's quasi-static model [21]; the new variables X_x and X_y are introduced to eliminate these functions to simplify the solution process. Among them, X_x and X_y denote the distance between the ball center and the inner curvature center in the horizontal and vertical directions, respectively. As shown in Figure 2, The contact angles of the ball at the azimuth position φ_j are written as Equations (9)–(12):

$$\sin\alpha_{oj} = \frac{X_{xj}}{(f_o - 0.5)\cdot D_w + \delta_{oj}} \tag{9}$$

$$\cos\alpha_{oj} = \frac{X_{yj}}{(f_o - 0.5)\cdot D_w + \delta_{oj}} \tag{10}$$

$$\sin\alpha_{ij} = \frac{A_{xj} - X_{xj}}{(f_i - 0.5)\cdot D_w + \delta_{ij}} \tag{11}$$

$$\cos\alpha_{ij} = \frac{A_{yj} - X_{yj}}{(f_i - 0.5)\cdot D_w + \delta_{ij}}. \tag{12}$$

In Figure 2, the geometric compatibility equations of the ball at the φ_j azimuth are given as Equations (13) and (14):

$$\left(A_{xj} - X_{xj}\right)^2 + \left(A_{xj} - X_{xj}\right)^2 + \left((f_o - 0.5)\cdot D_w + \delta_{oj}\right)^2 = 0 \tag{13}$$

$$X_{xj}{}^2 + X_{yj}{}^2 - \left((f_o - 0.5)\cdot D_w + \delta_{oj}\right)^2 = 0. \tag{14}$$

2.1.2. Interaction Force of Bearing Parts

Under high-speed operating conditions, the mechanical analysis of bearing components is complicated. The ball is applied to the centrifugal force, causing dissimilar ball-inner and ball-outer contact angles. The friction between the ball and the raceway is caused by the gyroscopic torque. The interaction of the bearing parts is displayed in Figure 3; Figure 3a is the force analysis of the entire bearing while Figure 3b is the *j*-th ball. When a load F (F_x, F_y, F_z, M_y, M_z) and rotational speed n_i are applied to the inner ring, the corresponding displacement δ (δ_x, δ_y, δ_z, θ_y, θ_z) is generated. At this time, the inner ring is pressed against the ball, and a corresponding contact deformation occurs between them. The gyroscopic torque is generated because of the non-zero contact angle.

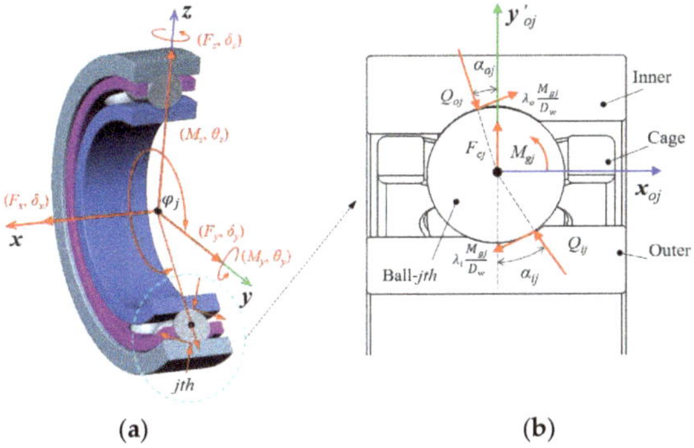

Figure 3. Force analysis of bearing parts. (**a**) Entire bearing; (**b**) *j*-th ball.

By observing Figures 2 and 3, the mechanical equilibrium equations of the ball at the azimuth position φ_j are as follows:

$$Q_{oj}\cdot\cos\alpha_{oj} - \lambda_o\frac{M_{gj}}{D_w}\cdot\sin\alpha_{oj} - Q_{ij}\cdot\cos\alpha_{ij} + \lambda_i\frac{M_{gj}}{D_w}\cdot\sin\alpha_{ij} - F_{cj} = 0 \tag{15}$$

$$Q_{oj}\cdot\sin\alpha_{oj} + \lambda_o\frac{M_{gj}}{D_w}\cdot\cos\alpha_{oj} - Q_{ij}\cdot\sin\alpha_{ij} - \lambda_i\frac{M_{gj}}{D_w}\cdot\cos\alpha_{ij} = 0, \tag{16}$$

where $Q_{i/oj} = K_{i/oj} \cdot \delta_{i/oj}{}^{1.5}$, $K_{i/oj}$ is the contact deformation coefficient between the ball and the inner/outer ring, which can be calculated from [22]. $\lambda_{i/o}$ represents the inner/outer raceway control coefficient. According to reference [23], if it is inner raceway control, $\lambda_i = 1$, $\lambda_o = 1$, otherwise it is outer raceway control, $\lambda_i = 0$, $\lambda_o = 2$.

Besides, the gyroscopic torque and centrifugal force of the ball are calculated with Equations (17) and (18):

$$M_{gj} = J_b \cdot \left(\frac{w_R}{w_i}\right)_j \cdot \left(\frac{w_m}{w_i}\right)_j \cdot w_i^2 \cdot \sin\beta_j \tag{17}$$

$$F_{cj} = 0.5 \cdot m_b \cdot d_m^2 \cdot w_i^2 \cdot \left(\frac{w_m}{w_i}\right)_j^2, \tag{18}$$

where J_b is the ball moment of inertia, w_i means the angular velocity of the inner ring, and m_b denotes the ball mass.

According to Jones' model theory [24], w_m/w_i, w_R/w_i, and $\tan\beta_j$ are related to the rotation speed and attitude of the ball, which can be expressed as the following formula:

$$\frac{w_m}{w_i} = \frac{1}{1 + \left(\frac{\cos\alpha_{oj} + \tan\beta \cdot \sin\alpha_{oj}}{\cos\alpha_{ij} + \tan\beta \cdot \sin\alpha_{ij}} \times \frac{1+\gamma\cdot\cos\alpha_{oj}}{1-\gamma\cdot\cos\alpha_{ij}}\right)} \tag{19}$$

$$\frac{w_R}{w_i} = \frac{-1}{\left(\frac{\cos\alpha_{oj} + \tan\beta \cdot \sin\alpha_{oj}}{1+\gamma\cdot\cos\alpha_{oj}} + \frac{\cos\alpha_{ij} + \tan\beta \cdot \sin\alpha_{ij}}{1-\gamma\cdot\cos\alpha_{ij}}\right) \cdot \gamma \cdot \cos\beta} \tag{20}$$

$$\beta = \arctan\left(\frac{\sin\alpha_{ij}}{\cos\alpha_{ij} + \gamma}\right), \tag{21}$$

where γ represents the dimensionless constant, $\gamma = D_w/d_m$.

Under high-speed operating conditions, the angular contact ball bearing can be regarded as a whole, and the load acting on the bearing should be balanced. In this paper, the inner ring is used as the carrier, and the load applied by the outside should be balanced with the load applied by the ball. The equilibrium equations of the inner can be established:

$$F_x - \sum_{j=1}^{Z}\left(Q_{ij} \cdot \sin\alpha_{ij} + \lambda_i \frac{M_{gj}}{D_w} \cdot \cos\alpha_{ij}\right) = 0 \tag{22}$$

$$F_y - \sum_{j=1}^{Z}\left(Q_{ij} \cdot \cos\alpha_{ij} - \lambda_i \frac{M_{gj}}{D_w} \cdot \sin\alpha_{ij}\right)\cos\varphi_j = 0 \tag{23}$$

$$F_z - \sum_{j=1}^{Z}\left(Q_{ij} \cdot \cos\alpha_{ij} - \lambda_i \frac{M_{gj}}{D_w} \cdot \sin\alpha_{ij}\right)\sin\varphi_j = 0 \tag{24}$$

$$M_y - \sum_{j=1}^{Z}\left(\left(Q_{ij} \cdot \sin\alpha_{ij} + \lambda_i \frac{M_{gj}}{D_w} \cdot \cos\alpha_{ij}\right) \cdot R_i - \lambda_i \cdot f_i \cdot M_{gj}\right)\sin\varphi_j = 0 \tag{25}$$

$$M_z - \sum_{j=1}^{Z}\left(\left(Q_{ij} \cdot \sin\alpha_{ij} + \lambda_i \frac{M_{gj}}{D_w} \cdot \cos\alpha_{ij}\right) \cdot R_i - \lambda_i \cdot f_i \cdot M_{gj}\right)\cos\varphi_j = 0, \tag{26}$$

where Z means the number of balls.

2.2. Model Solution

The quasi-static model of an angular contact ball bearing is composed of nonlinear equations, including the geometric compatibility Equations (13) and (14) of the ball, the force balance Equations (15) and (16), and the balance equation of the inner ring

Equations (22)–(26). The model has $4Z + 5$ nonlinear equations and $4Z + 5$ unknowns, such as X_{xj}, X_{yj}, δ_{ij}, δ_{oj}, δ_x, δ_y, δ_z, θ_y, θ_z ($j = 1 \cdots z$). Considering the solution accuracy and efficiency of non-linear equations, the Newton–Raphson iterative algorithm is adopted to solve. Because the initial value affects the accuracy of the iterative algorithm, and a large number of solution parameters is included in this model. The results of the statics model [21] are used as the initial value of the quasi-static and are transferred to the quasi-static model. The solution process of the above model is shown in Figure 4, and the iteration convergence accuracy (eps) sets to 10^{-5}.

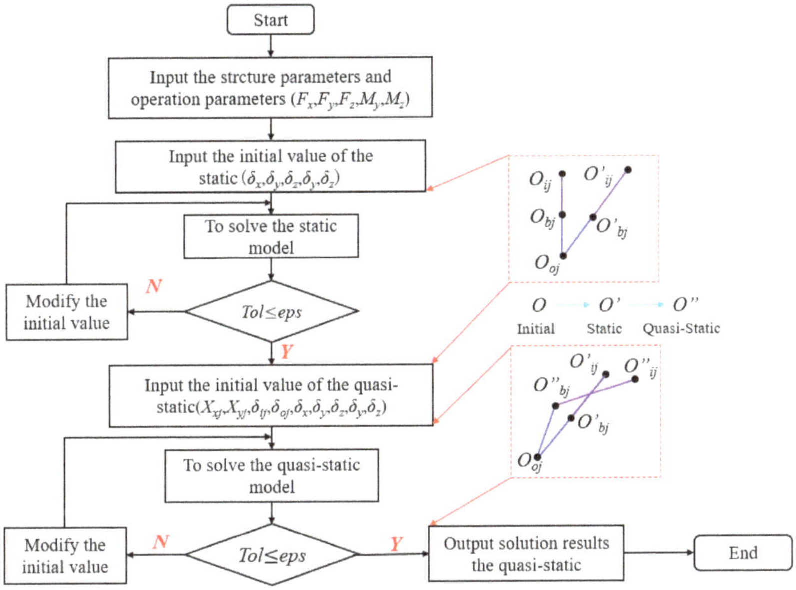

Figure 4. Solution process of the quasi-static model.

3. Model Verification and Sliding Loss Analysis

3.1. Model Versatility Verification

In this section, the mechanical model established above is verified by comparing the relationship between the force and displacement of the bearing and the contact characteristic. The displacement relation (stiffness) is verified by experiments, and the contact characteristic (contact angle) is compared with the published reference.

3.1.1. The Relationship between Force and Displacement

First, the bearing stiffness test system built by our team was used to verify the mechanical model above. As shown in Figure 5, the inner ring of bearing rotates with the precision mandrel 5 while the outer is fixed. The non-contact axial load is applied to the bearing through the air bearing plate 3, and the displacement sensor 6 and the force sensor 2 are used to test the relative displacement of the inner and outer rings and the axial force. The force and displacement signals are collected through the data acquisition system to measure the axial stiffness of the bearing.

Based on the stiffness test system, the stiffness of a B7008C/P4 angular contact ball bearing is measured. The specific structural parameters of B7008C/P4 are shown in Table 1. We applied an axial force of 150 N and 300 N respectively, the rotational speed was gradually increased from 100 r/min to 1900 r/min. The change in axial stiffness of the bearing is shown in Figure 6.

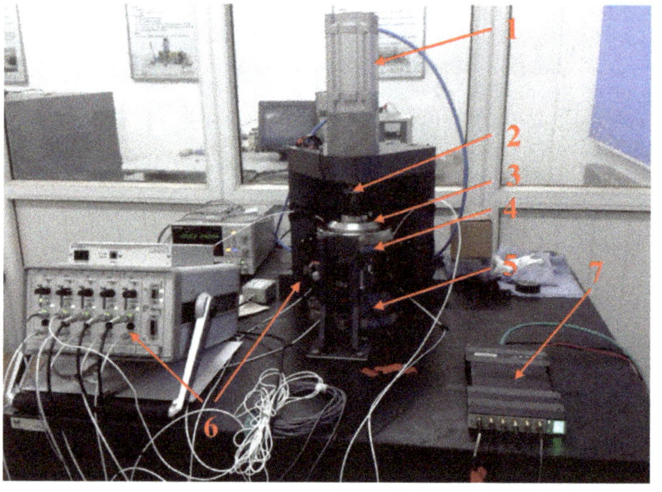

Figure 5. The Bearing stiffness test system. **1**—FESTO cylinder; **2**—The axial force sensor; **3**—Air bearing plate; **4**—Bearing gland; **5**—Precision mandrel; **6**—Capacitive displacement sensor; **7**—B&K.

Table 1. The B7008C/P4 structural parameters.

Parameters	Values
Inner raceway curvature radius r_i/mm	4
Outer raceway curvature radius r_o/mm	3.79
Ball diameter D_w/mm	7.144
Number of balls Z	19
Pitch diameter d_m/mm	54.007
Initial contact angle α^0/°	15

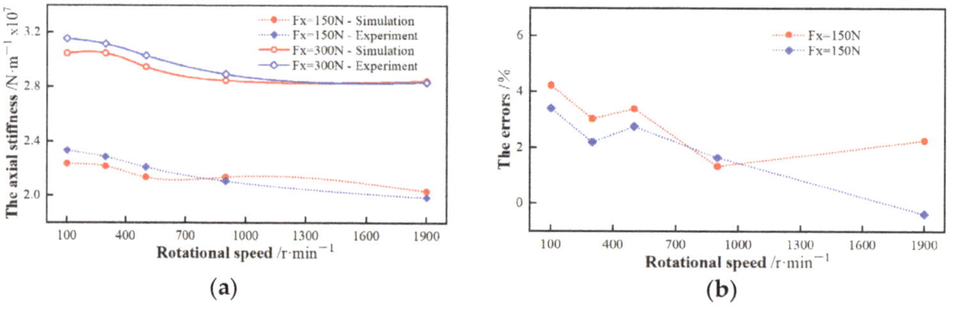

Figure 6. Comparison of experiment and model. (**a**) Stiffness between experiment and model; (**b**) Error between experiment and model.

By observing Figure 6, it can be seen that the bearing stiffness decreases as the rotational speed increases. This is caused by the change in contact angle due to the variable rotational speed. The axial force has an obvious influence on the bearing stiffness, and the stiffness increases as the axial force grows at the same speed. By comparing the theoretical simulation with the experiment, the errors do not exceed 5%, which proves the model. The errors may be caused by the axial runout of the spindle.

3.1.2. The Contact Characteristic

Second, by comparing the deviation of the solution results of the model with that of Harris et al. [24], the model is proved. A 218 angular contact ball bearing was taken as the research object and the model was solved under given conditions. From reference [24], the initial contact angle of the 218 was 40°, and the rotational speed respectively was set to 6000, 10,000, 15,000 r/min, while the axial force was gradually increased from 17,500 N to 44,500 N since the focus of this study was the contact angle, which was the object of verification. Figure 7 is the result of the comparison.

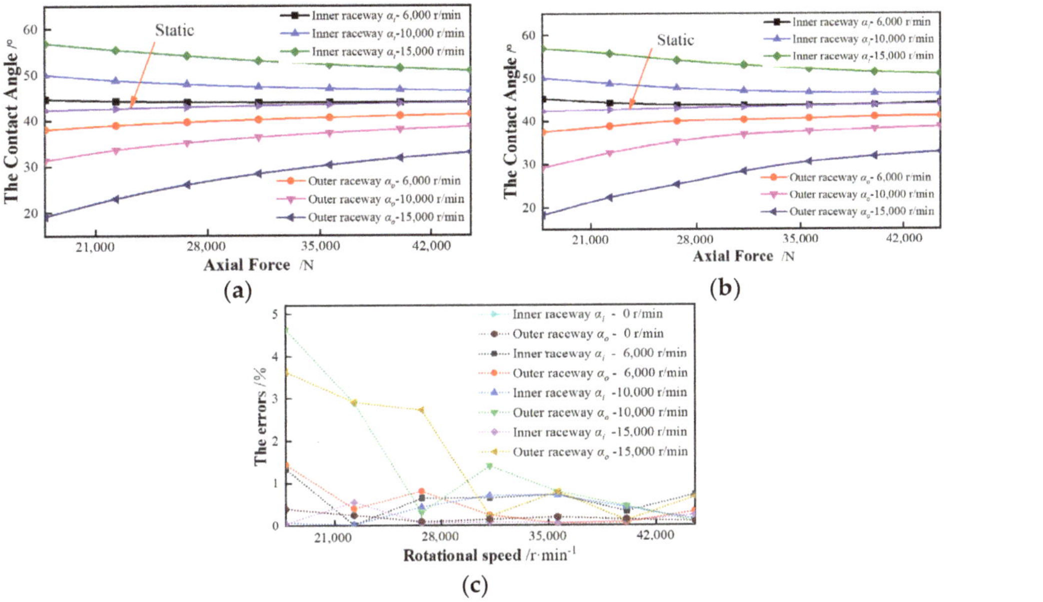

Figure 7. Comparison of the results [24] with the model. (**a**) Results of the model; (**b**) Results of the Ref. [24]; (**c**) Error between the reference and the model.

From Figure 7, the contact angle is greatly affected by the axial force and the rotational speed. The ball-inner raceway contact angle gradually reduces as the axial force increases, while the contact angle of the outer becomes larger. With the increasing rotational speed, the ball-inner raceway contact angle continuously increases and the contact angle of the outer decreases. By comparison, the deviation between the results of this model and the results [24] is within 5%, which is considered to be caused by unit conversion, and the correct and reliable of the model was obtained. The error may be caused by unit conversion.

3.2. Analysis of Sliding Loss Caused by Contact Angle Changes

Based on the model above, the B7008C/P4 bearing is taken as the research object to investigate the sliding. Firstly, the contact characteristics among the bearing components under the combined load are analyzed.

The dynamic characteristics of the 7008 can be obtained by solving the quasi-static model. The constant combined load (F_x = 500 N, F_z = −300 N) and rotational speed (n_i = 10,000 r/min) are considered. The force applied to each ball is not equal because of the radial load, which causes their contact angle values to be different. The variability of the contact angle will cause sliding between the ball and the raceway, as shown in Figure 8.

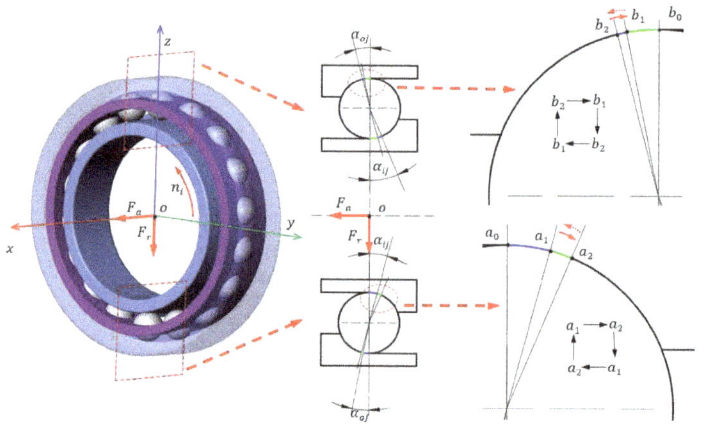

Figure 8. Analysis of sliding under a combined load.

Owing to the variable of the contact angle, the ball slides along the major axis of the contact ellipse relative to the raceway and the sliding length of the ball relative to the inner raceway at the azimuth position φ_j is:

$$s_{ij} = r_i \cdot (\alpha_{ij} - \alpha_{imin}). \tag{27}$$

The sliding length of the ball at this azimuth relative to the outer raceway is:

$$s_{oj} = r_o \cdot (\alpha_{oj} - \alpha_{omin}), \tag{28}$$

where α_{ij} is the contact angle of the ball at the azimuth φ_j with the inner ring and the angle of the ball with the outer is α_{ij}. α_{imin} expresses the minimum contact angle of the ball with the inner ring when it runs for a cycle while α_{omin} means the minimum angle of the ball with the outer ring.

To clearly express the ball information of each azimuth angle, the ball is labeled. The ball mark in the opposite direction to the z axis is designated as 0, where the azimuth angle is 0°. The counterclockwise rotation around the x-axis is specified as the positive direction.

By observing Figures 8 and 9, when the ball azimuth is 0°, the contact angle between the ball and the inner ring or the outer is a minimum. At this location, the contact point between the ball and the inner ring is a_1, and the contact point with the outer ring is b_1. At the azimuth position 180°, the contact angle between the ball and the inner ring or the outer is a maximum. The contact point between the ball and the inner ring at the position is a_2, and the contact point with the outer ring is b_2. When the ball moves from 0° to 180°, the contact angle between the ball and the ring gradually becomes bigger and the contact point between the ball and the ring moves from 1 to 2 along the major axis of the contact ellipse. In contrast, the contact angle between the ball and the ring reduces, when the ball moves from 180° to 360° (0°). The contact point between the ball and the ring moves from 2 to 1 along the major axis of the contact ellipse. In the dynamic operation process, when the ball runs for a week, the contact angle changes cyclically once (from small to large, and then from large to small). The contact point moves back and forth between 1 and 2 points. Figure 9 is the relationship between the change of the contact angle and the sliding length of the ball in a period. Under the working conditions, the maximum sliding length between the ball and the inner ring is 0.6987 mm for 7008, and the maximum sliding length of the outer ring is 0.5381 mm.

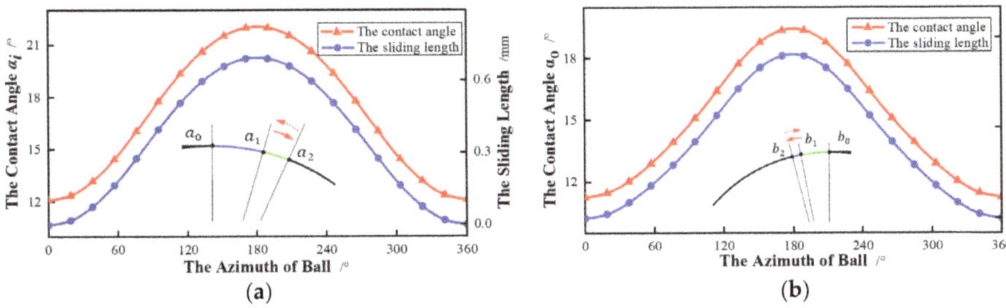

Figure 9. Relationship between sliding length and contact angle. (**a**) Ball-Inner raceway contact angle; (**b**) Ball-Outer raceway contact angle.

During operation, the motion of the bearing parts is shown in Figure 10. Assuming no sliding occurs on the bearing, the driving rotational speed applied by the inner ring only rotates around its own x-axis. Thus, the velocity component of the ball is concentrated in the ellipse contact minor axis direction between the ball and the inner ring, while the component in the major axis direction is smaller under the circumstances. The motion in the minor axis direction is rolling, and the sliding occurs in the major axis direction. Since the rolling friction coefficient is very small, there is little heat generation. However, the sliding friction coefficient is large, and the friction heat is serious. The sliding in major axis direction is the focus of the investigation.

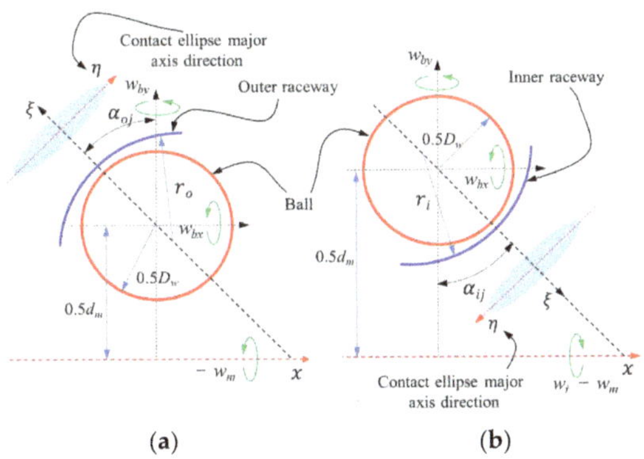

Figure 10. The relative relationship between ball and raceway. (**a**) Inner raceway; (**b**) Outer raceway.

The bearing friction loss above occurs due to sliding caused by changes in the contact angle. The micro-element method is used to calculate the loss. Firstly, the ball is divided into finite parts for a week and the results are added up to obtain the average value, which is expressed as:

$$H = \frac{1}{n}\sum_{j=1}^{n} Ff_j \cdot \frac{\Delta s_j}{\Delta t}, \tag{29}$$

where n means that the operation week of the ball is divided into n stages. Ff_j is the average friction force experienced by the ball in stage j. Δs_j represents the slip increment of the ball in stage j. Δt denotes the time taken to pass stage j for the ball.

For the frictional force Ff_j, assuming ignoring the slight rolling, the sliding is only considered in the major axis direction. Such force can be determined as follows:

$$Ff_j = \frac{\mu(Q_j + Q_{j+1})}{2}, \tag{30}$$

where μ is the sliding friction coefficient, which can be obtained in [25].

Based on the above theory, a combined-loaded bearing (F_x = 500 N, F_z = −300 N, n_i = 10,000 r/min) is solved. Because of the ball sliding along the major axis, the heat generated on the inner raceway is 9.456 W, while the outer is 8.624 W. The integral heating of the bearing obtained by empirical equation [26] is 132.2792 W. The power loss by sliding approximately accounts for 13.67% of the total heat. Ignoring the sliding loss, the results of frictional heat will be seriously affected and the heat generated by the sliding loss is almost proportional to the sliding length of the ball along the major axis direction on the raceway. Thus, the sliding length is taken as the evaluation criterion to investigate the friction, and the effects of working conditions and curvature ratio are analyzed.

4. Results and Discussion

Based on the model in Section 2, a 7008 angular contact ball bearing is taken as the object to be solved. By the sliding length of the ball on the raceway, the friction loss of the bearing with different initial contact angles (15°, 25°) is studied. The efforts of the working conditions and the structural parameters are discussed.

4.1. Radial Loaded Bearing

The influence of the radial force on the ball sliding is first studied, including the direction and magnitude. Without the torque ($M_y = M_z$ = 0 N·mm), a constant axial force (F_x = 500 N) and rotational speed (n_i = 10,000 r/min) is considered. By changing the radial force, the trend of sliding with different contact angles is analyzed.

Firstly, the y-axis is applied to the radial force, while the force of the z-axis is 0. With the rising of radial force (F_z) on the y-axis, the sliding length of the ball with different contact angles is shown in Figure 11.

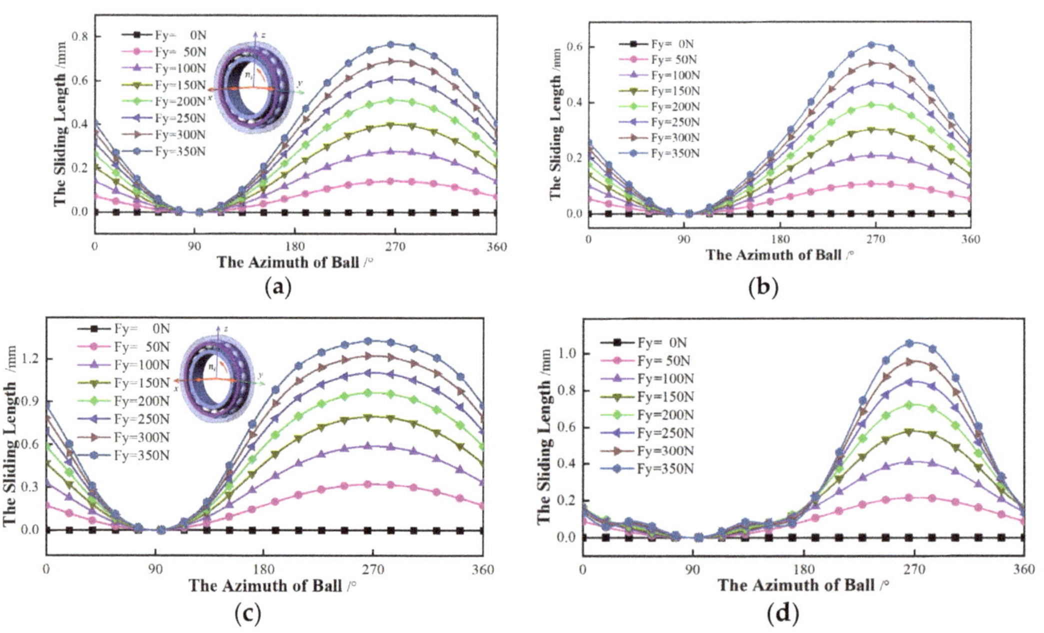

Figure 11. F_y's influence on the sliding of the inner ring and outer ring. (**a**) The sliding of inner raceway-15°; (**b**) The sliding of outer raceway-15°; (**c**) The sliding of inner raceway-25°; (**d**) The sliding of outer raceway-25°.

The relationship between the sliding length and the radial force is shown in Figure 9. For bearings with the same initial contact angle, the effect of force is similar. The minimum contact angle (α_{min}) appears when the ball is 90°, where the sliding length is 0. The maximum contact angle (α_{max}) is located at the ball azimuth of 270°, which is the largest sliding. As the radial force of F_y grows, the sliding length increases. The variation trend of the sliding is relatively slight when the contact angle of the ball near the α_{min} or α_{max}. The friction loss is quite small. On the contrary, the trend is relatively sharp in other positions, and the loss becomes large. As for bearings of different initial contact angles, the azimuth angles of α_{min} or α_{max} are the same. However, the sliding trend of the bearing with 25 ° is larger than that of the 15° under the same force, and the friction loss is also greater. When the ball azimuth closes 45° and 135°, the variation of sliding changes rapidly for the inner, while the larger change for outer appears near the azimuth of 225° and 315°. This means there is more friction.

Secondly, setting F_y to 0, the z-axis is applied to the radial force. As shown in Figure 12, the relationship between the sliding with different contact angles and the radial force is analyzed. No matter whether the initial contact angle is 15° or 25°, the α_{min} appears at the position where the azimuth is 0°, and the azimuth of the α_{max} is 180°. The other effects of the z-axis are similar to the force applied in the y-axis.

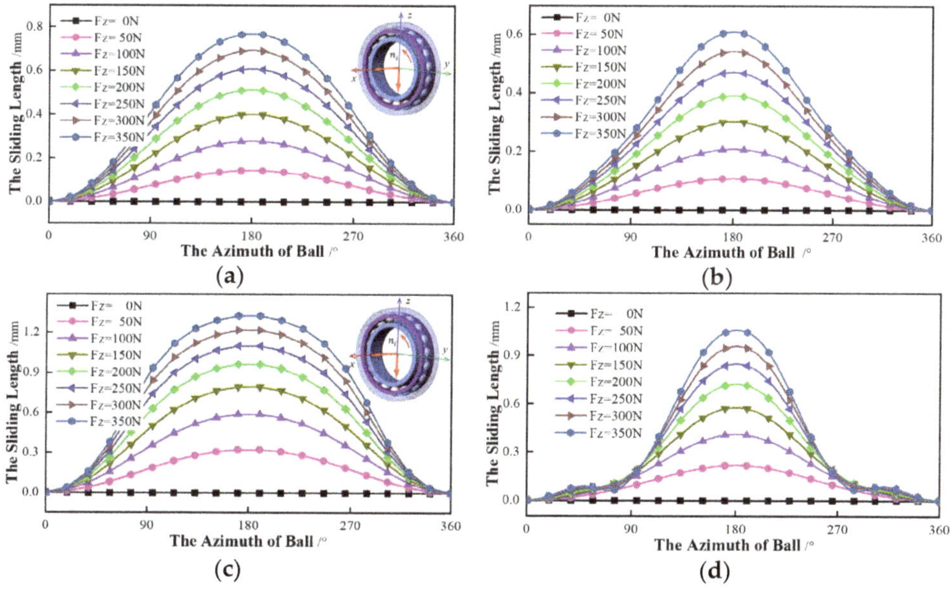

Figure 12. F_z's influence on the sliding of the inner ring and outer ring. (**a**) The sliding of inner raceway-15°; (**b**) The sliding of outer raceway-15°; (**c**) The sliding of inner raceway-25°; (**d**) The sliding of outer raceway-25°.

Finally, the y-axis and the z-axis are applied to the radial force at the same time, which is gradually growing. Figure 13 shows the relationship between the radial force and the sliding with different initial contact angles and the radial force. It can be found that, regardless of the initial contact angle, the azimuths of α_{min} and α_{max} are constant and the α_{min} appears at the azimuth of 45°, while the 225° azimuth is the α_{max}. Other changes have similar effects to the radial force applied to the y-axis or the z-axis individually.

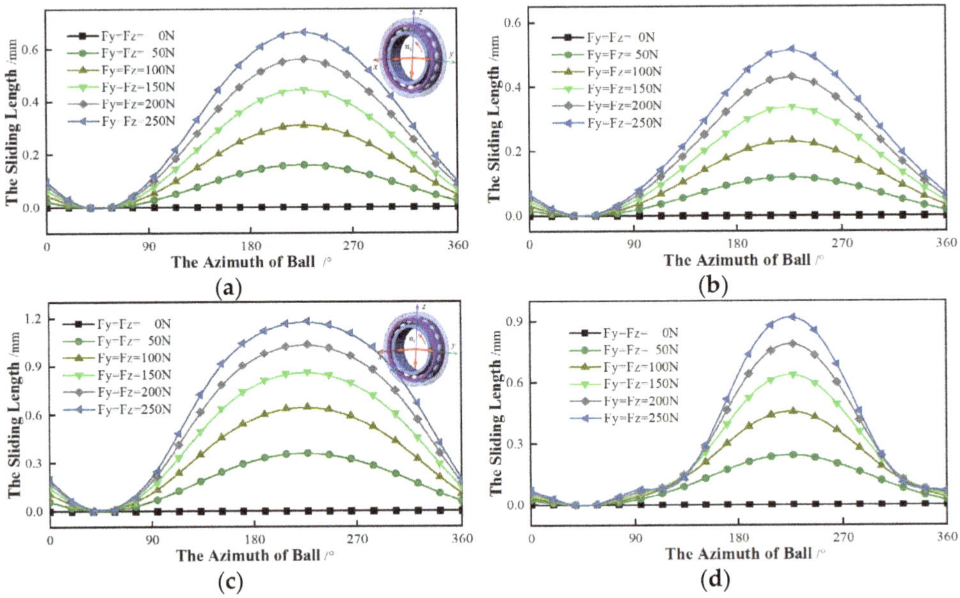

Figure 13. The combined effect of F_y and F_z on the sliding of the inner ring and outer ring. (**a**) The sliding of inner raceway-15°; (**b**) The sliding of outer raceway-15°; (**c**) The sliding of inner raceway-25°; (**d**) The sliding of outer raceway-25°.

To summarize, the sliding length is only affected by the magnitude of the radial force, independent of its direction. With the increasing of the radial force, the variation trend of sliding is increases. However, the direction of the radial force has an influence on the azimuth angles of α_{min} and α_{max}.

4.2. Torque Component

In this section, the relationship between the torque and the sliding is investigated. The axial force (F_x = 500 N), radial force (F_y = 200 N, F_z = 0 N) and rotational speed (n_i = 10,000 r/min) are constant. The sliding of different initial contact angles is discussed by changing the torque.

Firstly, the y-axis is only applied to the torque, which increases from 0 N·mm to 250 N·mm. The torque of the z-axis is 0. This means that the radial force and the torque are exerted to the same axis. Figure 14 shows the relationship between the torque and the sliding of different initial contact angles.

From Figure 14, the difference in sliding of different contact angles is small. As the torque component increases, the azimuth angles of α_{min} and α_{max} are same and the sliding lengths of different torques are almost equal. Although the sliding rate grows with the increase of torque, the effect is small. For bearings with different initial contact angles, the influence of torque on the sliding is not obvious. The sliding of an initial contact angle of 25 ° is slightly less sensitive to this torque than that of 15 °. From Figure 14, we can see that the total sliding length of a larger initial contact angle is greater than the sliding of a small contact angle.

Secondly, the z-axis applied to the torque is considered, which is gradually rising. In other words, the radial force is applied to the y-axis while the z-axis exerts to the torque. The relationship between the torque and the sliding length of different initial contact angles is shown in Figure 15.

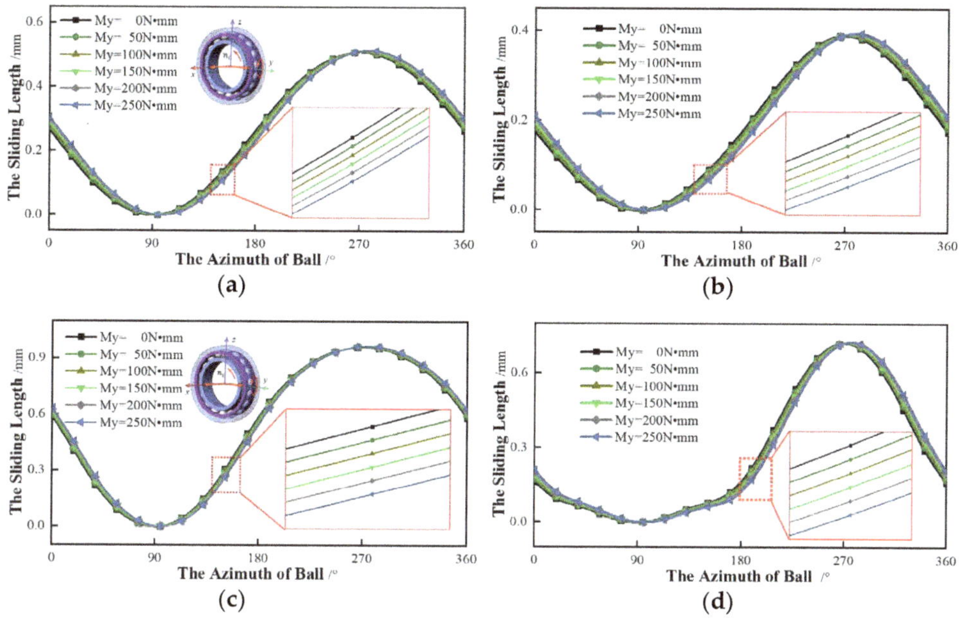

Figure 14. M_y's influence on the sliding of the inner ring and outer ring. (**a**) The sliding of inner raceway-15°; (**b**) The sliding of outer raceway-15°; (**c**) The sliding of inner raceway-25°; (**d**) The sliding of outer raceway-25°.

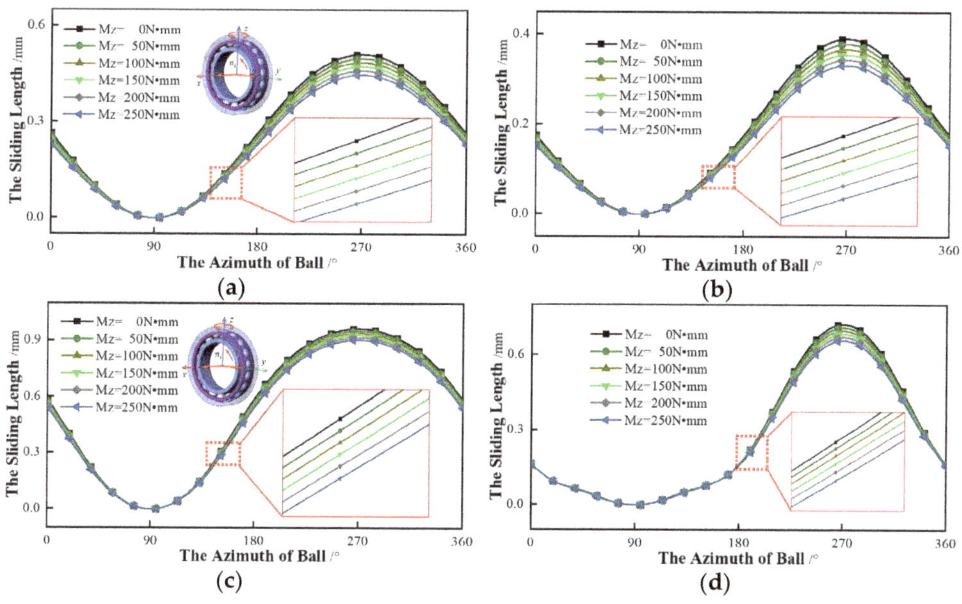

Figure 15. M_z's influence on the sliding of the inner ring and outer ring. (**a**) The sliding of inner raceway-15°; (**b**) The sliding of outer raceway-15°; (**c**) The sliding of inner raceway-25°; (**d**) The sliding of outer raceway-25°.

As shown in Figure 15, this type of torque has an influence on the sliding length of the bearing with the same initial contact angle. As the torque rises, the sliding length gradually decreases and the azimuth angles of α_{min} and α_{max} are constant. As for the variation rate of sliding, it also gradually reduces. With regard to bearings with different initial contact angles, the variation rates are distinct. The rate of bearings with an initial contact angle of 15° is more sensitive to torque than the rate of 25°. However, the total sliding length of the 25° is generally greater than that of the 15°.

In summary, when the torque and radial load are applied to the same axis, the effect of torque on sliding is very weak. On the contrary, applying the torque and radial load to the different axes, the torque has a corresponding effect on the sliding. With the torque increasing, the sliding length gradually lessons. From Figures 14 and 15, regardless of whether the torque and the radial force are on the same axis, the sliding lengths of different initial contact angles are unequal and the sliding with an initial contact angle of 25° is always greater than that of 15°.

4.3. Rotational Speed Bearings

To reveal the influence of rotational speed on sliding length, speed gradually increases. Regardless of the torque ($M_y = M_z = 0$ N·mm), the axial force ($F_x = 500$ N) and the radial force ($F_y = 200$ N, $F_z = 0$ N) are considered. Figure 16 shows the relationship between the sliding length and the rotational speed.

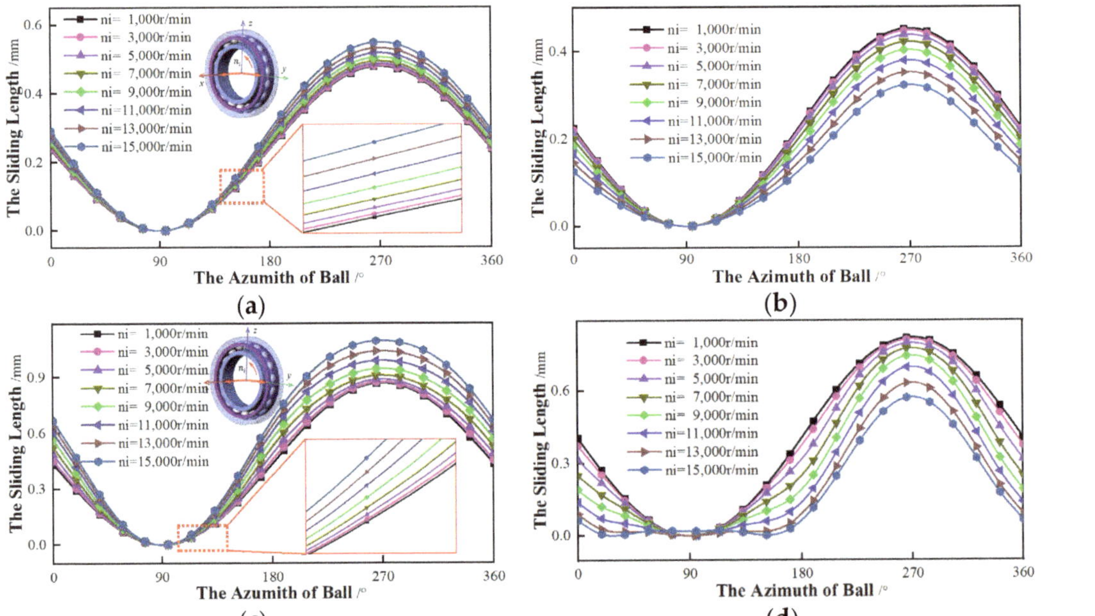

Figure 16. The effect of rotational speed on the sliding of the inner ring and outer ring. (**a**) The sliding of inner raceway-15°; (**b**) The sliding of outer raceway-15°; (**c**) The sliding of inner raceway-25°; (**d**) The sliding of outer raceway-25°.

Figure 16 shows the curves of the sliding length changing with the rotational speed for the bearing. It can be seen that, as the rotational speed increases, the azimuth angles of α_{min} and α_{max} remain unchanged. For the inner ring, the higher the speed, the greater the sliding length, and the faster the variation rate of sliding. On the contrary, with the rising rotational speed, the sliding length of the outer ring reduces and the change rate of outer sliding gradually grows. Compared with the outer ring, the inner ring is more sensitive to the speed. As for the bearing with different initial contact angles, the sliding with an initial contact angle of 25° changes more than that of 15°. Besides, for the bearing with an initial contact angle of 25°, the non-sliding area appears near the α_{min}.

4.4. Groove Curvature Ratio

In addition to working conditions, the structural parameters may also affect the sliding. The curvature ratio of the raceway has an important influence on the friction loss. Ignoring the torque ($M_y = M_z = 0$ N·mm), a constant combined force ($F_x = 500$ N, $F_y = 200$ N, $F_z = 0$ N) and constant speed ($n_i = 10{,}000$ r/min) are applied to the bearings. The effects of the curvature ratio of the raceway on the sliding are investigated.

Firstly, the influence of the curvature ratio of the inner raceway on the sliding is considered. The curvature ratio of the outer raceway is constant ($f_o = 0.535$), while the ratio of the inner (f_i) gradually increases from 0.515 to 0.565. The relationship between the ratio of the inner and the sliding lengths is shown in Figure 17.

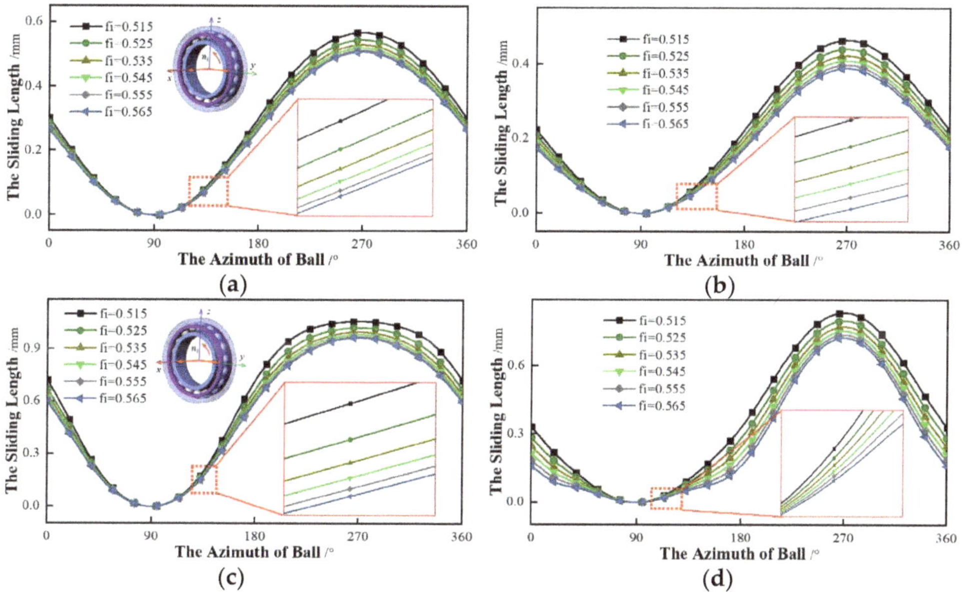

Figure 17. The effect of curvature ratios on the sliding of the inner ring and outer ring. (**a**) The sliding of inner raceway-15°; (**b**) The sliding of outer raceway-15°; (**c**) The sliding of inner raceway-25°; (**d**) The sliding of outer raceway-25°.

By observing Figure 17, as the curvature ratio of the inner raceway increases, the azimuths of α_{min} and α_{max} are constant and the sliding length of rings gradually decreases. However, once the ratio reaches a certain value, the sliding changes will be not obvious. The initial contact angle of the bearing is different, and the influence of the ratio on the sliding is different. The sliding length of a bearing with an initial contact angle of 15° is smaller than that of 25°.

Then the relationship between the curvature ratio of the outer raceway and the sliding is analyzed. Keeping the ratio of the inner constant ($f_i = 0.535$), the outer ratio (f_o) gradually increases. The influence of the outer ratio on the sliding is presented in Figure 18. The outer ratio has a weak effect on the sliding length, but the overall trend gradually increases. When the outer ratio exceeds a certain value and continues to increase, the sliding length is almost constant. For bearings with different initial contact angles, the effects of the outer ratio on the sliding are disparate. For the bearing with an initial contact angle of 15°, the effect of the outer ratio on sliding is not distinct. However, the outer ratio has a significant impact on the sliding for the bearing with 25°. As the outer ratio increases, not only the sliding length but also the variation rate of the sliding gradually grows.

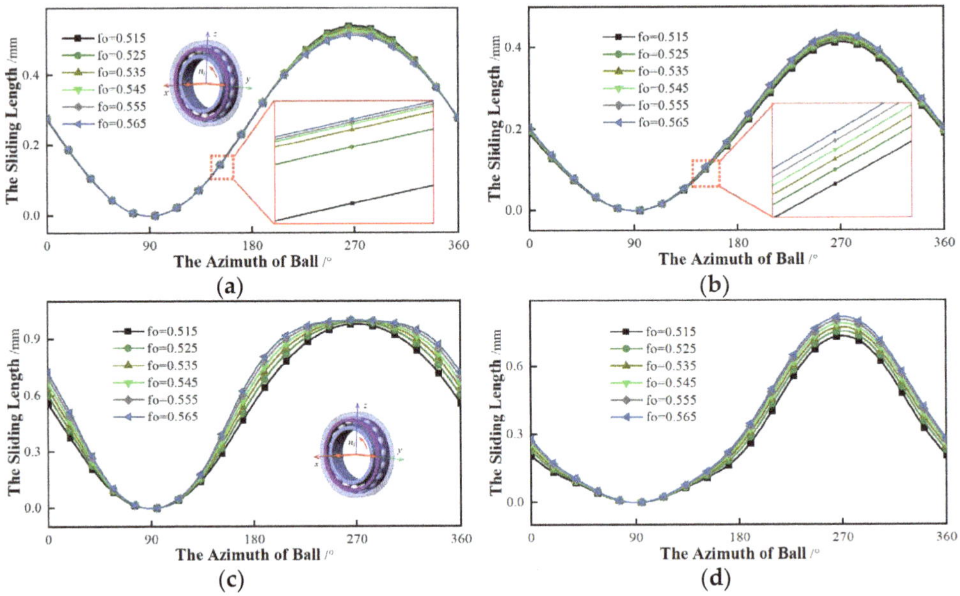

Figure 18. The effect of curvature ratios on the sliding of the inner ring and outer ring. (**a**) The sliding of inner raceway-15°; (**b**) The sliding of outer raceway-15°; (**c**) The sliding of inner raceway-25°; (**d**) The sliding of outer raceway-25°.

To summarize, compared with the curvature ratio of the outer raceway, the sliding is more sensitive to the inner ratio. In addition, the larger the inner ratio, the smaller the sliding length. The influence of the outer ratio on the sliding is just the opposite; with an increasing outer ratio, the sliding gradually decreases.

5. Conclusions

Taking the sliding length as the evaluation object, the influences of working conditions and structural parameters on bearing friction are investigated, and the conclusions are as follows:

(1) The effect of radial force on the sliding length is great. As the radial force rises, the sliding length between the ball and raceway grows, especially with a larger initial contact angle;

(2) The direction of torque also has an important influence on the sliding length. When the radial force and torque are applied on the same coordinate axis, the magnitude of the torque hardly affects the sliding, while when applied on the different axes, the influence is obvious;

(3) With the increase of the rotational speed, the sliding between the ball and the inner raceway increases, while the sliding of the outer raceway decreases;

(4) The curvature ratios of both inner and outer rings affect the sliding, while the influence trend is the opposite. As the ratio increases, the sliding of the inner increases, while that of the outer decreases.

Author Contributions: Conceptualization, S.M., K.Y. and J.H.; data curation, S.M. and M.L.; formal analysis, S.M. and Y.Z.; funding acquisition, J.H.; investigation, S.M., M.L. and K.Y.; methodology, S.M., M.L. and Y.Z.; project administration, K.Y. and J.H.; resources, J.H.; software, S.M.; supervision,

M.L. and K.Y.; validation, M.L. and Y.Z.; visualization, Y.Z., Y.Z. and J.H.; writing—original draft, S.M. and K.Y.; writing—review & editing, K.Y., Y.Z. and J.H. All authors have read and agreed to the published version of the manuscript.

Funding: This research received was founded by the National Outstanding Youth Science Fund Project of the National Science Foundation of China (52022077).

Data Availability Statement: Not applicable.

Acknowledgments: The authors are very grateful for the editors of *Lubricants* and the anonymous reviewers for their work in processing this article.

Conflicts of Interest: The authors declare no conflict of interest.

References

1. Liu, J.; Tang, C.; Wu, H.; Xu, Z.; Wang, L. An analytical calculation method of the load distribution and stiffness of an angular contact ball bearing. *Mech. Mach. Theory* **2019**, *142*, 103597. [CrossRef]
2. Ma, S.J.; Zhang, X.H.; Yan, K.; Zhu, Y.S.; Hong, J. A Study on Bearing Dynamic Features under the Condition of Multiball–Cage Collision. *Lubricants* **2022**, *10*, 9. [CrossRef]
3. Ge, L.; Wang, C.; Yan, K.; Zhu, Y.; Hong, J. Design of groove structures for bearing lubrication enhancement based on the flow mechanism analysis. *Tribol. Int.* **2021**, *158*, 106950. [CrossRef]
4. Palmgren, A. *Ball and Roller Bearing Engineering*, 3rd ed.; SKF Industries: Gothenburg, Sweden, 1959.
5. Kakuta, K. Generating mechanical of friction torque in a ball bearing I. *Mechanism* **1965**, *2*, 645–648.
6. Tong, V.C.; Hong, S.W. Improved formulation for running torque in angular contact ball bearings. *Int. J. Precis. Eng. Manuf.* **2018**, *19*, 47–56. [CrossRef]
7. Todd, M.J.; Stevens, K.T. *Frictional Torque of Angular Contact Ball Bearings with Different Conformities*; Technical Report ESA-CR(P)-1221; Risley: Warrington, UK, 1978.
8. Deng, S.E.; Hua, X.W.; Zhang, W.H. Analysis on friction torque fluctuation of angular contact ball bearing in gyro motor. *J. Aerosp. Power* **2018**, *33*, 1713–1724. [CrossRef]
9. Xu, Y.; Xia, X. Analysis of Parameter-non-Parameteric Fusion on Space Bearing Friction Moment. *J. Aerosp. Power* **2014**, *29*, 2129–3135. [CrossRef]
10. Zhang, Y.; Xia, X.T.; Wang, Z.Y.; Li, J.H. Study on the Prediction Method of Frictional Moment for Rolling Bearings. *J. Bear.* **2006**, *10*, 22–24.
11. Snare, B. Rolling resistance in lightly loaded bearings. *Ball Bear. J.* **1968**, *152*, 3–8.
12. Li, S.S.; Chen, J.; Lin, J.; Gu, J.M. Analytic and Experimental Study on Friction Torque for High-speed Micro Ball Bearing. *J. Lubr. Eng.* **2013**, *38*, 32–35. [CrossRef]
13. Wang, Y.; Lin, F.; Jiang, H.; Yuan, W. Investigation on frictional characteristic of deep-groove ball bearings subjected to radial loads. *Adv. Mech. Eng.* **2015**, *7*, 1–12. [CrossRef]
14. Ye, J.; Yang, L.F.; Ji, Z.J.; Guo, C.Y. Measurement and analysis of friction moment of frame bearing. *J. Bear.* **2001**, *4*, 7–9. [CrossRef]
15. Deng, S.E.; Li, X.L.; Wang, J.G. Frictional Torque Characteristic of Angle Contact Ball Bearings. *J. Mech. Eng.* **2011**, *47*, 114–120. [CrossRef]
16. Todd, M.J. *Selected ESTL Papers on Ball Bearings for Satellites*; ESA Scientific & Technical Publications Branch: Paris, France, 1980.
17. Houpert, L. Numerical and analytical calculations in ball bearing. In Proceedings of the 8th European Space Mechanisms and Tribology Symposium, Toulouse, France, 29 September–1 October 1999; pp. 283–290.
18. Cao, Y.; Altintas, Y. A general method for the modeling of spindle-bearing systems. *J. Mech. Des. Trans. ASME* **2004**, *126*, 1089–1104. [CrossRef]
19. Hao, L.; Deng, S.; Qian, D.; Hua, L. Accurate prediction method of initial value of high-speed ball bearing model and gyroscopic torque analysis. *J. Mech. Sci. Technol.* **2020**, *34*, 3745–3755. [CrossRef]
20. SKF Explorer. New Bearings for High-Speed Applications. Available online: https://evolution.skf.com/new-bearings-for-high-speed-applications/#related-articles (accessed on 9 January 2018).
21. Jones, A.B. A general theory for elastically constrained ball and radial roller bearings under arbitrary load and speed conditions. *J. Fluids Eng. Trans. ASME* **1960**, *82*, 309–320. [CrossRef]
22. Zhang, Y.; Fang, B.; Kong, L.; Li, Y. Effect of the ring misalignment on the service characteristics of ball bearing and rotor system. *Mech. Mach. Theory* **2020**, *151*, 103889. [CrossRef]
23. Zhang, J.H.; Fang, B.; Yan, K.; Hong, J. A novel model for high-speed angular contact ball bearing by considering variable contact angles. *J. Mech. Sci. Technol.* **2020**, *34*, 809–816. [CrossRef]
24. Harris, T.A. *Advanced Concepts of Bearing Technology, Rolling Bearing Analysis*, 5th ed.; John Wiley and Sons, Inc.: New York, NY, USA, 2006.
25. Wang, L.Q. *Design and Numerical Analysis of Rolling Element Bearing for extreme Applications*; Harbin Institute of Technology Press: Harbin, China, 2013.
26. Deng, S.E.; Jia, Q.Y.; Xue, J.X. *Design Principle of Rolling Bearing*; Standards Press of China: Beijing, China, 2014.

lubricants

Article

Research on the Mechanism of the Stiffness Performance of Rolling Bearings under Wrong Assembly State

Yanfei Zhang [1,2,*], Yunhao Li [1], Lingfei Kong [1], Zhenchao Yang [1,*] and Yue Si [1]

1 School of Mechanical and Precision Instrument Engineering, Xi'an University of Technology, Xi'an 710048, China; 2200221197@stu.xaut.edu.cn (Y.L.); lingfeikong@xaut.edu.cn (L.K.); siyue925@xaut.edu.cn (Y.S.)
2 Luoyang Bearing Science & Technology Co., Ltd., Luoyang 471039, China
* Correspondence: yfzhang@xaut.edu.cn (Y.Z.); zcyang@xaut.edu.cn (Z.Y.)

Abstract: In this paper, a quasi-static angular contact ball bearing model, considering assembly accuracy is constructed, while a numerical solution method for bearing stiffness under bad assembly state is established. A 7014C angular contact ball bearing is used as the research object and five groups of different spacer inclinations are designed to imitate the installation error of the spindle bearing. The bearing stiffness performance was comparatively analyzed, according to the five spacers. The effect of preload and rotation speed on bearing stiffness are systematically investigated, considering different parallelism errors, as induced by the spacers. The influence mechanism of the badly assembled bearing on the respective stiffness anisotropy is studied based on the proposed model. The results show that the variations of the inclination between the inner and outer rings of the bearing exhibit a very weak effect on the axial stiffness, while the influence on the radial and angular stiffness is more significant.

Keywords: angular contact ball bearing; spacer inclination; bearing stiffness

Citation: Zhang, Y.; Li, Y.; Kong, L.; Yang, Z.; Si, Y. Research on the Mechanism of the Stiffness Performance of Rolling Bearings under Wrong Assembly State. *Lubricants* **2022**, *10*, 116. https://doi.org/10.3390/lubricants10060116

Received: 11 April 2022
Accepted: 31 May 2022
Published: 5 June 2022

1. Introduction

The factors that affect the performance of a machine tool, over the course of a spindle system's whole life cycle, include not only the quality of the spindle components, but also the design of the spindle system, the component assembly process and the assembly technique. Angular contact ball bearings have a significant impact on the performance of the spindle system, as the core rotating support element, while bearing stiffness performance indicators have a direct impact on spindle system vibration, noise, rotational precision, end jump and service life [1–5]. Moreover, the accuracy of the bearing stiffness values is crucial in building a global model of the spindle system. Indeed, the frequency response function and, in particular, the critical eigenfrequencies are directly linked to the bearings' stiffness. As a result, additional quantitative study is required, to investigate the relationship between bearing assembly quality and bearing stiffness in all directions, in order to determine how the bearing assembly quality affects the stiffness performance of the bearing and even the spindle system.

Ball bearings, as a kind of rolling bearing, are one of the most important parts of spindle-bearing system. Many researchers [6–8] established that spindle performance changes dynamically due to the nonlinear effect of bearings stiffness. A great number of scholars have conducted studies on bearing stiffness, including bearing stiffness analysis and calculation, as well as bearing stiffness change affect factor analysis [9]. Jones, for example, was the first to develop the rolling bearing contact angle analysis mechanics model, which was the basis for the numerical assessment of the bearing load and change law, while Harris later enhanced the model and it was widely accepted by other researchers as Jones and Harris model [10,11].

J. Jedrzejewski [12] researched the relationship between the rotation speed and bearing stiffness, deriving the "bearing stiffness softening phenomenon", where the bearing stiffness continues to drop, as the bearing rotational speed rises. Matti Rantatalo [13] held the same view and the respective research also pointed that the radial stiffness would drop to 40%, when the rotational speed was up to 20,000 rpm, based on the presented calculations. However, both of them did not give a detailed calculation model to explain the speed varying stiffness, dependent on the analysis of bearing contact force. In the work of Sheng et al. [14]., the detail notion of rolling bearing speed-varying stiffness is introduced and explained, based on the relations of load-deflection, according to the bearing dynamic model, also based on Jones and Harris's model.

David Noel [15] proposed a new method for the computation of the stiffness matrix: a complete analytical expression is presented, including dynamic effects, in order to ensure accuracy at high shaft speed. This new method is particularly relevant in the case of lightly loaded bearings in critical applications (both high values of shaft speed and ball orbital diameter). Yi Guo [16] developed a finite element/contact mechanics model for rolling element bearings, with the focus on obtaining accurate bearing stiffness for a wide range of bearing types and parameters. The presented fully-populated stiffness matrix demonstrates the coupling between bearing radial, axial and tilting bearing deflections. However, this method did not consider the rotational speed effect on the bearing stiffness.

Cao [17] et al. tested the influence of bearing positioning preload and constant pressure preload on spindle-bearing dynamic stiffness, concluding that positioning preload is more effective than constant pressure preload in preserving spindle dynamic stiffness, under cutting loads at high speeds. Aramaki [18] studied bearing stiffness under the influence of preload force and confirmed the mapping link between preload force and bearing stiffness features. Yang [19] used the proposed static model of rolling bearing, to calculate the dynamic stiffness of the bearing and determined its operational stiffness, considering the influence of rotational speed, initial preload force, thermal preload force and oil film thickness on bearing stiffness. Liu [20] suggested a methodology, based on finite element analysis, in order to study the influence of preload on bearing stiffness and discover the bearing's equilibrium state while under load.

The experiment about bearing stiffness has also been conducted by many researchers. Walford et al. [21] developed a spindle-bearing test platform, to estimate the bearing's radial stiffness and damping, by measuring the response of the spindle, as well as to determine the effect of temperature on the stiffness of the bearings. Similarly, Kraus et al. [22] pointed out that, static bearing stiffness is very close to the stiffness measured while the bearing is running, based on stiffness and damping characteristics of a radial ball bearing, as derived by experimental modal tests. Marsh E R et al. [23] carried out an experimental measurement of precision bearing dynamic stiffness, where it was pointed out that an analysis of the method is sensitive to errors in sensor location, while several practical advantages of the proposed approach, over traditional static testing, were demonstrated. However, the majority of the above studies are based on a mechanical model of the bearing, under ideal operating conditions. Nonetheless, in engineering applications, various errors and even installation eccentricity in the bearing-spindle coupling system are inherent, while the bearing operation is affected by the additional moment.

Recently, ring misalignment occurrence has attracted much attention from the academia. Many researchers are devoted to studying the effect of ring misalignment error on the operating characteristics of rolling bearings. Zhang [24] studied the effect of ring misalignment on the service characteristics of ball bearing and rotor system; however, the analysis of bearing stiffness is not the key research content of this study. Xu [25] researched the effect of angular misalignment of inner ring on the surface contact characteristics and stiffness coefficients of duplex angular contact ball bearing. It was pointed out that, the ring misalignment considerably changes the surface contact characteristics of ball-raceway and causes uneven load distribution. As a result, Zhang [26] built a model that included the bearing's deflection angle and assessed the effect of axial and radial loads on the contact angle, but the change law of bearing stiffness performance was not investigated.

This issue is addressed in this paper, where the mechanism of bearing stiffness change is studied, in the context of poor assembly. Furthermore, the influence of rotational speed, preload and assembly quality on bearing stiffness, in all directions, is systematically analyzed (the relative tilt amount of the inner and outer ring of the bearing), while the quantitative description of the bearing, from the assembly error expression to bearing stiffness performance change, is realized.

2. Bearing Stiffness Solution Model Construction

2.1. Equivalent Transformation of Bearing Spacer Non-Parallel

A common way to adjust the initial assembly preload of spindle bearings is the use of spacers. In the case where the spacer end faces are not parallel, there is a greater impact on the bearing's initial assembly state, causing the bearing to operate in a relatively tilted attitude, between the inner and outer rings, which affects the bearing's mechanical and stiffness characteristics. This approach allows for the simulation of various bearing spacer tilting circumstances. Figure 1 shows the tilting of the spacer and the subsequent changes in the forces on the bearing's inner and outer rings.

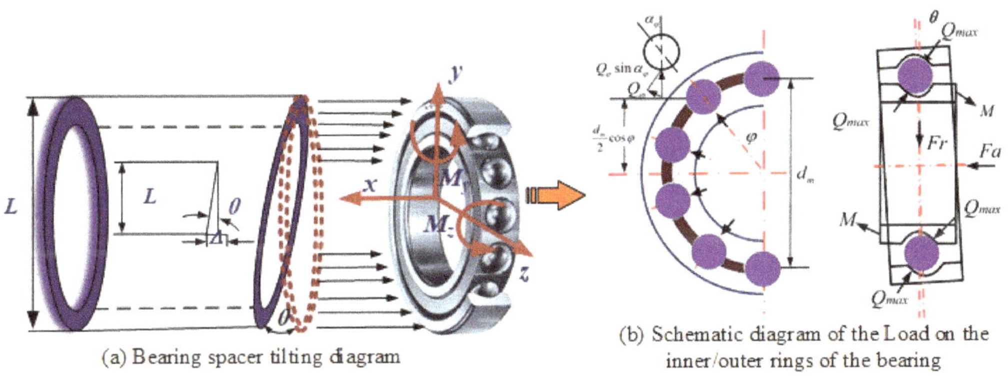

Figure 1. Diagram of the forces on the bearing assembly: (**a**) Bearing spacer titling diagram, (**b**) Schematic diagram of the Load on the inner/outer rings of the bearing.

The bearing outer ring force equilibrium equation is determined using the force analysis diagram in Figure 1:

$$\begin{bmatrix} F_a \\ F_r \end{bmatrix} = \begin{bmatrix} \sum_{j}^{Z}\left(Q_{oj}\sin\alpha_{oj} - \lambda_{oj}\frac{M_{gj}}{D_b}\cos\alpha_{oj}\right) \\ \sum_{j}^{Z}\left(Q_{oj}\sin\alpha_{oj} - \lambda_{oj}\frac{M_{gj}}{D_b}\cos\alpha_{oj}\right) \end{bmatrix} \quad (1)$$

where, F_a represents the axial load applied on the bearing (N); F_r is the radial load applied on bearing (N); Z is the number of the rolling balls; Q_{oj} is the contact load between jth ball and outer ring (); α_{oj} is the contact angle between the jth ball and the outer ring (rad); M_{gj} is the gyroscopic moment of the jth rolling element (N·mm); D_b is the diameter of the rolling ball; λ_{oj} is the load distribution coefficient of outer raceway.

Furthermore, in the case of bearing spacer non-parallelism, caused by the relative tilt of the bearing's inner and outer rings, the contact force created on the rolling body and raceway contact region produces a moment M, relative to the inner and outer ring of the bearing tilt, whose value is calculated as follows:

$$M = \sum_{j}^{Z} \left[\begin{array}{l} \left(Q_{oj} \sin \alpha_{oj} - \frac{\lambda_{oj} M_{gj}}{D_b} \cos \alpha_{oj} \right) R_o \\ + \frac{\lambda_{oj} M_{gj}}{D_b} r_o \end{array} \right] \cos \varphi_j \quad (2)$$

where, R_o is the radius of the outer raceway curvature center (mm); r is the distance from the point of load force application to the axis (mm); φ_j is the orientation of the jth rolling element (rad).

It is evident that, due to the non-parallelism of the spindle spacer and the resulting bearing inner ring, the outer ring experiences a relative tilt, causing non-uniformity in the rolling body and raceway contact area, while under load, which triggers the additional bending moment and has an impact on the bearing stiffness performance.

2.2. Bearing Stiffness Solution

The 5th order square matrix, created by taking the partial derivative of the external load f applied to the bearing, against the relative displacement d of the inner and outer rings of the bearing, is the analytical way of solving the bearing stiffness matrix. The Jones model is used to calculate the three-degree-of-freedom hydrostatic parameters, used as the initial value of the iterative algorithm to obtain the proposed three-degree-of-freedom hydrostatic solution parameters. Next, the derived three-degree-of-freedom hydrostatic parameters are used as the initial value of the Newton-Raphson iterative solution method, used to obtain the proposed five-degree-of-freedom hydrostatic parameters of the bearing and solve the 4Z + 5 equations. In the solution process, each rolling element equation is a local equation, whereas the overall bearing force equation is a global equation. The solution variables for each rolling element are: $\mathbf{X} = \{X_1, X_2, \delta_i, \delta_o\}_j$, $j = 1, 2 \ldots Z$ where Z is the number of scrolling bodies and is called the local solution variable. $\mathbf{f} = \{F_x, F_y, F_z, M_y, M_z\}$ is called a global load variable, d = $\{\delta_x, \delta_y, \delta_z, \theta_y, \theta_z\}$ is called a universal displacement variable. According to the global equations, the overall bearing force and each rolling body force are related, so when solving the overall bearing force equation, each rolling body equilibrium equation must also be solved, joining each rolling body solution formula and force equilibrium equation. The specific solution flow is shown in Figure 2.

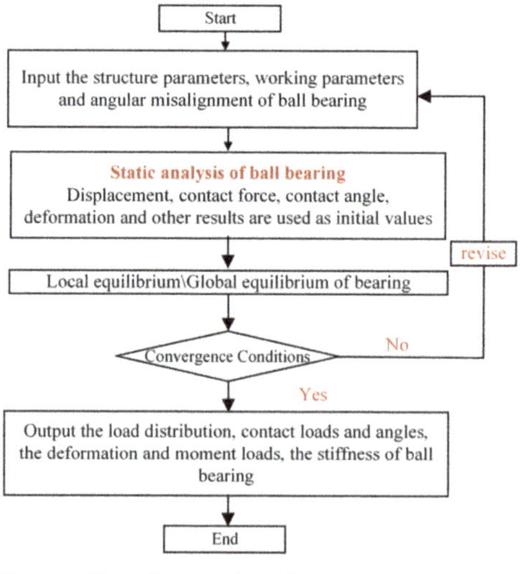

Figure 2. Flow diagram for calculation of bearing stiffness characteristics in all directions.

3. Bearing Stiffness Solution Condition Settings

In this paper, the well-known commercially available precision angular contact ball bearing 7014C is employed, with parameters as listed in Table 1.

Table 1. Basic bearing parameters.

Parameter	Value
D_b (mm)	11.9
f_i (mm)	0.52
f_o (mm)	0.52
Z	20
d_m (mm)	90.5
α_0 (°)	15
E_{ball} (Mpa)	2.06×10^5
ρ_{ball} (kg·mm^3)	7.8×10^{-6}
v_{ball}	0.3
E_{ring} (Mpa)	2.06×10^5
v_{ring}	0.3

The end face of the spacer is always machined with one end face as the reference surface and the other end face is ground. Due to manufacturing errors, the machined surface is not perfectly flat. In other words, there is a parallelism error between the machined end face of the spacer and the reference surface, which is depicted in Figure 2. To facilitate analysis and calculation, this paper uses the inclination angle θ of the spacer end face to characterize the non-parallelism of the spacer end face, expressed as:

$$\theta = \arctan\frac{\Delta}{L} \tag{3}$$

where L is the outer diameter dimension of the outer spacer (mm), in relation to the mating bearing. This paper is based on the analysis of the 7210C bearing, so the value of L is 90 mm; Δ: The height difference between the highest point and the lowest point of the spacer end face (μm) is shown in Figure 3.

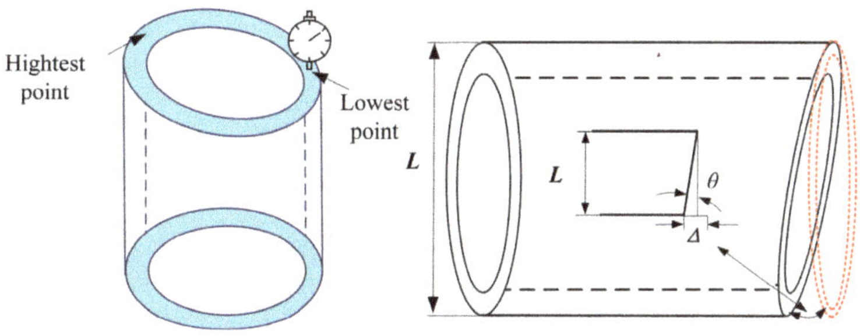

Figure 3. Graphical depiction of spacer end face non-parallelism.

For theoretical analysis purposes, several groups of spacers with parallelism errors are set up in this study, which are denoted as: Δ = 10 μm, 20 μm, 30 μm, 40 μm, while the corresponding tilt angle values of these four groups of outer spacers with parallelism errors are: $\theta_1 = 0.0052°$, $\theta_2 = 0.0104°$, $\theta_3 = 0.0156°$, $\theta_4 = 0.0208°$as listed in Table 2. According to the NSK bearing manual, the spacer face runout within 3 μm is within the design error tolerance, so the spacer mark within the design error range is: Δ = 0 μm, which indicates a well parallel spacer. The angle of tilt is noted as: $\theta_0 = 0°$. In contrast with the tilting of a bearing, caused by an individually applied bending moment, when the bearing is tilted due to a dimensional error in the outer spacer, the bearing is also tilted to a certain extent by the geometric constraints of the spacer end face.

Table 2. Description of bearing stiffness analysis conditions.

Preload	Extremely Light Preload (EL)	Light Preload (L)	Medium Preload (M)	Heavy Preload (H)
0 μm/0°	145 N	290 N	740 N	1470 N
10 μm/0.0052°	145 N	290 N	740 N	1470 N
20 μm/0.0104°	145 N	290 N	740 N	1470 N
30 μm/0.0156°	145 N	290 N	740 N	1470 N
40 μm/0.0208°	145 N	290 N	740 N	1470 N

The analysis shows that, in the assembly of the spindle, there will be a tilting gap between the outer spacer and the bearing outer ring, due to the existence of the tilting angle of the outer spacer end face. As shown in Figure 4, the bearing and housing is usually a clearance fit, resulting in clearance inside the bearing. There is a possibility of the bearing outer ring being tilted, during the transfer of force, under the action of the axial assembly force and the contact between the tilted end face of the outer spacer and the bearing outer ring.

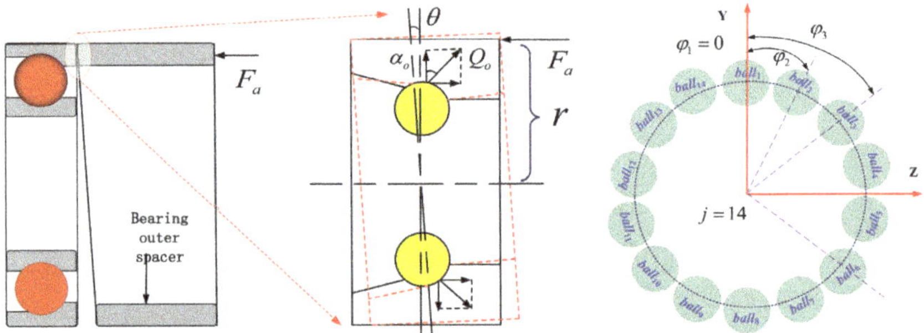

Figure 4. Force analysis diagram of bearing outer ring under inclined spacer ring: (**a**) contact between bearing and spacer, (**b**) bearing force, (**c**) ball orientation diagram.

The notion of bearing mounting quality, as opposed to the ideal load bearing stiffness change law, can more accurately reflect the actual engineering application of bearing stiffness performance, bearing mounting quality analysis, bearing joint parts quality on the bearing and even spindle stiffness service performance of the mechanism. In this view, it is the bearing mounting quality and recommended assembly parameters in real-world engineering applications that are considered in this paper; as well as a number of working scenarios for the investigation of bearing stiffness performance, with exact factors (Table 2). Different speed values (at 1000 rpm increments), preload conditions and bearing deflection circumstances are all used to compute stiffness variations.

4. Analysis of Factors Influencing Bearing Stiffness

4.1. Bearing Stiffness Variation Law under Different Working Conditions

The variation of axial, radial and angular stiffness values on a macro scale reflects the bearing stiffness characteristics. The diagram in Figure 5 demonstrates that there are five surfaces (five surfaces indicate five different tilt conditions: 0 μm −10 μm −20 μm −30 μm −40 μm). It is harder to determine that this is a superposition of five surfaces, based on the three-dimensional surface diagram. More specifically, the effect of tilting the inner and outer rings of the bearing on its overall axial stiffness K_{xx}, radial stiffness K_{yy}, radial stiffness K_{zz}, angular stiffness $K_{\theta y}$ and angular stiffness $K_{\theta z}$ does not cause significant value fluctuation. The reason for this could be that some of the local differences in surface changes are difficult to discern in 3D views, while the changes in bearing stiffness are so miniature that they are difficult to reflect on the overall stiffness change relationship graph.

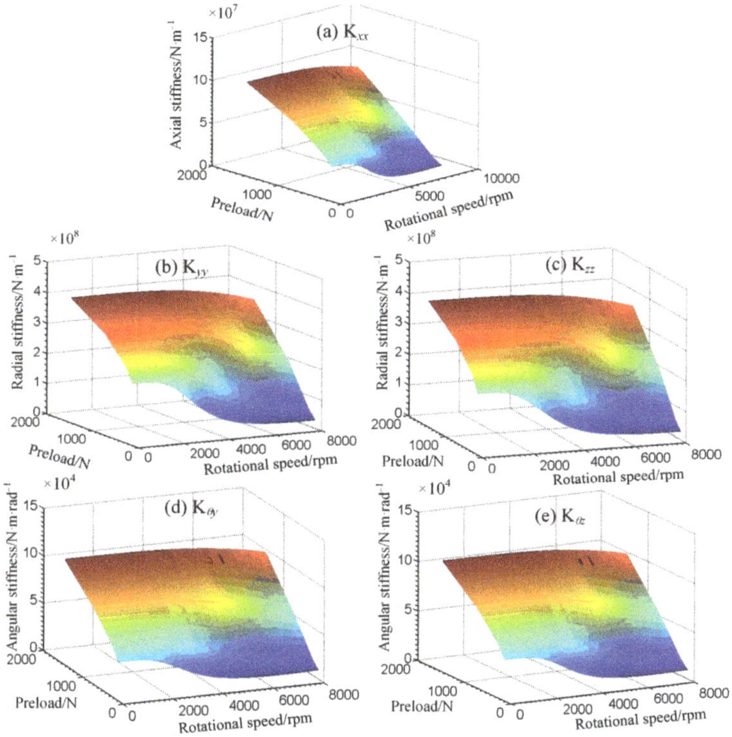

Figure 5. Variation of spindle bearing stiffness in all directions in relation to tilt angle.

As a result, a view is transformed to analyze the stiffness variation rule (Figure 6). It is observed that the bearing stiffness varies significantly under different preloads. Specifically, raising the bearing preload can effectively improve bearing stiffness performance, whereas as preload and speed increase, bearing stiffness reduces significantly (bearing heavy preload versus light preload conditions). In addition, the axial stiffness is one order of magnitude lower, compared to the radial stiffness, while no significant variation is found between the horizontal radial stiffness and the vertical radial stiffness, as well as between the two angular stiffness components $K_{\theta y}$. The bearing stiffness variations in all directions, in relation to its inner and outer ring tilt angle fluctuations, are not obvious but exhibit a certain regularity. That is, stiffness increases and decreases in relation to spindle speed and preload force, while it is not a linear relationship.

4.2. Analysis of the Variation of Bearing Stiffness in All Directions in Relation to Speed

The stiffness—speed—preload relationship graph, based on variations in the bearing's inner and outer ring tilt, does not adequately reflect the influence of imperfect mounting on bearing stiffness, as shown in Section 4.1. In engineering, the tilting of inner and outer rings inevitably leads to a change in the internal contact parameters of the bearing. If this tilting angle is known, the change in contact parameters will lead to a change in contact stiffness, until it is reflected in the isotropic stiffness of the bearing. As a result, the effect of the additional bending moment, induced by the relative tilting of the inner and outer rings, on bearing stiffness, will be investigated, in order to further confirm it, based on a two-dimensional plane perspective.

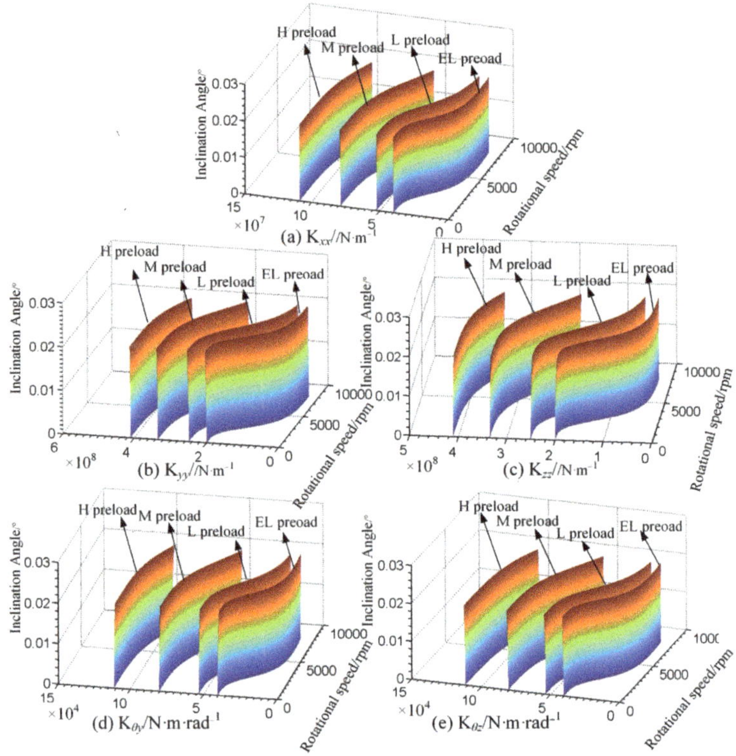

Figure 6. Variation of spindle bearing stiffness in all directions under preload conditions.

Figure 7 shows the change in the axial stiffness of the bearing, as a function of speed, under various preload forces. It also illustrates a comparison and analysis of the change in axial stiffness of the bearing, as a function of the inner and outer rings inclination value. Based on Figure 5, from a macroscopic point of view, bearing inner and outer ring relative tilt angle variations did not produce too high influence on the axial stiffness. In the operating conditions of Extremely Light preload (EL), Light preload (L), Medium preload (M), bearing stiffness change curve shows obvious consistency, as the five curves almost overlap. In the Heavy preload (H) operating conditions, the axial stiffness difference in the local magnification chart is slightly obvious. However, compared to the bearing axial stiffness value of 1–2 magnitude orders lower, this difference can be ignored. At the same time, another occurrence was specified, namely the considerable change in axial stiffness of the bearing, caused by the preload force. In the mild preload situation, when the speed approaches 1800 rpm, the axial rigidity gradually drops, until stabilizing at 5000 rpm. The slope of the drop is less than that, in the slight preload conditions. Similarly, in the bearing under preload conditions, the stiffness decreases more obviously after the speed reaches about 3000 rpm, while even more obviously in heavy preload conditions, when the axial stiffness only shows a significant decreasing trend, after the speed reaches about 4000 rpm. The investigation of the axial stiffness of bearings, at various degrees of inner and outer ring tilting, reveals that bearing mounting quality has only a minor impact on axial stiffness, as well as confirms the efficacy and significance of bearing preload in enhancing bearing stiffness.

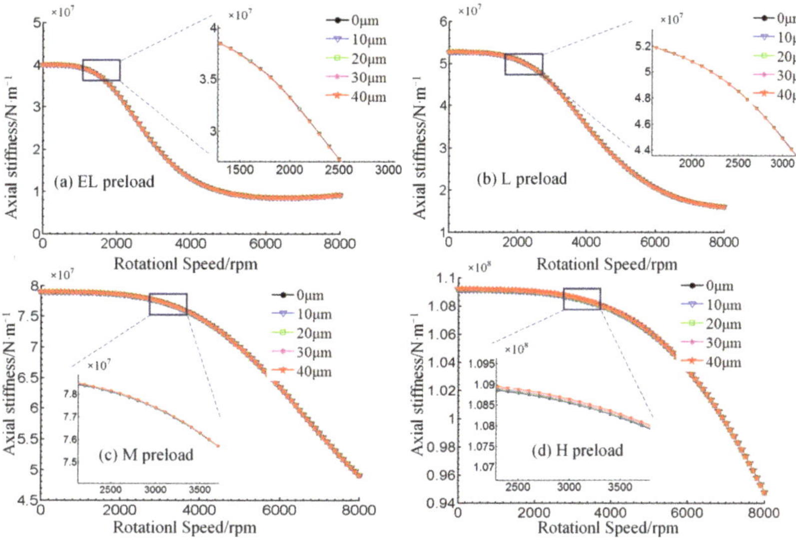

Figure 7. Variation of axial bearing stiffness K_{xx} in relation to speed.

Figure 8 shows how the bearing's horizontal radial stiffness varies, with relation to preload force, at various tilting angles, while systematically represented in four operating conditions: extremely light preload, light preload, medium preload and heavy preload. The radial stiffness of the bearing exhibits the same regularity as the axial stiffness, while the influence of the preload on the radial stiffness is also significant. At a higher preload, the bearing stiffness becomes more resistant to the "softening phenomenon", as the speed increases, whereas the radial stiffness is an order of magnitude higher than the axial stiffness. It should be noted that changing the bearing's inner and outer ring tilt has a significantly greater impact on radial stiffness, i.e., under different preload circumstances, the bearing's radial stiffness increases with a tilt angle of roughly 2.5% to 3%. This probably occurs due to the bearing inner and outer ring tilt orientation (see Figure 1 for a schematic depiction). As a result, a thorough understanding of the tilt angle's influence on the radial stiffness, as well as the general and specific laws of bearing stiffness, is critical, not only for accurate bearing stiffness exploration, but also for providing a solid theoretical foundation for subsequent bearing service performance accuracy.

Similar to Figures 8 and 9 shows how the bearing's vertical radial stiffness varies with relation to preload force, at different tilting angles, represented consistently under four operating conditions: EL, L, M and H. It is proven that the vertical radial stiffness of the bearing shows almost the same regularity as the horizontal radial stiffness, while the difference is reflected in a closer local view, where the horizontal radial stiffness of the bearing appears slightly increased, in an analogy to the increase of the inner and outer ring tilt. In regards to the vertical radial stiffness, the exact opposite relationship holds; that is, the vertical radial stiffness of the bearing decreases with the increase of the tilt angle, at about 2~2.5%. The bearing spacer tilt orientation, generated by this common difference, tilts the bearing about the horizontal axis, according to the analysis. In order to accurately describe the law of tilt angle influence on bearing radial stiffness, besides the bearing tilt angle changes, the bearing tilt process of relative azimuth changes should be considered in a comprehensive analysis, which will provide results that are more in line with the actual engineering reality, while also more accurate.

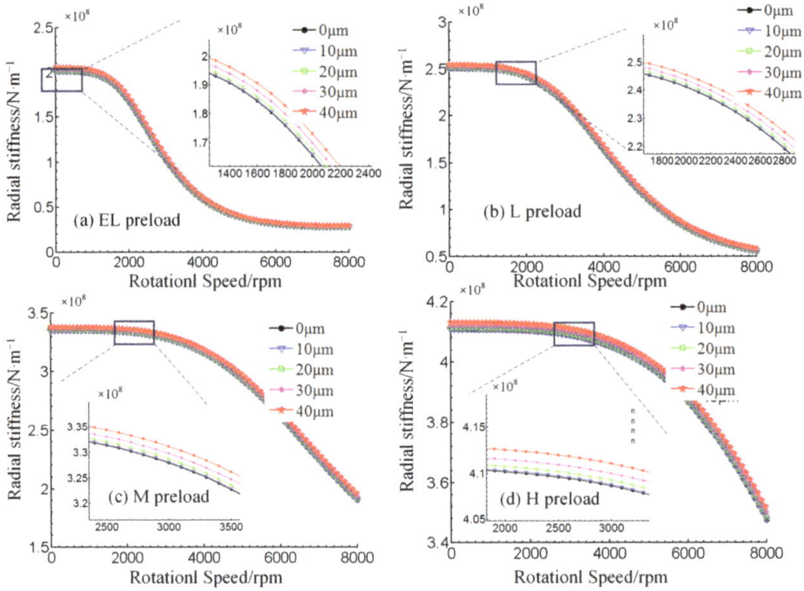

Figure 8. Variation of horizontal radial bearing stiffness K_{yy} with relation to speed.

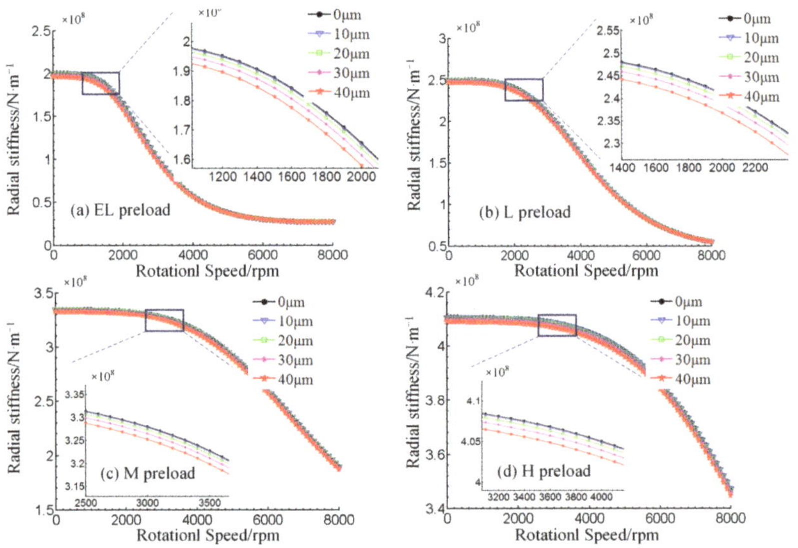

Figure 9. Bearing vertical radial stiffness K_{zz} variation with relation to speed.

Figures 7–9 illustrate the three directions of the bearing plain stiffness change law. Since angular contact ball bearings also have an angular stiffness, an in-depth analysis should explore whether the bearing inner and outer ring relative tilt will have what effect on the angular stiffness and whether it is still similar to the plain stiffness change law, so as to accurately grasp the overall stiffness change mechanism.

Figure 10 shows that, the angular stiffness of the bearing changes in a similar way to the flatness of the bearing. From a macroscopic perspective, bearing speed and initial preload are the most important factors in determining bearing stiffness. Furthermore, angular stiffness and plain stiffness decrease as frequency increases, and they rise as preload decreases. The fact that the tilt angle of the inner and outer rings vary around the z axis, makes the angular stiffness of the bearing changes, with relation to the tilt angle of the inner and outer rings, a little more visible (Figure 1), so the angular stiffness $K_{\theta y}$ decreases as the tilt angle increases.

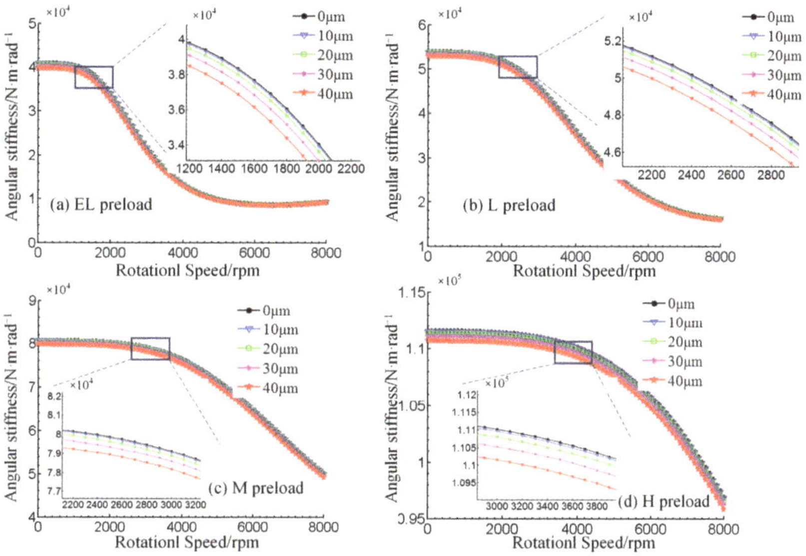

Figure 10. Variation of angular bearing stiffness $K_{\theta y}$ with relation to speed.

Based on Figure 11, the angular stiffness $K_{\theta z}$ variation characteristics with relation to speed are almost identical to the respective angular stiffness $K_{\theta y}$ variation characteristics, whereas minor differences are only reflected in the local view enlargements; however, the effect of the bearing tilt angle is evident in this difference in variation. A comparison of the local enlargements in Figure 8 shows that the angular stiffness $K_{\theta y}$ tends to decrease as the angle of inclination increases, while the angular stiffness $K_{\theta z}$ tends to increase as the angle of inclination increases, while the greater the preload the more significant the difference is.

Throughout the bearing stiffness curve with relation to speed and the preload force change law, it was discovered that the bearing's initial preload force, working speed and stiffness performance play dominant roles. However, the bearing inner and outer ring tilt angle, as well as the bearing stiffness in all directions show a significant correlation and a significant regularity difference. Although this variability may have a minor effect on the performance of rough machining spindles, it cannot be overlooked in precision machining spindles. Consequently, while mounting spindle bearings, special attention should be paid to the quality of the installation, ensuring that their stiffness characteristics and even the service performance of the spindle system are carefully controlled.

Due to the numerous elements affecting bearing stiffness fluctuation, a study focused just on the total stiffness under these five tilting scenarios would still yield traditional regularity results. Thus, the bearing stiffness influence must be decomposed into the local area of each rolling element and raceway contact, i.e., the overall rigidity. The bearing's performance is determined by the specific rolling element, while the stiffness of the inner and outer ring raceways follow a certain law combination, which requires a more complete examination of the local contact unit area.

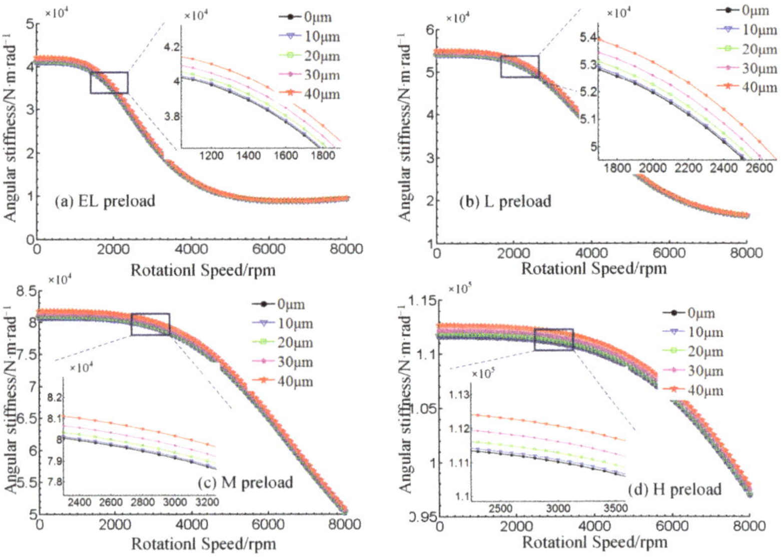

Figure 11. Variation of angular bearing stiffness $K_{\theta z}$ with relation to speed.

4.3. Analysis of Bearing Stiffness Variability in All Directions

The bearing stiffness in all directions was studied, as well as bearing speed, preload, inner and outer ring tilt working circumstances, which derived that the bearing stiffness is predominantly represented in the radial stiffness. This paper further analyses the inner contact angle αi_i, outer contact angle α_{oo}, inner ring contact force Q_i and outer ring contact force Qo of the rolling element, in addition to the raceway in the local area of the bearing, carrying out numerical research on the internal contact mechanical properties of the bearing. Since the general law of the bearing is the same for different preload values, only the medium preload condition, which is typically employed in engineering, is selected as a representative case for the specific analysis, whereas the variation of the bearing's internal contact characteristics is also investigated.

The characteristics of the 7014C bearing, under the influence of the inner and outer ring tilt rolling inner contact angle change, are shown in Figure 12. The inner contact angle gradually increases, as the speed rises, which is mostly due to the bearing's centrifugal force gradually increasing along the speed rise. At a constant speed, the analysis of the curve can provide certain results. Specifically, when the bearing inner and outer ring tilt is 0, each ball contact angle is a constant value (the impact of the bearing rolling body own weight is not considered). When the inner and outer rings show a certain tilt, the contact angle is then changed and some rolling body at the contact angle increased and some become smaller, while with the bearing inner and outer ring tilt degree of intensification, the difference of inner contact angle on individual balls is also increased. As the contact angle variations will affect the bearing performance, these will inevitably have a certain impact on the overall bearing rigidity.

Figure 13 shows the characteristics of the 7014C shaft at four different speed values, demonstrating the effect of the inner and outer ring tilt on the rolling element's outer contact angle. It becomes evident that the outer contact angle of the bearing decreases as the speed increases; a phenomenon due to the centrifugal effect, caused by the speed change. Moreover, the difference between the outer contact angles of each rolling element shows an obvious tendency.

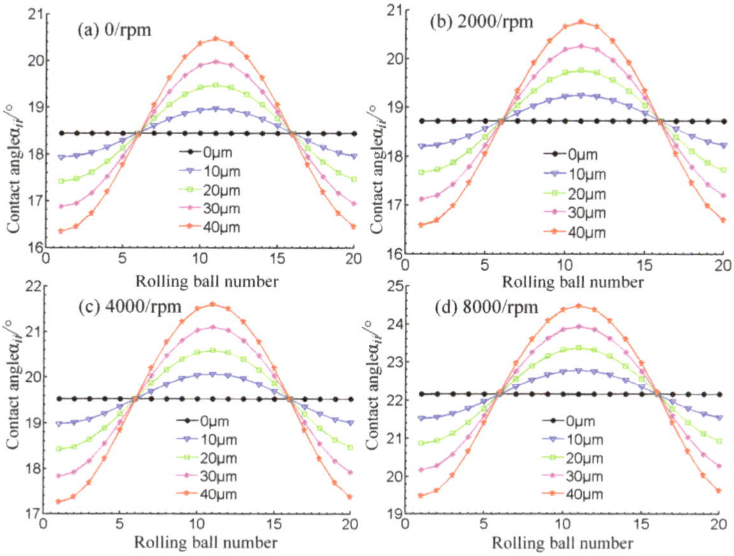

Figure 12. 7014C bearing rolling body and raceway internal contact angle α_{ii}.

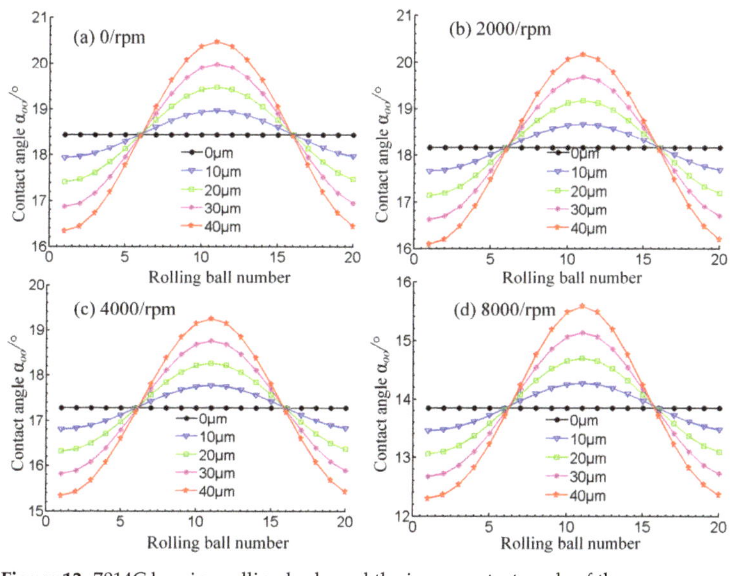

Figure 13. 7014C bearing rolling body and the inner contact angle of the raceway α_{oo}.

In Figure 14, the variation of the local contact force, between the rolling element and raceway of the inner ring of the bearing, is depicted. It can be seen that when the bearing speed increases, the inner ring local maximum contact force changes with the inner ring contact angle, while with the increase of speed the inner ring maximum local contact force significantly decreases. The main reason is that the contact angle increases along with the speed, which reduces the maximum local contact force. Under the same speed, the maximum local contact force changes more significantly with the increase of the inner and outer ring tilt angle. This indicates that when the tilt amount increases, the impact on the internal contact parameters of the bearing is more significant, intensifying the uneven load on the contact area of each rolling element of the bearing, whereas the unevenly distributed contact force on the circumferential radial is one of the reasons for the non-uniformity of the radial stiffness of the bearing, which may have a great impact on the accuracy and even the service life of the bearing.

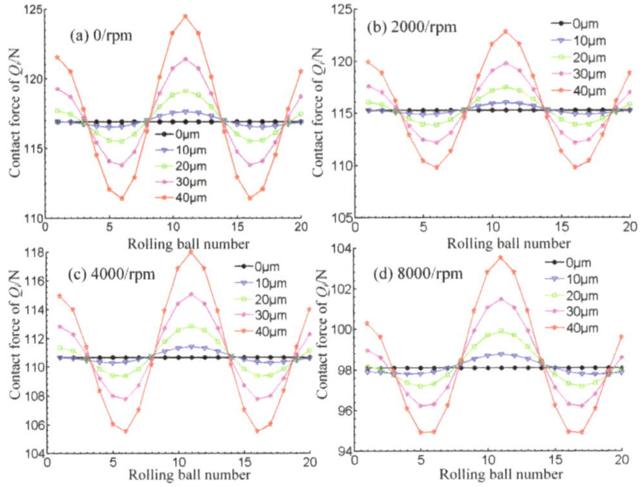

Figure 14. 7014C bearing rolling body and raceway internal contact force Q_i.

Figure 15 illustrates the variation pattern of the bearing outer ring contact force. It can be seen that, the bearing outer ring contact force exhibits a similar pattern to the bearing inner ring contact force. The difference lies in the speed rise causing the contact force on the outer ring of the bearing to increase. The change is still related to the increasing speed, caused by the effect of centrifugal force. At a certain speed, the outer ring local contact force changes. This means that the contact force on the outer ring of the bearing is non-uniform, along the circumferential direction, while this non-uniformity increases with the amount of skew.

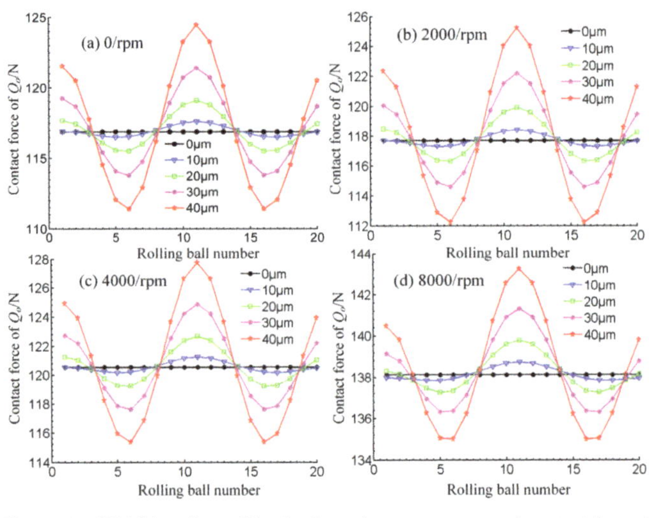

Figure 15. 7014C bearing rolling body and raceway external contact force Q_o.

5. Conclusions

In view of the common problems of bearing mounting in actual working conditions, the bad assembly state is simplified and represented by the relative tilting of the bearing spacer. Then, the influence of bearing preload, rotation speed and tilting of the spacer on the stiffness, along all directions, is examined the bearing stiffness (radial, axial and angular) is comparatively discussed under different assembly conditions, as well as the bearing preload and rotation speed. The main conclusions of this paper are as follows:

- Bearing inner and outer ring tilt angle has a greater impact on bearing radial stiffness than on axial stiffness.
- The inner and outer rings of the bearing changes lead to variations of the inner ring contact load, revealing the reasons for the bearing's anisotropic stiffness.
- Among the three influence factors of bearing preload, rotational speed and spacer inclination, although the inclination has the smallest influence on the macroscopic stiffness of the bearing, it will cause uneven radial stiffness of the bearing, which is very important for a precision spindle.
- Moreover, the anisotropy of radial stiffness, caused by the inclination of the bearing spacer, will be verified through experiments in the follow-up research. In the future, the study on the precision spindle system stiffness, induced by bearing anisotropy of radial stiffness, will be exhaustive.

Author Contributions: Conceptualization, Y.Z. and Y.L.; methodology, Y.L.; validation, Y.Z., Y.L. and L.K.; formal analysis, Z.Y. and Y.S.; investigation, L.K. and Y.Z.; data curation, Y.L.; writing—original draft preparation, Y.S.; writing—review and editing, Y.L.; visualization, Y.S.; supervision, Y.S.; project administration, Z.Y.; and funding acquisition, Y.S. All authors have read and agreed to the published version of the manuscript.

Funding: This research was funded by National Natural Science Foundation of China, grant number 52005405, General project of Shaanxi Natural Science Foundation, grant number 2022JM-244 and Natural Science Basic Research Program of Shaanxi Province, grant number 2020JQ-639.

Institutional Review Board Statement: Not applicable.

Informed Consent Statement: Not applicable.

Data Availability Statement: Detailed data are contained within the article.

Conflicts of Interest: The authors declared no potential conflict of interest with respect to the research, authorship, and/or publication of this article.

References

1. Wan, C.S. *Analysis Method of Rolling Bearing*; China Machanical Press: Beijing, China, 1995. (In Chinese)
2. Yang, Z.; Li, C.; Zhou, N.; Zhang, J. Research on the cage stability of high-precision ball bearing with image acquirement and error separation. *Measurement* **2021**, *186*, 110149. [CrossRef]
3. Sier, D.; Xinglin, L.; Jiugen, W. Frictional torque characteristics of angular contact ball bearings. *J. Mech. Eng.* **2011**, *47*, 114–120.
4. Yang, Z.; Niu, X.; Li, C. Experimental study on cage dynamic behavior of long-life high-precision ball bearing with trajectory deviation. *IEEE Trans. Instrum. Meas.* **2022**, *71*, 5011511. [CrossRef]
5. Liu, J.; Tang, C.; Wu, H.; Xu, Z.; Wang, L. An analytical calculation method of the load distribution and stiffness of an angular contact ball bearing. *Mech. Mach. Theory* **2019**, *142*, 103597. [CrossRef]
6. Abele, E.; Altintas, Y.; Brecher, C. Machine tool spindle units. *CIRP Ann.* **2010**, *59*, 781–802. [CrossRef]
7. Lin, C.-W.; Lin, Y.-K.; Chu, C.-H. Dynamic models and design of spindle-bearing systems of machine tools: A review. *Int. J. Precis. Eng. Manuf.* **2013**, *14*, 513–521. [CrossRef]
8. Ma, S.; Zhang, X.; Yan, K.; Zhu, Y.; Hong, J. A Study on Bearing Dynamic Features under the Condition of Multiball–Cage Collision. *Lubricants* **2022**, *10*, 9. [CrossRef]
9. Gao, S.H.; Meng, G.; Long, X.H. Study of milling stability with Hertz contact stiffness of ball bearings. *Arch. Appl. Mech.* **2011**, *81*, 1141–1151. [CrossRef]
10. Jones, A.B. A general theory for elastically constrained ball and radial roller bearings under arbitrary laod and speed conditions. *J. Basic Eng.* **1960**, *82*, 309–320. [CrossRef]
11. Harris, T.A. *Rolling Bearing Analysis*, 4th ed.; John Wiley and Sons, Inc.: New York, NY, USA, 2000.

12. Jedrzejewski, J.; Kwasny, W. Modelling of angular contact ball bearings and axial displacements for high-speed spindles. *CIRP Ann.* **2010**, *59*, 377–382. [CrossRef]
13. Rantatalo, M.; Aidanpaa, J.-O.; Göransson, B. Milling machine spindle analysis using fem and non-contact spindle excitation and response meaure-ment. *Int. J. Mach. Tools Manuf.* **2007**, *47*, 1034–1045. [CrossRef]
14. Sheng, X.; Li, B.; Wu, Z.; Li, H. Calculation of ball bearing speed-varying stiffness. *Mech. Mach. Theory* **2014**, *81*, 166–180. [CrossRef]
15. Noel, D.; Ritou, M.; Furet, B.; Le Loch, S. Complete Analytical Expression of the Stiffness Matrix of Angular Contact Ball Bearings. *J. Tribol.* **2013**, *135*, 041101. [CrossRef]
16. Guo, Y.; Parker, R.G. Stiffness matrix calculation of rolling element bearings using a finite element contact mechanics model. *Mech. Mach. Theory* **2012**, *51*, 32–45. [CrossRef]
17. Cao, H.R.; Tomas, H.; Yusuf, A. A comparative study on the dynamics of high speed spindles with respect to different preload mechanisms. *Int. J. Adv. Manuf. Technol.* **2011**, *57*, 871–883. [CrossRef]
18. Aramaki, H.; Shoda, Y.; Morishita, Y.; Sawamoto, T. The performance of ball bearing with silicon nitride ceramic balls in high speed spindles for machine tools. *ASME J. Tribol.* **1988**, *10*, 693–698. [CrossRef]
19. Yang, Z.W.; Yin, G.F.; Sang, X.; Jiang, H.; Zhong, K.Y. Coupling Analysis Model of Thermal and Dynamic Characteristics for High-speed Motorized Spindle. *J. Jilin Univ. (Eng. Technol. Ed.)* **2011**, *41*, 205–210.
20. Liu, X.; Hong, J.; Zhu, Y.S.; Liu, Z. Iterative Method to Solve Bearing's Force and Stiffness for a Multi-support Spindle System Based on Finite Element Analysis. *J. Xi'an Jiaotong Univ.* **2010**, *44*, 41–45.
21. Stone, B.J.; Walford, T. The measurement of the radial stiffness of rolling element bearings under oscillating conditions. *J. Mech. Eng. Sci.* **1980**, *22*, 175–181.
22. Kraus, J.; Blech, J.; Braun, S. In situ determination of rolling bearing stiffness and damping by modal analysis. *J. Vib. Acoust. Stress Reliab. Des.* **1987**, *109*, 235–240. [CrossRef]
23. Marsh, E.R.; Yantek, D.S. Experimental Measurement of Precision Bearing Dynamic Stiffness. *J. Sound Vib.* **1997**, *202*, 55–66. [CrossRef]
24. Zhang, Y.; Fang, B.; Kong, L.; Li, Y. Effect of the ring misalignment on the service characteristics of ball bearing and rotor system. *Mech. Mach. Theory* **2020**, *151*, 103889. [CrossRef]
25. Xu, T.; Yang, L.; Wu, W.; Wang, K. Effect of angular misalignment of inner ring on the contact characteristics and stiffness coefficients of duplex angular contact ball bearings. *Mech. Mach. Theory* **2021**, *157*, 104178. [CrossRef]
26. Zhang, X.; Han, Q.; Chu, F. Contact angle of ball bearings based on a simplified Jones-Harris method. *J. Vib. Shock* **2013**, *170*, 175.

lubricants

MDPI

Article

An Investigation for the Friction Torque of a Tapered Roller Bearing Considering the Geometric Homogeneity of Rollers

Yuwei Liu *, Xingyu Fan, Jia Wang and Xiayi Liu

School of Mechanical and Precision Instrument Engineering, Xi'an University of Technology, Xi'an 710049, China; 105266@xaut.edu.cn (X.F.); wangxiaojia@xaut.edu.cn (J.W.); liuxyee@xaut.edu.cn (X.L.)
* Correspondence: liuyw_mg@xaut.edu.cn

Abstract: The geometric homogeneity of rollers, namely the dimension and shape deviations among rollers in a roller bearing, is one of the most important manufacturing errors. However, to the best of the authors' knowledge, no specified investigation has been carried out on the effects of the geometric homogeneity of rollers on the friction torque of tapered roller bearings (TRBs). By introducing the diameter deviation of rollers and the distribution form of rollers with a diameter deviation, this study presents a mathematic model to reveal the effects of the geometric homogeneity of rollers on the friction torque of TRBs. The geometric homogeneity of the rollers, although having only a minimal influence on the overall friction torque acting on rings, can lead to a significant increase in the slide friction force between the individual rollers and the inner raceway. By comparing the distribution form of rollers with a diameter deviation, the diameter deviation value of the roller shows a significant influence on the maximum sliding friction between the roller and the inner raceway. The impact of the geometric homogeneity of rollers on the sliding friction between the roller and the inner raceway is more pronounced under light load conditions. The above-mentioned comparisons and conclusions can be used in formulating machining error criteria for TRB rollers.

Keywords: geometric homogeneity of rollers; tapered roller bearings; friction torque; contact force distribution

Citation: Liu, Y.; Fan, X.; Wang, J.; Liu, X. An Investigation for the Friction Torque of a Tapered Roller Bearing Considering the Geometric Homogeneity of Rollers. *Lubricants* **2022**, *10*, 154. https://doi.org/10.3390/lubricants10070154

Received: 10 May 2022
Accepted: 9 July 2022
Published: 12 July 2022

1. Introduction

Tapered roller bearings (TRBs) are generally used in heavy equipment such as wind power generators and high-speed trains. The operating temperature and limited rotating speed of the machinery are determined by the friction torques of their interior TRBs [1]. Thus, the behavior of the friction torque on TRBs of high-speed industrial machinery has continually attracted the attention of researchers and field engineers in the last few decades.

Throughout the years of development and investigation, efforts have been made in theoretical and experimental analyses to calculate TRBs' friction torque. Witte [2] analyzed the friction between each roller and raceway and presented an analytical model to calculate the friction torque of TRBs. However, the model did not consider the mixed lubrication state between bearing components. Consequently, the simulated results of the starting torque of TRBs showed a non-negligible error. With the development of elastohydrodynamic lubrication (EHL) theory, Karna [3] and Aihara [4] further modified and perfected the friction torque model of TRBs to meet engineering accuracy requirements. Zhou [5] explored a novel friction torque model to analyze the influences of a starved condition and a contaminant on the friction torques of TRBs. The Svenska Kullager Fabriken (SKF) Group [6] proposed a concise empirical formulation for predicting the friction torques of the TRBs based on the experimental results.

Despite numerous previous studies focusing on analyzing the friction torque of TRBs with the designed geometric dimensions, in practical engineering applications, manufacturing errors are unavoidable, which can affect the contact state between mating bearing components [7–10] and thus the friction torque of TRBs. Aschenbrenner [11] presented a variational simulation framework for the analysis of the load distribution of cylindrical roller bearings with component geometric deviations, providing a basis for investigating the frictional torque of tapered roller bearings with component geometric deviations. Deng [12] developed a theoretical method to investigate the effects of the surface waviness of bearing components on the friction torque of ball bearings. Heras [13] proposed a finite element model to calculate the friction torque of four contact point slewing bearings and analyzed the effects of raceway geometric errors on bearing friction torque. Halminen [14] and Xu [15] established a multibody model of ball bearings involving the surface waviness of the components. In their study, the impact of surface waviness on the dynamic performance of ball bearings was explored. Liu [16] utilized a time-varying calculation method to estimate the friction moments of angular contact ball bearings and analyzed the effects of the waviness amplitude and order on bearing friction torque under different operating conditions. Liu [17] analyzed the effects of roundness errors on the friction torque of bearings and found that both the magnitude and order of the roundness error have a significant impact on the bearing friction torque.

In addition to raceway machining errors, the geometric homogeneity of rollers, namely the dimension deviations between each roller, also plays a significant role in the evaluation of manufacturing errors in roller bearings [18,19]. Similar to a localized defect, the geometric homogeneity of rollers first affects the contact state between the roller and the raceway [20,21], thereby changing the overall bearing performance. However, the effects of the geometric homogeneity of rollers on the friction torque of TRBs are rarely reported in the literature. It is therefore timely to present this research to fill the knowledge gap.

In this paper, a quasi-statics model and a friction torque model for TRBs concerning the geometric homogeneity of rollers have been proposed, in which the geometric homogeneity of rollers was represented by the diameter deviation value of rollers and the distribution form of rollers with a diameter deviation. Based on the proposed models, the effects of the geometric homogeneity of rollers on the contact force distribution and friction characteristics of TRBs have been analyzed under different axial external loads, rotating speeds, and cage slip rates.

2. Materials and Methods

2.1. Quasi-Statics Analysis of TRB Considering the Geometric Homogeneity of Rollers

In order to estimate TRB friction torque accurately, the contact forces between each roller and raceway should be obtained in advance. As shown in Figure 1a, when the inner ring is constrained by displacement, an external load $\{F\}^T = \{F_x, F_y, F_z\}$ acting on the outer ring causes the corresponding displacement of the outer ring $\{d\}^T = \{d_x, d_y, d_z\}$. As presented in Figure 1b, for the cross-section at the jth roller with the location angle φ_j, the displacement of the outer ring along the r_j-axis and z-axis can be represented as:

$$\begin{cases} d_{rj} = d_x \cos\varphi_j + d_y \sin\varphi_j \\ d_{zj} = d_z \end{cases} \tag{1}$$

The displacement of the outer ring at the jth roller $\{d_j\}^T = \{d_{rj}, d_{zj}\}$ determines the contact deformations of the jth roller-raceways and jth roller-flange $\{\delta_j\}^T = \{\delta_{ij}, \delta_{ej}, \delta_{fj}\}$, as well as the contact forces acting on the jth roller $\{Q_j\}^T = \{Q_{ej}, Q_{ij}, Q_{fj}\}$. Here, the subscripts i, e, and f denote the inner raceway, outer raceway, and flange, respectively. The above contact deformation process and contact forces can be represented as [22]:

$$\begin{cases} \delta_{ij} = X_{rj} \cos\alpha_i + X_{zj} \sin\alpha_i \\ \delta_{ej} = (d_{rj} - X_{rj}) \cos\alpha_e + (d_{zj} - X_{zj}) \sin\alpha_e \\ \delta_{fj} = (d_{rj} - X_{rj}) \cos\alpha_f + (d_{zj} - X_{zj}) \sin\alpha_f \end{cases} \tag{2}$$

$$\begin{cases} Q_a = K_a \delta_a^{10/9}, \ a = i, e \\ Q_f = K_f \delta_f^{3/2} \end{cases} \quad (3)$$

where X_{rj}, X_{zj} indicate the displacement of the jth roller along the r_j-axis and z-axis, respectively; α_i, α_e, α_f indicate the contact angles of the roller-inner raceway, roller-outer raceway, and roller-flange, respectively; δ_a, δ_f indicate the contact deformations of the roller-raceways and roller-flange, respectively; K_a, K_f indicate the load-deformation factors of the roller-raceways and roller-flange, respectively, which depend on the material and geometry at the contact.

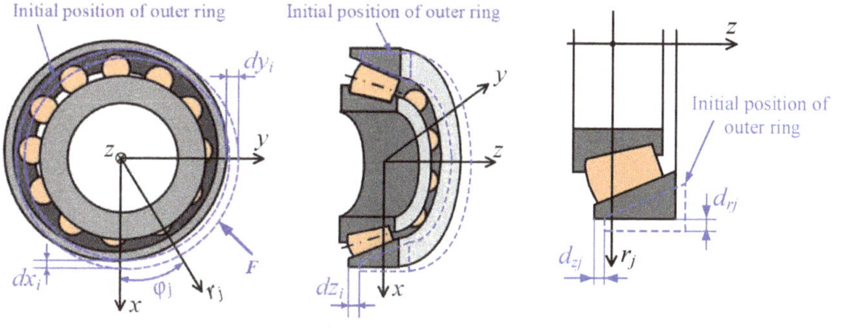

Figure 1. The displacement of the outer ring under the external load: (**a**) a global view of TRBs; (**b**) a cross-section at the jth roller.

Since the surface topography deviation of rollers is ignored in this study, the geometric homogeneity of rollers can be simplified as the diameter dimension deviation of rollers. When the diameter dimension deviation occurs on rollers, the contact force between the roller and its mating components will be changed. As a result, Equation (3) should be modified based on Taylor's series as

$$\begin{cases} Q_a = K_a \delta_a^{10/9} + \frac{10K_a}{9} \delta_a^{1/9} \frac{\Delta D \cos \varepsilon}{2} \ a = i, e \\ Q_f = K_f \delta_f^{3/2} \end{cases} \quad (4)$$

where ΔD is the diameter deviation value of the roller; ε is the half roller angle.

As shown in Figure 2, from the force equilibrium with respect to the ξ and ζ axes, the roller equilibrium equations are established considering the centrifugal force as Equation (4). Equations (2), (4) and (5) describe a local equilibrium system that can be solved by numerical methods to obtain the contact forces acting on the jth roller $\{Q_j\}^T = \{Q_{ej}, Q_{ij}, Q_{fj}\}$ under a certain displacement of the outer ring at the jth roller $\{d_j\}^T = \{d_{rj}, d_{zj}\}$.

$$\begin{cases} (Q_i - Q_e) \cos \varepsilon + Q_f \sin \mu + F_c \cos \kappa = 0 \\ (Q_i + Q_e) \sin \varepsilon + Q_f \cos \mu - F_c \sin \kappa = 0 \end{cases} \quad (5)$$

where F_c is the roller centrifugal force; κ is the angle between the roller center line and TRB center line; and β is the angle between the roller center line and the roller-flange contact line.

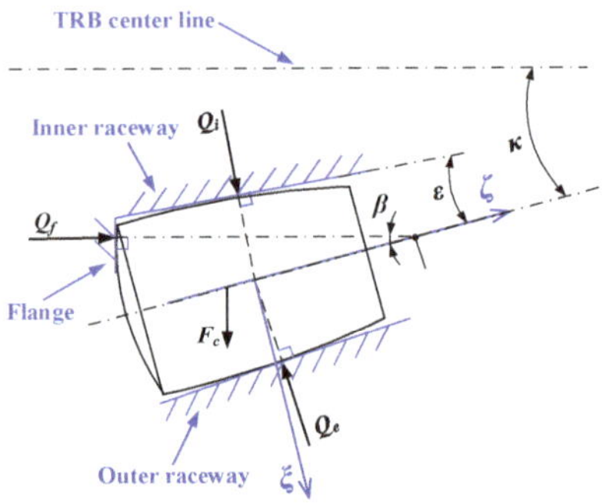

Figure 2. Contact forces acting on the roller.

The global equilibrium system of external loads acting on the outer ring and the contact forces of the roller-outer raceway can be represented as

$$\begin{cases} \sum\limits_{j=1}^{Z} Q_{ej} \cos\alpha_e \cos\varphi_j + F_x = 0 \\ \sum\limits_{j=1}^{Z} Q_{ej} \cos\alpha_e \sin\varphi_j + F_y = 0 \\ \sum\limits_{j=1}^{Z} Q_{ej} \sin\alpha_e + F_z = 0 \end{cases} \tag{6}$$

By giving a certain external load, $\{F\}^T = \{F_x, F_y, F_z\}$, the unknown contact forces of each roller can be obtained from the coupled solution of the global equilibrium system and the local equilibrium system, which is usually called the quasi-statics analysis of TRB. The detailed calculation procedure for the above coupled solution is shown in Figure 3. Since the global equilibrium equations and local equilibrium equations are nonlinear, the iterative Newton–Raphson method is adopted in this study.

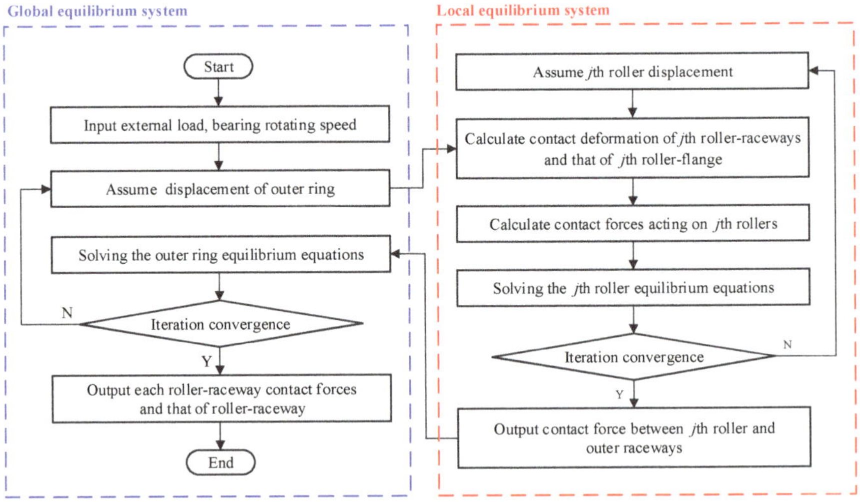

Figure 3. The coupled solution procedure of the global equilibrium system and the local equilibrium system.

2.2. Friction Torque Analysis of TRBs Considering the Geometric Homogeneity of Rollers

The friction in TRBs is mainly composed of the rolling friction between the roller and raceways F_{ro}, the sliding friction between the roller and raceways F_{rs} and the sliding friction between the roller and flange F_{fs}. According to theoretical analyses and experimental studies, the above components of friction in TRBs can be represented as in [23]:

$$F_{ro} = \frac{0.88 \times 10^2}{\alpha_0}(GU)^{0.658}W^{0.31}Rl_w \tag{7}$$

$$F_{rs} = 0.168 v_s \eta_0 l_w I U^{-0.74} G^{-0.4} W^{0.2} R^{-1} \tag{8}$$

$$F_{fs} = Q_f \mu_0 e^{-1.8\Lambda_r^{1.2}} \tag{9}$$

where α_0 is the viscosity-pressure coefficient; G, U, and W are the dimensionless material parameter, the dimensionless velocity parameter, and the dimensionless load parameter, respectively [24]; R is the equivalent radius between the roller and raceway; l_w is the effective contact length between the roller and raceway; v_s is the slide speed between the roller and raceway; I is the integrals used to describe the tractive effect; η_0 is the viscosity at atmospheric pressure; μ_0 is the Coulomb friction coefficient; and Λ_r is the oil film parameter.

According to an ideal Hertz line contact pressure distribution, the integrals of tractive effect I can be approximated as in [5]:

$$I = \int_0^b \exp\Big\{(\ln \eta_0 + 9.67)[(1 + 5.1 \times 10^{-9} p)0.601 - 1]\Big\} dx \tag{10}$$

where b is the semi-width of the Hertz line contact; and p is the contact pressure.

Based on the above analyses, the total friction torque acting on the outer ring M_{er} and that acting on the inner ring M_{ir} can be represented as

$$M_{er} = \sum_{j=1}^{z} (F_{ro,ej} - F_{rs,ej}) R_e \tag{11}$$

$$M_{ir} = \sum_{j=1}^{z} \Big[(F_{ro,ij} + F_{rs,ij}) R_{\mathrm{i}} + F_{fs,fj}(R_i + e_r) \Big] \tag{12}$$

where R_e, R_i are the radius of the roller-outer raceway contact point and the radius of the roller-inner raceway contact point on the roller mean diameter, respectively; and e_r is the height of the roller end and flange contact. Here the subscripts *ej*, *ij*, and *fj* denote the roller-outer raceway contact, roller-inner raceway contact, and roller-flange contact, respectively.

As shown in Equation (8), in addition to the contact force, the sliding friction between the roller and raceway F_{rs} is also affected by the sliding speed between the roller and raceway v_s. The kinematic relationships of the bearing components are assumed to follow the outer raceway control hypothesis [25,26] in this article; therefore, the sliding speed between the roller and raceways can be represented as:

$$\begin{cases} v_{se} = 0 \\ v_{si} = \omega_i R_i S_c \end{cases} \tag{13}$$

where S_c is the cage slip rate; and ω_i is the angular velocity of the inner ring.

2.3. Geometrical and Material Parameters

Based on the theoretical analyses described above, the friction torque of TRBs considering the geometric homogeneity of rollers is analyzed and discussed in this section. Take

TRB 30228J as an example to construct the corresponding analysis model, and its geometric characteristics are given in Table 1.

Table 1. The geometric and material properties of TRB 32008J.

Geometrical Characteristics	Value
Small diameter of taper roller (mm)	6.131
Large diameter of taper roller (mm)	6.846
Length of roller (mm)	13.66
Number of rollers	23
Outer raceway angle (rad)	0.2473
Inner raceway angle (rad)	0.1949
Roller angle (rad)	0.0262
Flange angle (rad)	1.5621
Elasticity modulus (N/mm^2)	2.1×10^5
Poisson's ratio	0.278

To simulate the geometric homogeneity of rollers caused by machining, the diameter deviation value of roller ΔD is assumed to be a standard Gaussian distribution and can be represented as

$$f(\frac{\Delta D}{S_d}) = \frac{1}{\sqrt{2\pi}} \exp\left(-(\frac{\Delta D}{S_d})^2/2\right) \tag{14}$$

where S_d is the deviation magnitude determined by the roller diameter and the machining methods.

Multiple series of ΔD were generated according to the above distribution characteristics, which consider different deviation magnitudes and distribution forms of the rollers with diameter deviations (random distribution and distribution in descending order of deviation value). As shown in Table 2, Series 1 corresponds to an S_d value of 0.5 μm and random distribution, Series 2 corresponds to an S_d value of 0.5 μm and distribution in descending order of deviation value, Series 3 corresponds to an S_d value of 1 μm and random distribution, Series 4 corresponds to an S_d value of 1 μm and distribution in descending order of deviation value, Series 5 corresponds to an S_d value of 1.5 μm and random distribution, and Series 6 corresponds to an S_d value of 1.5 μm and distribution in descending order of deviation value.

Table 2. The generating data of diameter deviation value ΔD.

Roller Number		1	2	3	4	5	6	7	8
ΔD (μm)	Series 1	0.28	0.52	−0.56	0.63	0.33	−0.03	−0.10	−0.11
	Series 2	0.76	0.63	0.52	0.47	0.41	0.33	0.31	0.28
	Series 3	0.55	1.04	−1.12	1.26	0.66	−0.07	−0.20	−0.22
	Series 4	1.53	1.26	1.04	0.95	0.83	0.66	0.63	0.55
	Series 5	0.83	1.56	−1.68	1.89	0.99	−0.10	−0.29	−0.33
	Series 6	2.29	1.89	1.56	1.42	1.24	0.99	0.94	0.83
Roller Number		**9**	**10**	**11**	**12**	**13**	**14**	**15**	**16**
ΔD (μm)	Series 1	−0.15	0.01	0.03	0.41	0.76	0.23	−0.10	0.31
	Series 2	0.26	0.23	0.15	0.13	0.09	0.07	0.03	0.01
	Series 3	−0.30	0.02	0.05	0.83	1.53	0.47	−0.21	0.63
	Series 4	0.52	0.47	0.31	0.26	0.18	0.14	0.05	0.02
	Series 5	−0.45	0.03	0.08	1.24	2.29	0.70	−0.31	0.94
	Series 6	0.77	0.70	0.46	0.39	0.27	0.20	0.08	0.03
Roller Number		**17**	**18**	**19**	**20**	**21**	**22**	**23**	
ΔD (μm)	Series 1	0.09	−0.51	0.47	0.15	0.07	0.26	0.13	
	Series 2	−0.03	−0.10	−0.10	−0.11	−0.15	−0.51	−0.56	
	Series 3	0.18	−1.03	0.95	0.31	0.14	0.52	0.26	
	Series 4	−0.07	−0.20	−0.21	−0.22	−0.30	−1.03	−1.12	
	Series 5	0.27	−1.54	1.42	0.46	0.20	0.77	0.39	
	Series 6	−0.10	−0.29	−0.31	−0.33	−0.45	−1.54	−1.68	

3. Results and Discussions

3.1. Effects of the Geometric Homogeneity of Rollers on Contact Force Distribution

According to the diameter deviation values of the rollers shown in Table 2, the internal contact force distribution of TRBs considering the geometric homogeneity of rollers is analyzed.

The contact force between each roller and the outer raceway is shown in Figure 4, in which the axial external load is 10 kN and the bearing rotating speed is 1000 rpm. In this figure, the contact force distribution of a TRB with an ideal roller diameter was used as a benchmark to reflect the effects of the geometric homogeneity of rollers. Changes in the diameter deviations of the rollers, whether positive or negative, will cause corresponding changes in the contact force between the roller and the outer raceway. Comparing Figure 4a,b, it can be seen that the distribution form of rollers with a diameter deviation also affects the internal contact force distribution of the TRB. It should be noted that the centrifugal force of rollers is much smaller than the roller-raceway contact force in the limited speed range of the TRBs; therefore, the influence of the bearings' rotating speeds on the contact force distribution is not further analyzed [20].

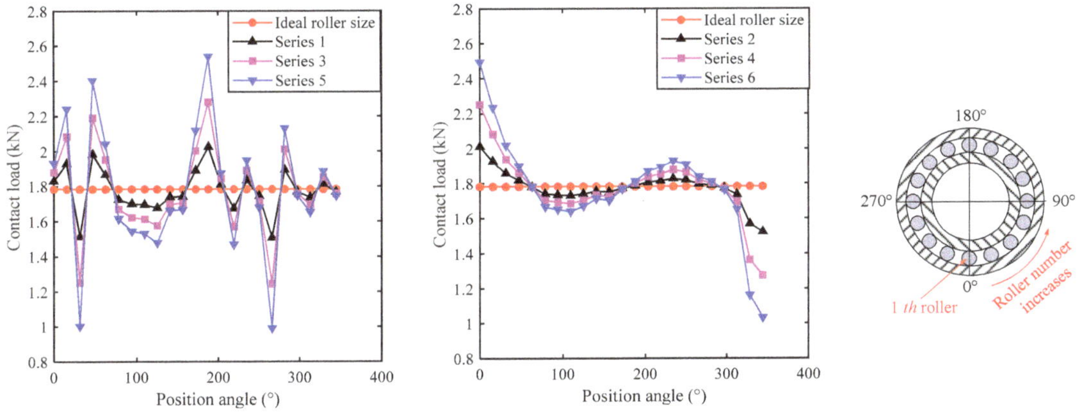

(a) (b) (c)

Figure 4. Effects of the geometric homogeneity of rollers on contact force distribution of TRBs: (**a**) Rollers with a diameter deviation are randomly distributed; (**b**) Rollers with a diameter deviation are distributed by deviation size; (**c**) Schematic diagram of the position angle of the jth roller.

According to the analysis results shown in Figure 4, the maximum value and the variance of contact forces are extracted and shown in Table 3. Comparing the maximum value and the variance of contact force corresponding to different series, it can be found that the maximum contact force is mainly determined by the diameter deviation value of the roller. However, both the diameter deviation value of the roller and the distribution form of rollers with a diameter deviation will have a significant effect on the uniformity of contact force distribution (reflected by variance values).

Table 3. Maximum contact force and variance of contact force.

	Series 1	Series 2	Series 3	Series 4	Series 5	Series 6
Maximum value (kN)	2.026	2.013	2.280	2.253	2.539	2.495
Variance (kN^2)	0.0162	0.0096	0.0645	0.0383	0.1441	0.0856

3.2. Effects of the Geometric Homogeneity of Rollers on Friction Force and Torque

According to the contact force distribution and Equations (11) and (12), we can obtain the friction torque acting on the outer ring and inner ring. The friction torque acting on the inner = and outer rings is shown in Figure 5, in which the axial external load is 10 kN and the rollers have an ideal diameter.

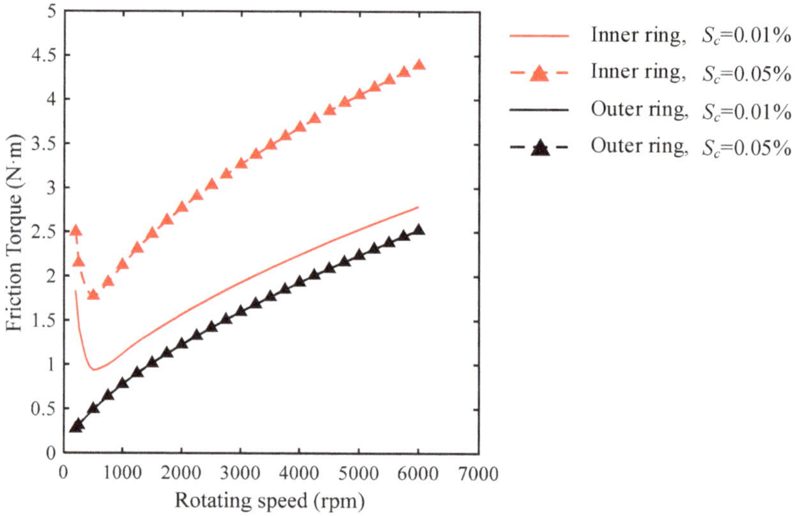

Figure 5. Friction torque acting on rings as a function of cage slip rate.

Based on the outer raceway control hypothesis, the friction torque acting on the outer ring only involves the rolling friction between the rollers and the raceway; however, the friction torque acting on the inner ring involves the rolling friction and sliding friction between the rollers and the raceway as well as the slide friction between the rollers and the flange. Therefore, the cage slip rate, which determines the slide speed between the rollers and the raceway, only has a significant effect on the friction torque acting on the inner ring as shown in Figure 5. In addition, as the TRBs' rotating speed increases, the lubrication state between the rollers and the flange changes from mixed lubrication to full-film lubrication, which leads to an initial decrease in the friction torque acting on the inner ring. The variation trends in friction torque obtained from the proposed model are consistent with those in the literature [4,5], thus validating the rationality of the proposed model.

The effects of the geometric homogeneity of rollers on friction torque acting on the rings are shown in Figure 6, in which the axial external load is 10 kN and the cage slip rate is 0.01%. As shown in Figure 6, the geometric homogeneity of rollers causes a decrease in the friction torque acting on the rings. Compared to the distribution form of the rollers with a diameter deviation, the diameter deviation value of the roller has a greater influence on the friction torque acting on the rings. When the TRBs' rotating speed is low, the roller-flange is in a mixed lubrication state, which causes the percentage change in the friction torque acting on the inner ring to be affected by the TRBs' rotating speed, as shown in Figure 6a. After the roller-flange lubrication enters the full oil film lubrication state, the percentage change in the friction torque tends to be constant. Since the friction torque acting on the outer ring only involves the rolling friction between the rollers and the raceway, the percentage change in the friction torque acting on the outer ring is not affected by the changes in the lubrication state between the roller and the flange. It should be noted that the above friction torque c reduction caused by the geometric homogeneity of rollers is very small at only a 10^{-2}% order of magnitude, which means that the geometric homogeneity of rollers will not have a significant influence on the overall energy loss or heat generation of TRBs.

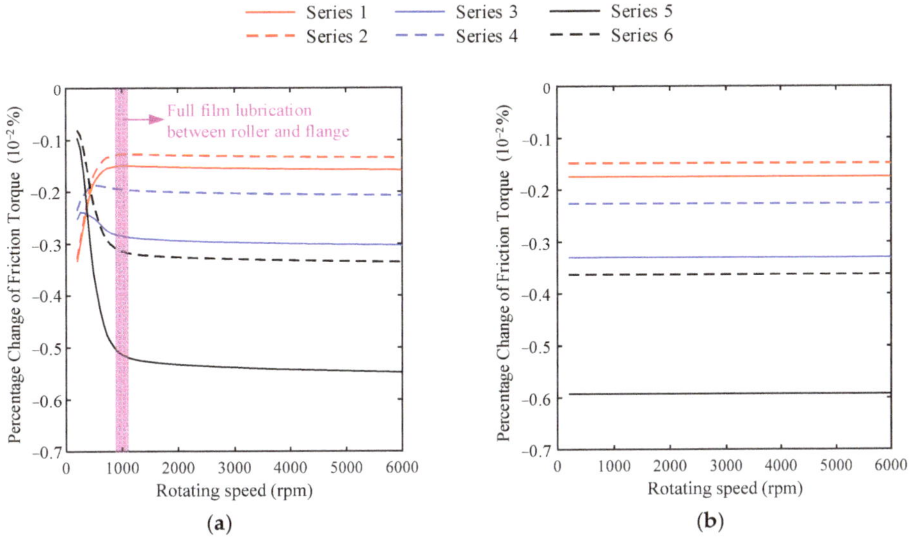

Figure 6. Effects of the geometric homogeneity of rollers on friction torque acting on rings: (**a**) Friction torque acting on inner ring; (**b**) Friction torque acting on outer ring.

Excessive sliding friction between individual rollers and raceways may induce excessive localized high temperatures inside the TRBs. The maximum increment in the slide friction between the roller and inner raceway caused by the geometric homogeneity of rollers (occurring at the roller with the greatest contact force) is shown in Figure 7. As shown in Figure 7a, the diameter deviation value of the roller results in a significant increase in the sliding friction between the roller and the inner raceway, whereas the distribution form of the rollers with a diameter deviation has little influence on the slide friction between the roller and the inner raceway.

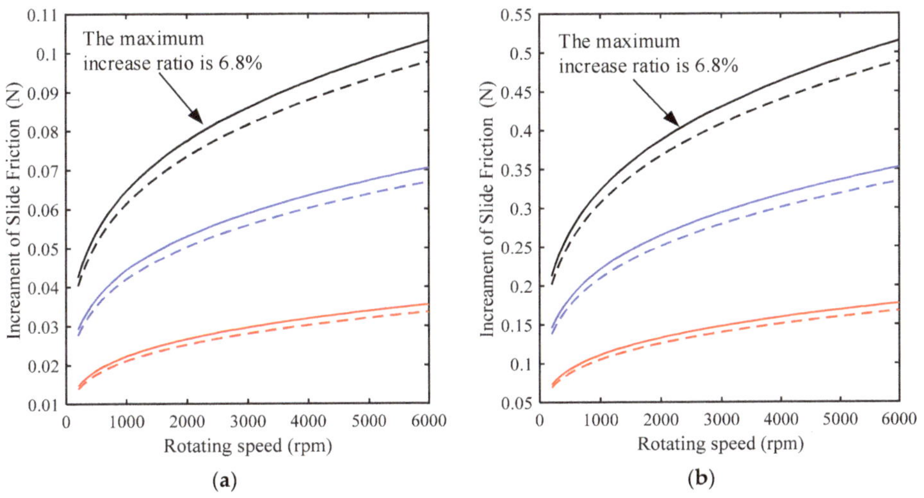

Figure 7. *Cont.*

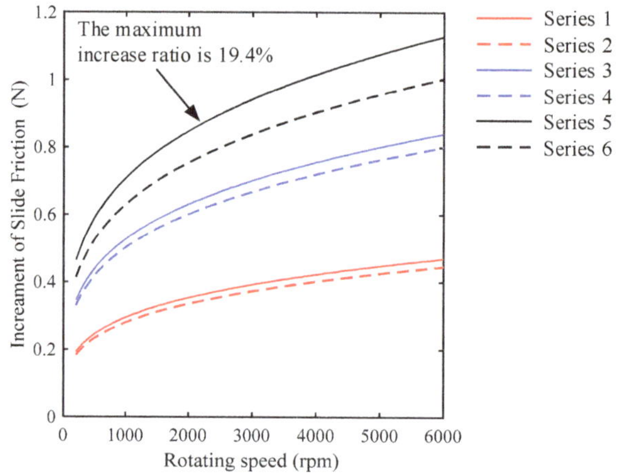

(c)

Figure 7. Effects of geometric homogeneity of rollers on roller-inner raceway maximum sliding friction force: (**a**) Axial external load is 10 kN and cage slip rate is 0.01%; (**b**) Axial external load is 10 kN and cage slip rate is 0.05%; (**c**) Axial external load is 2 kN and cage slip rate is 0.05%.

Comparing Figure 7a,b, it can be seen that the increments in the sliding friction caused by the geometric homogeneity of rollers increase as the cage slip rate increases, but the corresponding increase in the ratio remains constant. In addition, as shown in Figure 7b,c, the increments in the sliding friction caused by the geometric homogeneity of rollers and the corresponding increase in the ratio increase when the axial external load increases from 2 kN to 10 kN.

4. Conclusions

In this paper, a quasi-statics model and a friction torque model for TRBs considering the geometric homogeneity of rollers were proposed, and the effects of the geometric homogeneity of rollers on the contact force distribution and friction torque of TRBs are analyzed thoroughly. In the proposed model, the diameter deviation value of rollers and the distribution form of rollers with a diameter deviation were fully considered to improve the accuracy of the model. From the results, the following conclusions have been obtained.

Both the diameter deviation value of the roller and the distribution form of rollers with a diameter deviation have a significant effect on the uniformity of contact force distribution and the maximum contact force. The geometric homogeneity of the rollers, although having only a minimal influence on the overall friction torque acting on the rings, can lead to a significant increase in the slide friction force between the individual rollers and the inner raceway. Compared to the distribution form of rollers with a diameter deviation, the diameter deviation value of the roller has a greater influence on the maximum sliding friction between the roller and the inner raceway. As the cage slip rate increases, the slide friction force increment caused by the geometric homogeneity of the rollers will obviously increase. The effects of the geometric homogeneity of rollers on the sliding friction between the roller and the inner raceway are more pronounced under light axial external load conditions.

This paper provides a mathematical model for the friction characteristic analysis for TRBs considering the geometric homogeneity of rollers as well as a theoretical basis for formulating machining error criteria for TRB rollers. It should be noted that although this paper only analyzes the case of pure axial loading, the proposed model is also applicable to the case of axial and radial combined loading. Under axial and radial combined loading, there are load-bearing and non-load-bearing areas inside the TRBs. Therefore, the contact load of the individual roller is significantly affected by its position angle at some point in the

TRBs' operation. To make the analysis more reasonable, for TRBs under radial loading and combined loading, the friction torque analysis should consider the time-varying angular position of rollers.

Author Contributions: Conceptualization, Y.L. and X.F.; methodology, Y.L.; software, Y.L. and X.F.; validation, X.F. and J.W.; investigation, Y.L.; writing—original draft preparation, Y.L. and X.L.; writing—review and editing, X.F. and J.W.; visualization, Y.L. and J.W.; supervision, X.F.; project administration, X.F. and J.W.; funding acquisition, X.L. All authors have read and agreed to the published version of the manuscript.

Funding: This research was funded by the Province Key Research and Development Program of Shaanxi (2022KW-18).

Institutional Review Board Statement: Not applicable.

Informed Consent Statement: Not applicable.

Data Availability Statement: Detailed data are contained within the article.

Conflicts of Interest: The authors declare no conflict of interest.

References

1. Harris, T.A.; Kotzalas, M.N. Chapter 7: Rolling bearing temperatures. In *Advanced Concepts of Bearing Technology: Rolling Bearing Analysis*, 5th ed.; Taylor & Francis Group: Boca Raton, FL, USA, 2006.
2. Witte, D.C. Operating torque of tapered roller bearings. *ASLE Trans.* **1973**, *16*, 61–67. [CrossRef]
3. Karna, C.L. Performance characteristics at the rib roller end contact in tapered roller bearings. *ASLE Trans.* **1974**, *17*, 14–21. [CrossRef]
4. Aihara, S. A new running torque formula for tapered roller bearings under axial load. *J. Tribol.* **1987**, *109*, 471–478. [CrossRef]
5. Zhou, R.S.; Hoeprich, M.R. Torque of Tapered Roller Bearings. *J. Tribol.* **1991**, *113*, 590–597. [CrossRef]
6. SKF. Friction. In *General Catalogue 4000*; SKF: Gothenburg, Sweden, 2003; Available online: https://imparayaycia.com/SKF%20CATALOGO%20GENERAL.pdf (accessed on 11 July 2022).
7. Ma, S.; Zhang, X.; Yan, K.; Zhu, Y.; Hong, J. A Study on Bearing Dynamic Features under the Condition of Multiball–Cage Collision. *Lubricants* **2022**, *10*, 9. [CrossRef]
8. Wu, H.; Liu, J.; Shao, Y. Vibration characteristics of a roller bearing with the waviness error. In Proceedings of the 2019 58th Annual Conference of the Society of Instrument and Control Engineers of Japan, Higashi-Hiroshima, Japan, 10–13 September 2019. [CrossRef]
9. Savalia, R.; Ghosh, M.K.; Pandey, R.K. Vibration Analysis of Lubricated Angular Contact Ball Bearing of Rigid Rotor Considering Waviness of Ball and Races. *Tribol. Online* **2008**, *3*, 322–327. [CrossRef]
10. Nan, G.; Zhang, Y.; Zhu, Y.; Guo, W. Nonlinear dynamics of rotor system supported by bearing with waviness. *Sci. Prog.* **2020**, *103*, 003685042094409. [CrossRef]
11. Aschenbrenner, A.; Schleich, B.; Tremmel, S.; Sandro, W. A variational simulation framework for the analysis of load distribution and radial displacement of cylindrical roller bearings. *Mech. Mach. Theory* **2020**, *147*, 103769. [CrossRef]
12. Deng, S. Analysis on the Friction Torque Fluctuation of Angular Contact Ball Bearings. *J. Mech. Eng.* **2011**, *47*, 104. [CrossRef]
13. Heras, I.; Aguirrebeitia, J.; Abasolo, M. Friction torque in four contact point slewing bearings: Effect of manufacturing errors and ring stiffness. *Mech. Mach. Theory* **2017**, *112*, 145–154. [CrossRef]
14. Halminen, O.; Aceituno, J.F.; Escalona, J.L.; Sopanen, J.; Mikkola, A. A touchdown bearing with surface waviness: Friction loss analysis. *Mech. Mach. Theory* **2017**, *110*, 73–84. [CrossRef]
15. Xu, L.; Li, Y. Modeling of a deep-groove ball bearing with waviness defects in planar multibody system. *Multibody Syst. Dyn.* **2014**, *33*, 229–258. [CrossRef]
16. Liu, J.; Li, X.; Ding, S.; Pang, R. A time-varying friction moment calculation method of an angular contact ball bearing with the waviness error. *Mech. Mach. Theory* **2020**, *148*, 103799. [CrossRef]
17. Liu, J.; Yan, Z.; Shao, Y. An investigation for the friction torque of a needle roller bearing with the roundness error. *Mech. Mach. Theory* **2018**, *121*, 259–272. [CrossRef]
18. Yao, W.; Yuan, J.; Zhou, F.; Chen, Z.; Zhao, T.; Zhong, M. Trajectory analysis and experiments of both-sides cylindrical lapping in eccentric rotation. *Int. J. Adv. Manuf. Technol.* **2017**, *88*, 2849–2859. [CrossRef]
19. Yuan, J.; Yao, W.; Zhao, P.; Lyu, B.; Chen, Z.; Zhong, M. Kinematics and trajectory of both-sides cylindrical lapping process in planetary motion type. *Int. J. Mach. Tools Manuf.* **2015**, *92*, 60–71. [CrossRef]
20. Liu, Y.; Zhu, Y.; Yan, K.; Wang, F.; Hong, J. A novel method to model effects of natural defect on roller bearing. *Tribol. Int.* **2018**, *122*, 169–178. [CrossRef]
21. Kang, W.; Zhu, Y.; Yan, K.; Ren, Z.; Gao, D.; Hong, J. Research on extracting weak repetitive transients of fault rolling element bearing. *ISA Trans.* **2021**, *123*, 381–397. [CrossRef] [PubMed]

22. Tong, V.; Hong, S. The effect of angular misalignment on the running torques of tapered roller bearings. *Tribol. Int.* **2016**, *95*, 76–85. [CrossRef]
23. Harris, T.A. An analytical method to predict skidding in high speed roller bearings. *ASLE Trans.* **1966**, *9*, 229–241. [CrossRef]
24. Marian, M.; Bartz, M.; Wartzack, S.; Andreas, R. Non-dimensional groups, film thickness equations and correction factors for elastohydrodynamic lubrication: A review. *Lubricants* **2020**, *8*, 95. [CrossRef]
25. Harris, T.A.; Kotzalas, M.N. Chapter 7: Distributions of internal loading in statically loaded bearings. In *Essential Concepts of Bearing Technology: Rolling Bearing Analysis*, 5th ed.; Taylor & Francis Group: Boca Raton, FL, USA, 2006.
26. Zhang, Y.; Fang, B.; Kong, L.; Li, Y. Effect of the ring misalignment on the service characteristics of ball bearing and rotor system. *Mech. Mach. Theory* **2020**, *151*, 103889. [CrossRef]

lubricants MDPI

Article

Study on Distribution of Lubricating Oil Film in Contact Micro-Zone of Full Ceramic Ball Bearings and the Influence Mechanism on Service Performance

Jinmei Yao [1], Yuhou Wu [1], Jiaxing Yang [1], Jian Sun [1,2,*], Zhongxian Xia [1], Junxing Tian [1], Zhigang Bao [1] and Longfei Gao [1]

[1] School of Mechanical Engineering, Shenyang Jianzhu University, Shenyang 110168, China; yaojinmei06@126.com (J.Y.); wuyh@sjzu.edu.cn (Y.W.); a18946295255@163.com (J.Y.); xiazx17@sjzu.edu.cn (Z.X.); tianjunxingge@163.com (J.T.); bzg0123@126.com (Z.B.); gaolongfei005@126.com (L.G.)

[2] School of Mechanical Engineering, Tsinghua University, Beijing 100084, China

* Correspondence: sunjian@sjzu.edu.cn

Abstract: Compared with metal ball bearings, full ceramic ball bearings have more outstanding service performance under extreme working conditions. In order to reveal the lubrication mechanism and improve the operation performance and service life of full ceramic ball bearings, in this paper, the friction, vibration, and temperature rise characteristics of 6208 silicon nitride full ceramic deep groove ball bearing, under the condition of oil lubrication, are studied experimentally. Based on the test results, and through theoretical calculation and simulation analysis, the distribution of the lubricating oil film in bearing contact micro-zone under different working conditions was simulated. After that, the surface of contact micro-zone of full ceramic ball bearing was analyzed. It was found that there is an optimal oil supply for full ceramic ball bearing oil lubrication in service. Under the optimal oil supply lubrication, full film lubrication can be achieved, and the bearing exhibits the best characteristics of friction, vibration, and temperature rise. Compared with the load, the rotational speed of the bearing has a decisive influence on the optimal oil supply. When the rotational speed and load are constant, the minimum oil film thickness and oil film pressure in the contact area of the rolling body decrease with the increase of angle ψ from the minimum stress point of the rolling body. Under the action of high contact stress, thin oil film will be formed in the bearing outer ring raceway. In the field of full ceramic ball bearings, the research content of this paper is innovative. The research results of this paper have an important guiding significance for revealing the oil lubrication mechanism of full ceramic ball bearing and enriching its lubrication theory and methods.

Keywords: full ceramic ball bearing; lubricant oil film; service performance; simulation model

Citation: Yao, J.; Wu, Y.; Yang, J.; Sun, J.; Xia, Z.; Tian, J.; Bao, Z.; Gao, L. Study on Distribution of Lubricating Oil Film in Contact Micro-Zone of Full Ceramic Ball Bearings and the Influence Mechanism on Service Performance. *Lubricants* **2022**, *10*, 174. https://doi.org/10.3390/lubricants10080174

Received: 29 May 2022
Accepted: 26 July 2022
Published: 1 August 2022

1. Introduction

Full ceramic ball bearings refer to high-tech bearing products whose rings and rolling bodies are made of ceramic materials. It has excellent performance at high speed, compression resistance, high/low temperature resistance, wear resistance, corrosion resistance, and electromagnetic insulation, as well as other aspects, and can be widely used in aerospace, navigation, metallurgy, chemical and national defense, and military fields [1–3]. A good lubrication state is the prerequisite for the normal operation of full ceramic ball bearings. It is also an important factor affecting the performance of the bearings, such as friction, vibration, and temperature rise under complex working conditions [4]. At present, based on the lubrication theory of metal ball bearing, experts and scholars at home and abroad have carried out relevant research on the operation performance of ceramic ball bearings under oil lubrication conditions.

Meyer L. D [5] studied the analytical method for the continuous and periodic changes of the ball contact force when the bearing rotates, based on the Lagrange equation for solving time-varying displacement of the bearing. N. Lynagh [6] established a detailed bearing vibration model by considering the influence of rolling body surface corrugation and ball size on radial clearance and deduced the vibration force and torque generated by bearing movement by using the formula. It provides a great reference for the subsequent research. Mohammed Alfares [7] studied the influence of heat generation on the performance of ball bearings during service, as well as the influence of this heat effect on system dynamics, by using transient thermal model. A set of differential equations was obtained by using thermal equilibrium. Jalali-vahid D. [8] studied and obtained the multi-stage multi-grid solution for isothermal elastohydrodynamic circular contact problem. By deploying a large number of units, the accuracy of the numerical solution is improved. The conclusion presented in this paper is of great prospective significance and has guiding value for the analysis of isothermal elastic hydrodynamics. As a pioneer in this field, Dowson D. [9] conducted a numerical evaluation on the analysis of point contact of isothermal elastohydrodynamic lubrication. In the elastic analysis, the contact area is divided into equal rectangular areas, and it is assumed that a uniform pressure is applied to each element to analyze the oil film thickness. This paper also carries out a more detailed study on this basis. Ioannides, E. [10] put forward an idea a long time ago to study an analytical model for predicting the life of rolling bearings under the consideration of the fatigue life, fatigue criterion, and fatigue limit of bearings. The viewpoints proposed at that time were forward-looking and provided great significance for the research of bearing field. Zhu Weibing [11] established a calculation model of oil injection lubrication and under-race lubrication for angular contact ball bearings. The results showed that: when the speed and load were low, it was more appropriate to use oil injection lubrication. When the speed and load are high, it is more advantageous to use under-race lubrication. Nagare [12] experimentally analyzed and investigated the effects of performance parameters, such as speed, eccentricity, load, and friction torque, on the performance of overloaded low speed bearings. Liming Lu [13] designed a set of independent devices to explore the impact of impact load on the lubrication performance of roller sliding bearings. The results showed that: the greater the impact load frequency, the greater the minimum oil film thickness, greater the impact load amplitude, and thinner the roller oil film thickness. The sliding of the rolling element may lead to the failure of the rolling bearing. Kang Jianxiong [14] considered the interaction between the ball and raceway, interaction between the cage and raceway, and elastohydrodynamic lubrication and other factors, in order to effectively analyze the sliding characteristics of the rolling bearing using the dynamic response. Antonio [15] studied the influence of lubricant film thickness on bearing service performance under hydrodynamic lubrication conditions. Cho [16] improved the Elrod algorithm based on the mass conservation boundary conditions and obtained the governing equation and the lubrication equation of the complete oil film region that can automatically determine the dynamic boundary. Biswas [17] studied the lubrication performance of medium-low speed bearings under different load conditions, as well as the changes regarding oil film thickness, oil film pressure, and oil film velocity. Dmitrichenko [18] established the dynamic model of ball bearings and studied the influence of different distribution models of lubricants and fluid dynamic pressure on the dynamic characteristics of bearings. Brizmer [19] studied the mechanism of micro-pitting resistance of hybrid ceramic bearings under reduced lubrication conditions, thus providing a new idea for tribology and performance of silicon nitride ceramic rolling bearings. Kang Li [20] analyzed the tribological properties of GCr15-GCr15/Si_3N_4-GCr15 materials under the condition of oil lubrication by using pin-disc friction and wear testing machine. Xiang Guo [21] studied and investigated the dynamic mixed elastohydrodynamic lubrication behavior of water-lubricated bearings with unbalanced rotors in the starting process.

From the above, it can be seen that most studies were based on metal ball bearings, only part of the research is suitable for the hybrid ceramic ball bearing. Additionally, a small number of reports on the lubrication mechanism and method of full ceramic ball bearings are only described for the objective phenomena in the lubrication process. The objective laws and scientific problems behind it have not been further revealed. Therefore, this paper takes silicon nitride full ceramic ball bearing as the research object and studies its service performance, regarding friction, vibration, and temperature rise under the condition of oil lubrication. The effects of lubrication, rotational speed, and load on the performance of full ceramic ball bearings were revealed by means of experiments, and the optimal oil supply was determined. Based on the optimal oil supply, the distribution characteristics of lubricating oil film and its influence on service performance of full ceramic ball bearings were revealed by means of calculation and simulation. Finally, the surface characteristics of contact microzone after service of full ceramic ball bearing under oil lubrication condition were analyzed. The research results have important guiding significance for forming a lubrication theory and method suitable for full ceramic ball bearing, as well as improving its service performance and life.

2. Experimental Study on Oil Lubrication Characteristics of Full Ceramic Ball Bearing

2.1. Test Bearings and Components

The test took 6208 silicon nitride full ceramic deep groove ball bearing with P4 accuracy as the research object, carried out the service performance test under oil lubrication condition, and revealed the distribution law of the full ceramic ball bearing lubricating oil film, as well as its influence mechanism on friction performance. The structure of the test bearing is shown in Figure 1. The clearance of the test bearing was C_N standard clearance, and the cage guide mode was an outer ring guide. The lubricating oil used in the test was principal axis oil, with a viscosity value of 30 mm^2/s. The material used for the inner and outer rings and rolling bodies was silicon nitride ceramic, and the powder was produced by Ube Group in Japan and formed by hot isostatic pressing sintering process. The material performance test results are shown in Table 1. The structural parameters of the test bearing are shown in Table 2.

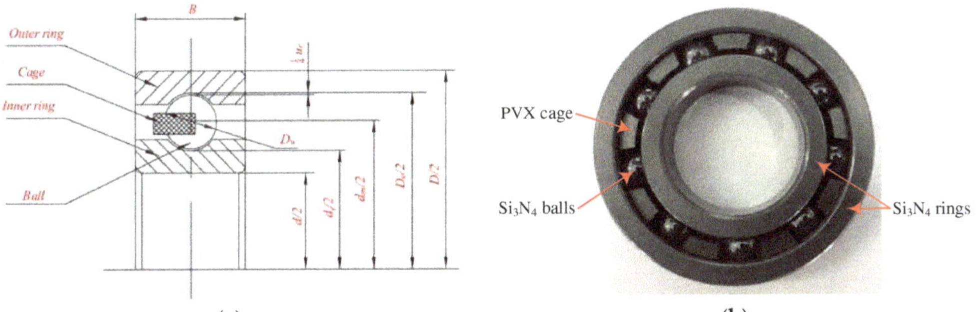

(a) (b)

Figure 1. Full ceramic ball bearing for test. (**a**) Structure diagram (**b**) Test bearing diagram.

Table 1. Physical properties of Si_3N_4 for full ceramic ball bearings.

Elastic Modulus (Gpa)	Poisson Ratio	Density (g/cm^3)	Thermal Expansion Coefficient (10^{-6}/K)	Thermal Conductivity (W/m∗K)	Hardness (kg/mm^2)
300–320	0.26	3.2–3.3	3.1–3.5	20–29	1520–1800

Table 2. Structure parameters of 6208 full ceramic ball bearings.

Structure Parameters	Values
Inner diameter d/mm	40
Outer diameter D/mm	80
Pitch diameter d_m/mm	60
Bearing width B/mm	18
Diameter of ball D_w/mm	12
Raceway radius of inner ring r_i/mm	6.17
Raceway radius of outer ring r_e/mm	6.29
Number of rolling elements z	9

The test bearing cage material was carbon dairy produce, PTFE, graphite reinforced PVX-based composite material, which was produced by Ensinger Group in Berlin, Germany. The specific properties of the material are shown in Table 3. The experimental verification and application experience show that the material can be applied to the wide temperature range of −70–200 °C. In addition, the material has the appropriate strength and toughness to withstand a certain load and impact, small friction coefficient, and good wear resistance. It also has smaller specific gravity and similar expansion coefficient to the rolling body, which is suitable for the test project.

Table 3. PVX composite enhanced cage material properties.

Material Properties	Values
Modulus of elasticity/MPa	5500–6000
Tensile strength/MPa	84
Flexural strength/MPa	142
Compression strength/MPa	22–102
Compression modulus/MPa	4000
Impact strength/kJm^{-2}	28
Ball indentation hardness/MPa	250
Glass transition temperature/°C	146
Service temperature/°C	+260−−200
Thermal expansion/10^{-5} K^{-1}	3–4
Specific heat/J/(g·K)	1.1
Thermal conductivity/W/(K·m)	0.82

2.2. Introduction to the Test Equipment

Figure 2 shows the JH-200E rolling bearing testing machine (Shenyang Jianzhu University, Shenyang, China) and its structure. The testing machine is a horizontal bearing performance life testing machine, which is mainly composed of bearing test chamber, axial loading system, radial loading system, test shafting, oil supply system, force sensor, temperature sensor, vibration sensor, and control system. The main test parameters were: the outer diameter of the test bearing was 30–200 mm, the inner diameter was 10–170 mm, and the width was 10–30 mm. Oil supply system flow range: 0.001–10 mL/min, and flow control accuracy was ±0.001 mL/min. The radial loading range of test bearings was 100–30,000 N, and the axial loading range was 50–10,000 N. The maximum spindle rotational speed was 30,000 r/min.

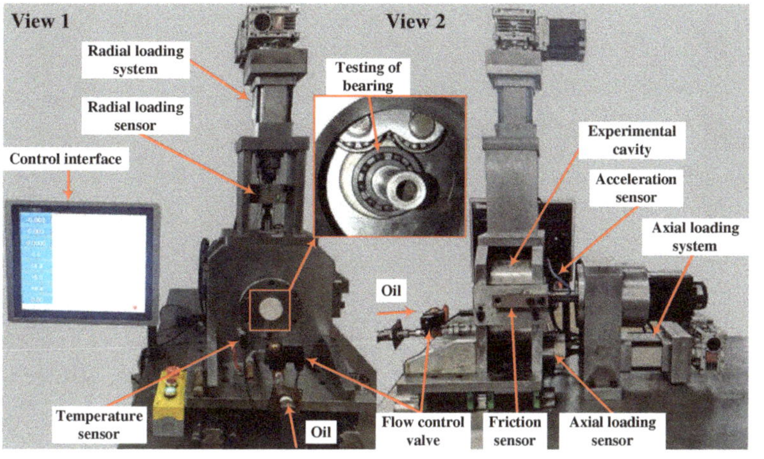

Figure 2. Lubrication test of full ceramic ball bearing oil based on JH-200E (Shenyang Jianzhu University, Shenyang, China).

2.3. Test Scheme Design

The oil lubrication test of the full ceramic ball bearing was carried out at room temperature. The target value is measured by changing the bearing oil supply, bearing radial load, bearing rotational speed, and other working conditions. The specific experimental process was as follows. (1) Before the bearing was started, oil was supplied to the test bearing according to the target set value. (2) Under no-load condition, the test bearing speed was increased from 0 r/min to the set value of the target speed within 3 min. (3) After the oil supply and speed of the test bearing were stable, the radial loading of the test bearing was carried out according to the set value of the target load. (4) After the test lasted for 30 min and the temperature of the test chamber and outer ring of the bearing, vibration value of the bearing, friction value, and other target measurement values were stable, the measurement and recording data were started. The test scheme is shown in Table 4.

Table 4. The 6208 full ceramic ball bearing oil lubrication test scheme.

Serial Number	Radial Load	Contact Stress		Other Index Parameters
		Inner Raceway	Outer Raceway	
1	400 N	1.65 Gpa	1.54 Gpa	Axial load: 0 N Test rotation speed: 5000/10,000 rpm Oil supply: 0.2–2.0 mL/min
2	900 N	2.16 Gpa	2.02 Gpa	
3	1750 N	2.70 Gpa	2.52 Gpa	
4	3000 N	3.23 Gpa	3.01 Gpa	

3. Test Results and Analysis

3.1. Variation Characteristics of Friction

The JH-200E rolling bearing testing machine was used to carry out the full ceramic ball bearing oil lubrication test. Based on the friction sensor, the friction change of the full ceramic ball bearing under variable working conditions was measured, as shown in Figure 3.

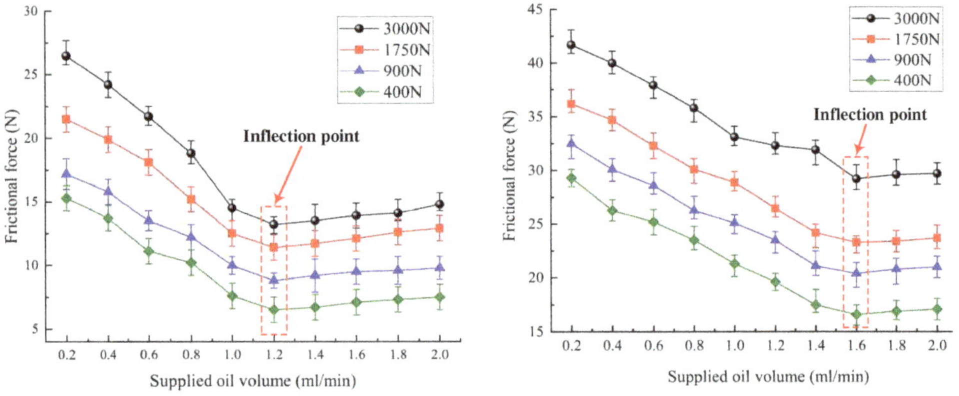

(**a**) (**b**)

Figure 3. Variation law of friction of full ceramic ball bearing under oil lubrication condition. (**a**) n = 5000 r/min, (**b**) n = 10,000 r/min.

It can be seen from Figure 3 that, when the rotational speed is constant with the increase of oil supply, the friction value of the full ceramic ball bearing decreases first and then increases. When the bearing speed was 5000 r/min, the friction force had a minimum inflection point value in the range of 0.2–2.0 mL/min oil supply, and the value was about 1.2 mL/min oil supply. This indicates that in this particular working condition, when the oil supply was 1.2 mL/min, the full ceramic ball bearing showed the best friction characteristics. When the oil supply was less than 1.2 mL/min, the full ceramic ball bearing was in the state of lack of oil lubrication, and the lubricating oil in the bearing contact micro-zone had not formed the state of full film lubrication, so the friction force was large. In this range, with the continuous increase of oil supply, the lubrication state improved, lubricating oil film gradually thickened, friction decreased, and change trend was very obvious. When the oil supply was greater than 1.2 mL/min, due to the large amount of lubricating oil, the excess lubricating oil produced viscous resistance to the operation of the bearing under the action of viscosity, thus leading to the gradual increase of the friction of the bearing. However, the influence of oil viscosity resistance on the change of friction was relatively small, so when the oil supply was greater than the optimal oil supply, and with the continuous increase of oil supply, the increase trend of friction was relatively moderate. By comparing Figure 3a,b, it can be seen that the oil supply at the inflection point of friction of the full ceramic ball bearings increased gradually with the increase of rotational speed. The change of load had no obvious effect on the oil supply at the inflection point of friction.

3.2. Variation Characteristics of Vibration Acceleration

The JH-200E rolling bearing testing machine was used to carry out the full ceramic ball bearing oil lubrication test, and the vibration variation of the full ceramic ball bearing outer ring under variable working conditions was measured based on the acceleration sensor, as shown in Figure 4.

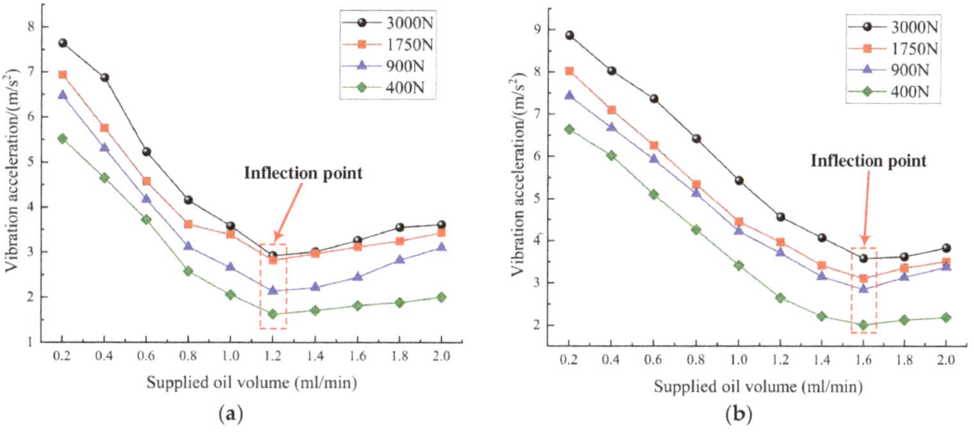

Figure 4. Vibration law of outer ring of full ceramic ball bearing under oil lubrication condition. (**a**) n = 5000 r/min, (**b**) n = 10,000 r/min.

According to Figure 4, when the rotational speed was constant, the vibration acceleration value of the outer ring of the full ceramic ball bearing decreased from large to small and then increased with the increase of oil supply. When the rotational speed was 5000 r/min, the vibration acceleration of outer ring had a minimum inflection point value, and the corresponding oil supply was also about 1.2 mL/min. This indicates that, when the rotational speed was 5000 r/min, the full ceramic ball bearing exhibited the best friction and vibration characteristics under the condition of 1.2 mL/min oil supply, which can be determined as its optimal oil supply. When the oil supply was less than the optimal oil supply, due to the influence of lack of oil lubrication, there was oil–solid mixed lubrication in the bearing contact micro-zone; the sliding roll ratio changed more frequently, and the vibration acceleration value of the bearing outer ring increased accordingly [22–24]. With the continuous increase of oil supply, the lubrication state improved; the vibration value of the bearing's outer ring decreased, and the change trend was very obvious. When the oil supply was greater than the optimal oil supply, the amount of lubricating oil was large, and the excess lubricating oil produced viscous resistance to the operation of the bearing under the action of viscosity, thus leading to a gradual increase in the vibration value of the bearing's outer ring, this phenomenon could be seen in Figure 4. However, the influence of oil viscosity resistance on the vibration change of the bearing's outer ring was relatively small, so when the oil supply was greater than the optimal oil supply, and with the continuous increase of oil supply, the increase trend of vibration acceleration value was relatively moderate [25,26]. In addition, when the radial load of the bearing increased, the vibration acceleration value of the bearing increased accordingly.

3.3. *Variation Characteristics of Outer Ring Temperature Rise*

The JH-200E rolling bearing testing machine was used to carry out the oil lubrication test of the full ceramic ball bearing, and the temperature rise change of the outer ring of the full ceramic ball bearing is shown in Figure 5, based on the temperature sensor measured under variable working conditions.

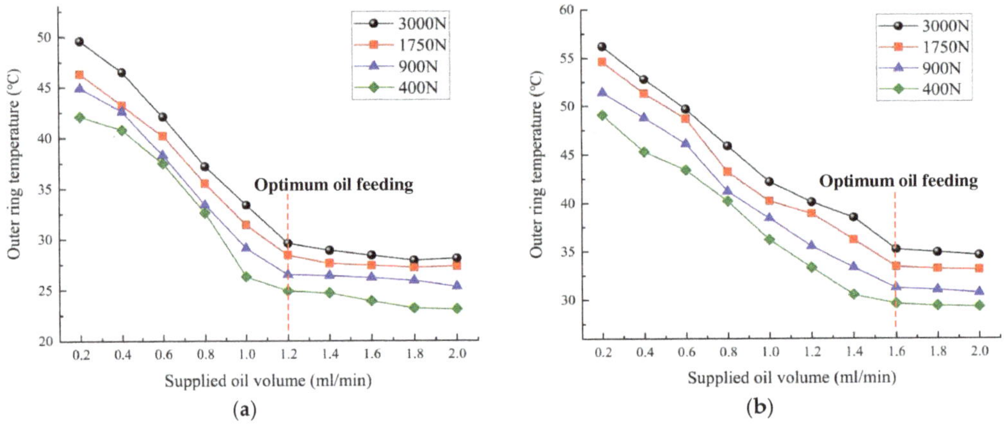

Figure 5. Temperature law of outer ring of full ceramic ball bearing under oil lubrication condition. (**a**) $n = 5000$ r/min, (**b**) $n = 10{,}000$ r/min.

According to the test data in Figure 5, with the increase of oil supply, the temperature rise of the outer ring gradually decreased. When the oil supply was lower than the optimal range, the temperature rise decreased obviously with the increase of oil supply. This is because, in the state of lack of oil lubrication, the lubrication effect of bearings is not good. The running state was mixed friction, and dry friction in some areas led to a significant rise in temperature. As the oil supply continued to increase, the lubrication state improved, and the temperature rise of bearing outer ring decreased. When the optimum oil supply was in place, the full ceramic ball bearing was in a state of full-film lubrication, which had a great improvement effect on the temperature rise of the bearing. When the oil supply continued to increase and was greater than the optimal oil supply, a large amount of lubricating oil removed the temperature rise generated by the operation of the bearing and caused a cooling effect, so the temperature rise of the bearing outer ring continued to decrease. The temperature rise of outer ring increased with the increase of bearing speed; it also increased with the increase of radial load. This is because, with the increase of the radial load and speed, the contact stress and contact frequency in the contact micro-zone of the full ceramic ball bearing became larger, and the heat generated by friction increased. The heat gradually accumulated in the outer ring of the bearing, and the temperature of the outer ring rose correspondingly.

4. Theoretical Calculation and Analysis

4.1. Oil Lubrication Dynamics Model of the Full Ceramic Ball Bearing

4.1.1. Establishment of Coordinate System of the Full Ceramic Ball Bearing

In the operation process of full ceramic ball bearings, the ceramic ball has the most complex force and contact with the inner and outer rings and cage, thus resulting in friction and impact. In addition, under the action of oil–gas lubrication, the ceramic ball was also affected by hydraulic force, which caused the ceramic ball to have a more complicated motion state [27,28]. In order to accurately describe the motion characteristics and interaction forces of the internal parts of the full ceramic ball bearing, a coordinate system was established, as shown in Figure 6.

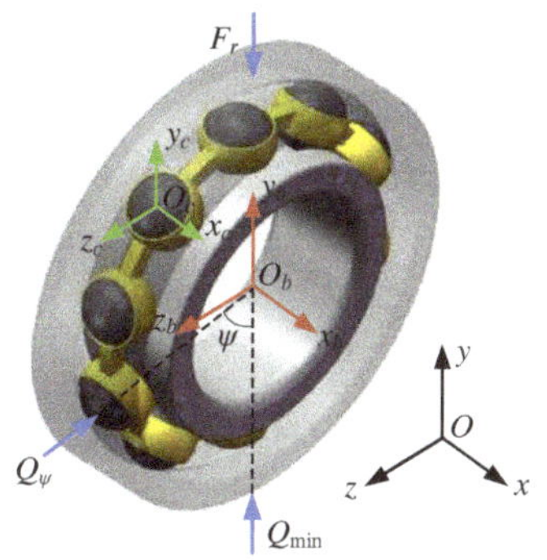

Figure 6. Coordinate system of deep groove ball bearing.

(I) The inertial coordinate system $Oxyz$ was established, with the bearing center O as the origin.

(II) The coordinate system $O_cx_cy_cz_c$ was established, with the spherical centroid O_c as the origin, y_C axis along the radial direction of the bearing, and ZC axis along the circumferential direction of the bearing.

(III) The inner circle centroid Ob was used as the origin to establish the coordinate system $O_bx_by_bz_b$.

The force between the balls, rings, and cage is shown in Figure 7, in which the subscripts *I* and *e* represent the inner and outer rings, respectively, *j* represents the *Jth* ball, Q represents the normal contact force between the ball and the ring raceway, T_η and T_ξ are the drag forces on the contact surface between the ball and the ring raceway, Q_c represents the force of the cage pocket on the ball, and the angle between the cage pocket hole and the three directions of $O_px_py_pz_p$ in the coordinate system, βx, βy, βz. $P_{R\eta(\xi)}$, and $P_{S\eta(\xi)}$, are the rolling friction force and sliding friction force of the fluid at the entrance of the contact surface between the ball and cage pocket, respectively.

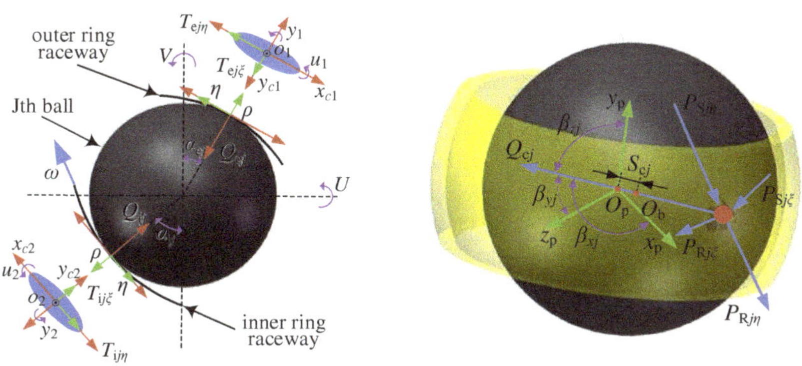

(a) (b)

Figure 7. Mechanical model of full ceramic deep groove ball bearing. (**a**) Ball-ring contact force model, (**b**) Ball-cage contact force model.

4.1.2. Elastohydrodynamic Model of Full Ceramic Ball Bearing under Oil Lubrication

Under the action of oil–gas lubrication, each rolling body had two velocity directions of contact ellipse, i.e., the long and short axes, and the density and viscosity of lubrication medium changed along the direction of lubrication film thickness. Its steady-state generalized Reynolds lubrication equation can be written as follows:

$$\begin{cases} \frac{\partial}{\partial x_c}\left[\left(\frac{\rho}{\eta}\right)h^3\frac{\partial p}{\partial x_c}\right]+\frac{\partial}{\partial y_c}\left[\left(\frac{\rho}{\eta}\right)h^3\frac{\partial p}{\partial y_c}\right]=12U\frac{\partial(\rho_x^* h)}{\partial x_c}+12V\frac{\partial(\rho_y^* h)}{\partial y_c} \\ U=\frac{1}{2}(u_1+u_2) \\ V=\frac{1}{2}(v_1+v_2) \end{cases} \tag{1}$$

where h is the oil film thickness, p is oil film pressure distribution, and U and V are the entrainment velocity in x_c and y_c directions, respectively (the coiling speed in y_c direction is much higher than that in x_c direction), which is related to the actual contact speed between the inner and outer ring and the rolling bodies. ρ and η are density and viscosity coefficients, respectively, and $\rho_x{}^*$ and $\rho_y{}^*$ are the equivalent densities related to the equivalent viscosity of contact extrusion.

The lubrication film thickness h considering elastic deformation can be expressed as:

$$h=h_0+\frac{x_c^2}{2R_x}+\frac{y_c^2}{2R_y}+\delta(x_c,y_c) \tag{2}$$

where h_0 is the central film thickness of the elastic contact area, and R_x and R_y are the equivalent radius of curvature in the x_c and y_c directions, respectively. $\delta(x_c, y_c)$ is the elastic deformation in the contact region, and its expression is:

$$\delta(x_c,y_c)=\frac{2}{\pi E}\iint_\Omega \frac{p(s,t)}{\sqrt{(x_c-s)^2+(y_c-t)^2}}\mathrm{d}s\mathrm{d}t \tag{3}$$

where E is the equivalent elastic model of two contact objects x.

4.1.3. Friction Torque

The rolling friction torque M_E generated when the ball rolls on the raceway was:

$$M_E=0.25D_{pw}\left[\left(1-\gamma_i^2\right)\sum_{j=1}^{z}\Phi_{ij}+\left(1-\gamma_e^2\right)\sum_{j=1}^{z}\Phi_{ej}\right]\beta_a \tag{4}$$

$$\gamma_{\mathrm{i}(e)}=D_{\mathrm{w}}\cos\alpha_{\mathrm{i}(e)}/D_{\mathrm{pw}} \tag{5}$$

where β_a is the elastic hysteresis coefficient. Φ can be calculated by referring to reference.

The friction torque M_{D} caused by differential sliding was:

$$M_{\mathrm{D}}=\frac{D_{\mathrm{pw}}}{2D_{\mathrm{w}}}\left[\left(1-\gamma_{\mathrm{i}}^2\right)\sum_{j=1}^{Z}M_{\mathrm{D}ij}+\left(1-\gamma_{\mathrm{e}}^2\right)\sum_{j=1}^{Z}M_{\mathrm{De}j}\right]f_{\mathrm{s}} \tag{6}$$

where f_{s} is the sliding friction factor between the ball and the raceway. The friction moment M_{s} caused by the spin sliding of the ball was:

$$M_{\mathrm{s}}=\frac{3}{8}f_{\mathrm{s}}\left[\sum_{j=1}^{Z}\left(E_{\mathrm{w}}E_{\mathrm{i}}a_{\mathrm{i}}Q_{\mathrm{ij}}\sin\alpha_{ij}\right)+\sum_{j=1}^{Z}\left(E_{\mathrm{w}}E_{\mathrm{e}}a_{\mathrm{e}}Q_{\mathrm{ej}}\sin\alpha_{ej}\right)\right] \tag{7}$$

where E_{w} is the elastic modulus of the ball material. E_{i} and E_{e} are the elastic moduli of inner and outer ring materials, respectively. a_{i} and a_e are the long half axes of the contact ellipse of the inner and outer rings and the balls, respectively.

The friction torque M_{c} caused by the friction between the ball and the cage was:

$$M_{\mathrm{c}}=0.25D_{\mathrm{pw}}\left(1-\gamma^2\right)\times\sin\left(\alpha_0+\arctan\frac{D_w\sin\alpha_0}{2\gamma_1}\right)m_d\mu_c \tag{8}$$

where M_c is the cage mass and μ_c is the sliding friction coefficient between the ball and cage.

The friction torque M_l caused by the viscous resistance of lubricating oil in the running process of the bearing was:

$$M_l = 6.53\alpha^{-1} S_1 D_{pw} \times \left\{ 2 \sum_{j=1}^{Z} \left[\frac{h_{ij} + h_{ej}}{2} (\alpha_{ij} + \alpha_{ej}) \right] S_2 \right\}^{-1} \quad (9)$$

where S_1 is the sufficient lubrication coefficient, and the lubrication coefficient of oil film was taken. h is the oil film thickness in the center of the bearing contact area. S_2 is the side leakage coefficient of lubrication, and the value was 1 in the calculation.

In the process of rotational service, the total friction torque M of the bearing was:

$$M = M_E + M_D + M_s + M_c + M_l \quad (10)$$

4.1.4. Influence of Temperature Rise on Structural Parameters of the Full Ceramic Ball Bearing under Oil Lubrication Condition

Due to the inconsistent deformation of bearing inner and outer rings and ball under the condition of temperature rise change, the clearance of deep groove ball bearings will change, and the specific calculation formula is as follows:

$$\Delta p = \Delta p_i - \Delta p_o + 2\Delta p_r \quad (11)$$

$$C_r = C_0 + \Delta p \quad (12)$$

where Δp is the change of bearing clearance. Δp_o is the deformation of outer ring. Δp_i is the deformation of inner ring. Δp_r is the deformation amount of ceramic sphere. C_0 is the initial clearance of bearing. C_r is bearing clearance.

The deformation of bearing ring and cage affected by temperature change can be expressed as:

$$\Delta u = \Gamma_s D_c \Delta T \quad (13)$$

where Δu refers to the inner and outer diameters of bearing ring and cage. Γ is the expansion coefficient of the corresponding material. D_c represents the bearing inner and outer rings, as well as the cage's inner and outer diameter sizes. ΔT is the temperature difference.

4.1.5. Boundary Conditions

In order to ensure the convergence and accuracy of the contact elastohydrodynamic model of the full ceramic deep groove ball bearing, the value range of boundary coordinates was $x_{\text{in}} = 2a$, $x_{\text{out}} = 2a$, $y_{\text{in}} = 3.5b$, $y_{\text{out}} = 1.5b$. Thus, the x_c and y_c directions to solve the area was defined as: $\{(x_c, y_c) \mid -2a \leq x_c \leq 2a, -3.5b \leq y_c \leq 1.5b\}$ or less or less. In this paper, the region is divided into 50×50 grids in two directions.

Boundary conditions of Reynolds lubrication equation: boundary pressure was 0. The pressure of the whole bearing area was greater than or equal to 0. Considering the type of lubricating oil was isothermal solution, Thermo elastohydrodynamic lubrication(TEHL) method should be used to solve the problem. The pressure gradient in the oil film rupture zone was 0, as follows:

$$\begin{cases} p(x_{\text{in}}, y_c) = p(x_{\text{out}}, y_c) = 0 \\ p(x_c, y_{\text{in}}) = p(x_c, y_{\text{out}}) = 0 \\ p(x_c, y_c) \geq 0 (x_{\text{in}} < x_c < x_{\text{out}}, y_{\text{in}} < y_c < y_{\text{out}}) \\ \frac{\partial p(x_{\text{out}}, y_c)}{\partial x_c} = \frac{\partial p(x_c, y_{\text{out}})}{\partial y_c} = 0 \end{cases} \quad (14)$$

4.2. Numerical Solution Process

Figure 8 shows the data coupling and solving process of the above equations.

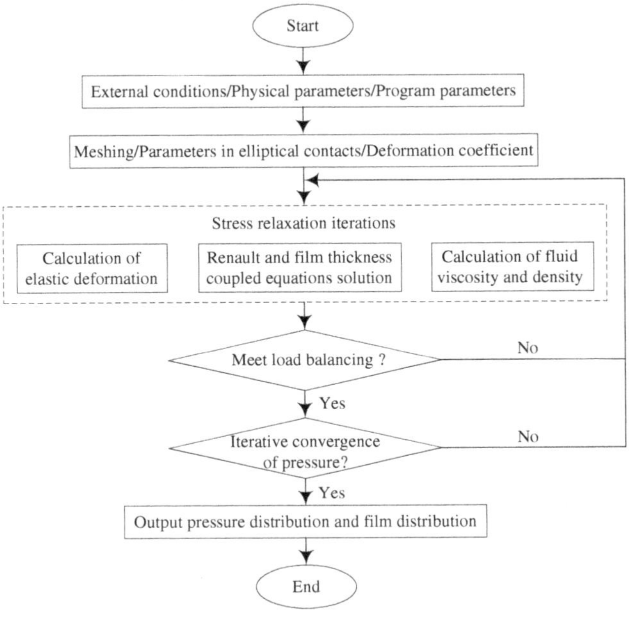

Figure 8. Flow chart of elastohydrodynamic coupling calculation of full ceramic ball bearings.

4.3. Calculation Results and Analysis

In 6208CE silicon nitride ceramic ball bearings, and combining the elastohydrodynamic lubrication equation shown in the Figure 2 flow calculation program, the test process to get the best oil for lubrication conditions, performing lubrication, and contact characteristics analysis were concluded, and the ceramic ball bearing lubrication under the condition of different speed and load of oil film thickness distribution of the simulation results are shown in Figure 9.

By comparing the simulation results in Figure 9a–f, it can be seen that the minimum liquid film thickness in the contact area of the rolling body decreased gradually with the increase of the angle ψ from the distance to the minimum stress point of the rolling body. This is because the squeezing effect of liquid film decreases with the increase of ψ. The minimum liquid film thickness of the rolling body at the same position increased with the increase of rotational speed and load. This was caused by the enhancement of dynamic pressure effect of oil film with the increase of rotational speed. Furthermore, the influence of rotational speed on the minimum liquid film thickness at different locations was similar, and the influence of rotational speed on the liquid film at different locations had no obvious change.

Under the condition of optimal oil supply lubrication for the full ceramic ball bearings, the pressure distribution of lubricating oil film of different rolling bodies under different rotational speeds and loads is shown in Figure 10.

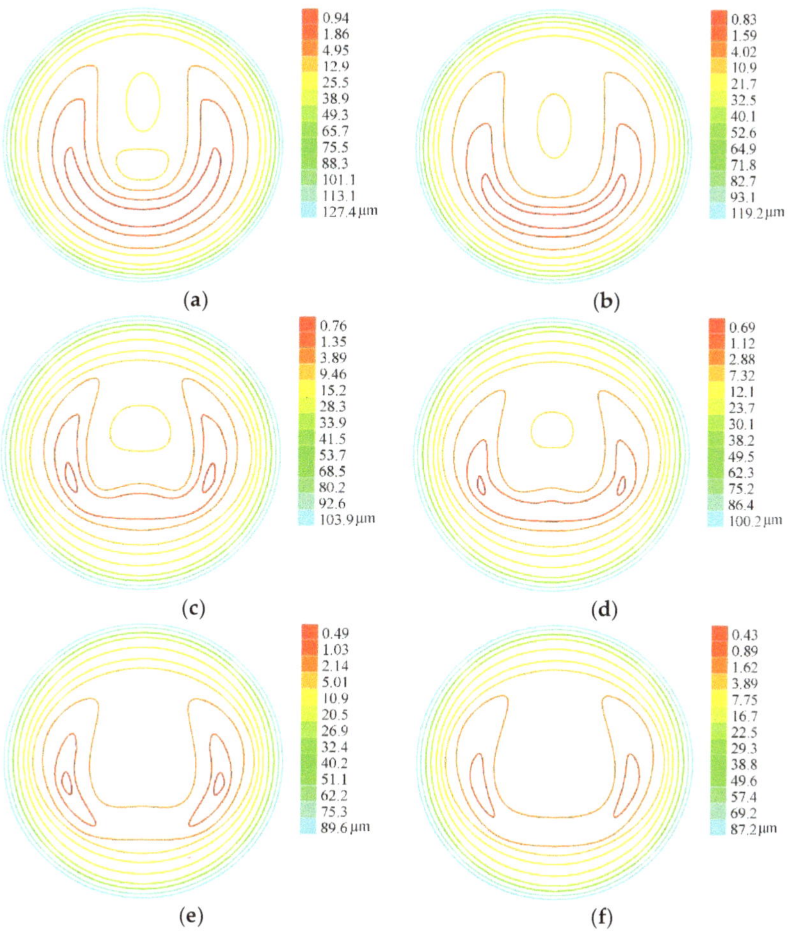

Figure 9. Oil film thickness distribution of ceramic ball bearings under different working conditions. (**a**) $\psi = 0°$, $F = 3000$ N, $n = 10{,}000$ r/min, (**b**) $\psi = 0°$, $F = 900$ N, $n = 5000$ r/min, (**c**) $\psi = 80°$, $F = 3000$ N, $n = 10{,}000$ r/min, (**d**) $\psi = 80°$, $F = 900$ N, $n = 5000$ r/min, (**e**) $\psi = 160°$, $F = 3000$ N, $n = 10{,}000$ r/min, (**f**) $\psi = 160°$, $F = 900$ N, $n = 5000$ r/min.

By comparing the simulation results of Figure 10a–f, it can be seen that the maximum pressure of the rolling body increased with the increment of ψ. This is because the maximum Hertz contact pressure of the rolling body increased with the increase of ψ. It can be seen from the figure that the maximum pressure of lubrication film decreased with the decrease of load contact pressure for different rolling bodies. The maximum oil film pressure corresponding to the rolling body at the same position increased with the increase of rotational speed and load. This was due to the enhanced dynamic pressure effect of oil film, caused by the increase of rotational speed. At the same speed, the angle between the rolling body and maximum stress position of the rolling body at different positions of the same bearing were larger, and the corresponding load was smaller, so the maximum oil film pressure caused by it also decreases correspondingly.

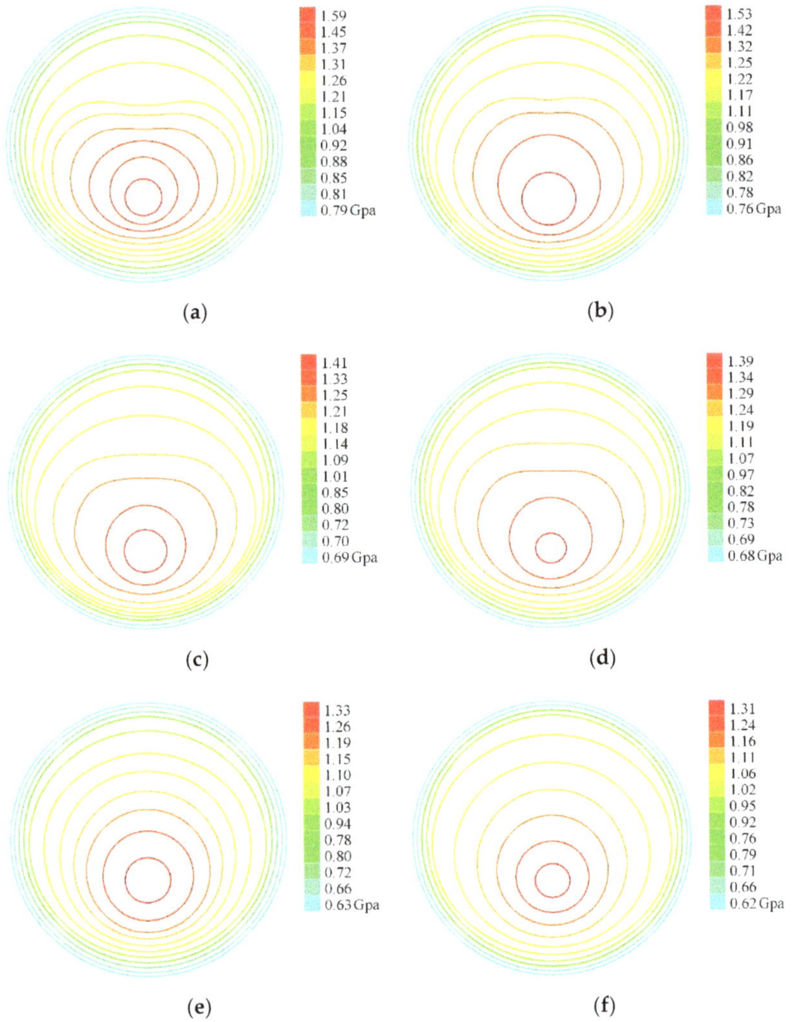

Figure 10. Oil film pressure distribution of ceramic ball bearings under different working conditions. (**a**) $\psi = 0°$, F = 3000N, n = 10,000 r/min, (**b**) $\psi = 0°$, $F = 900$ N, $n = 5000$ r/min, (**c**) $\psi = 80°$, $F = 3000$ N, $n = 10{,}000$ r/min, (**d**) $\psi = 80°$, $F = 900$ N, $n = 5000$ r/min, (**e**) $\psi = 160°$, $F = 3000$ N, $n = 10{,}000$ r/min, (**f**) $\psi = 160°$, $F = 900$ N, $n = 5000$ r/min.

By comparing the simulation and experimental results, it can be found that the thickness and pressure of the bearing oil film in the simulation model changed with the change of working conditions, and the change trend had a good consistency in the experimental results. In the case of large lubricating oil film thickness, the bearings showed excellent characteristics in the experiment. The reliability and correctness of the experimental data were verified.

5. Morphological Characteristics and Microstructure Properties of Contact Micro-Zone of Full Ceramic Ball Bearings under Oil Lubrication

5.1. Test Analysis of Full Ceramic Ball Bearings

The full ceramic ball bearings, tested under different lubrication conditions and working conditions, were tested and analyzed. Among them, the bearing in the figure below was tested under the condition of 900 N load, 5000 rpm, and 0.2 mL/min oil supply. After disassembling the tested bearing assembly, no obvious failure was found in the observation, as shown in Figure 11.

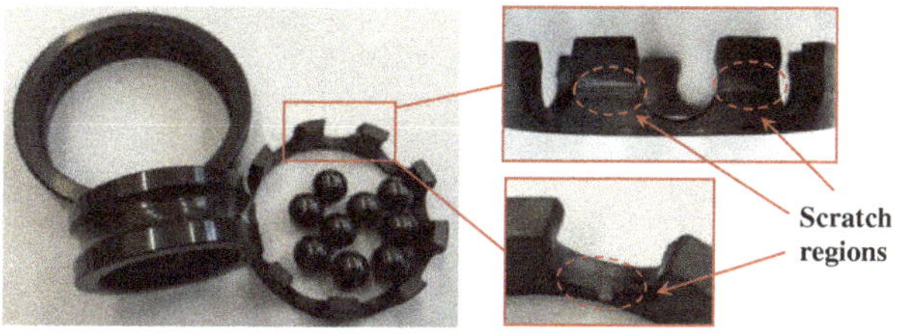

Figure 11. Removed silicon nitride full ceramic ball bearing and its components.

However, in Figure 11, the surface of the PVX cage side beam (the outer circular surface of the cage) showed obvious scratching and rubbing areas. According to the scratches, it can be judged that the surface of the outer ring and inner circle contact and collide when the bearing rotates. The reason for this phenomenon is that the cage was made of composite materials. According to the material properties in Tables 1 and 3, the thermal expansion coefficient of the cage was greater than that of silicon nitride. In the service process, due to the influence of temperature rise, the cage deformation was relatively large, and the cage was guided by the outer diameter. In this guidance mode, the coupling effects of temperature rise, load, and impact made the cage produce elliptic deformation, which then caused friction with the outer ring and inner circle, thus resulting in the cage's rub. In addition, PVX abrasive chips have certain lubricity, which helps to lubricate bearings in service conditions, to some extent.

5.2. Full Ceramic Ball Bearing Contact Area Surface and Surface Quality Testing

The contact area between the rolling body and the raceway was observed, as shown in Figure 12.

(**a**) (**b**)

Figure 12. Outer ring raceway and rolling bodies of full ceramic ball bearing after the test. (**a**) Bearing outer ring raceway, (**b**) Rolling bodies.

It can be seen from Figure 12 that there was no damage phenomenon on the surface of the contact area of the full ceramic ball bearing after the test. However, black film appeared in some areas of outer ring raceway and ball bearing surfaces. In order to further reveal the chemical composition of the film, the mechanism of the film generation and its influence on the surface friction and wear quality of the bearing contact micro-zone were analyzed from the microscopic point of view. We performed non-destructive cutting of test bearing ring and cage. They were tested together with ceramic balls by SEM, XRD, and other instruments.

Hitachi S-4800 scanning electron microscope (SEM), which was made by Hitachi in Japan, was used to detect the raceway and black film covered area on ball surface after the test in Figure 12, as shown in Figures 13 and 14. As can be seen from scanning electron microscopy, the black film area was the covered area in Figures 13 and 14. This indicates that a thin film was formed on the surface of the contact micro-zone between the raceway and ball bearings under the coupling effects of temperature rise, heavy load, sliding effect, and cage collision. By comparing the pictures in Figures 12–14, it can be seen that both the macro- and micro-observation results show that the film formed on the surface of outer ring raceway had a better and more significant effect. Meanwhile, observe the removed bearing inner ring in Figure 11—the black film was not visible on its raceway surface.

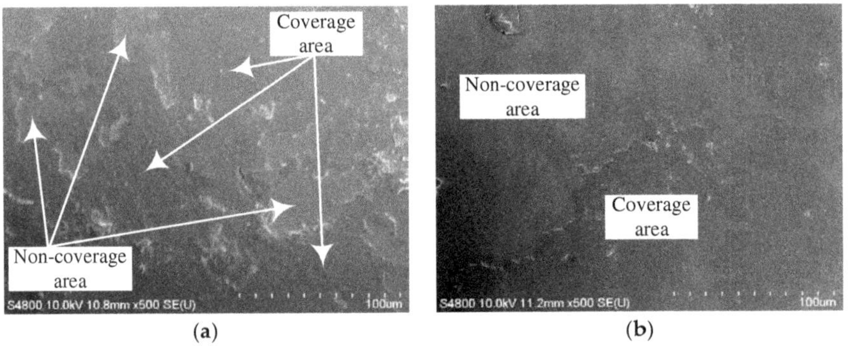

(**a**) (**b**)

Figure 13. SEM imaging of full ceramic ball bearing after the test. (**a**) Measuring point 1, (**b**) Measuring point 2.

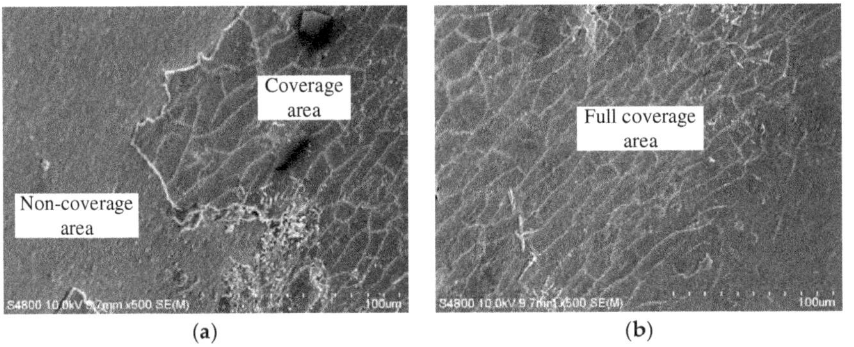

(**a**) (**b**)

Figure 14. SEM imaging of outer raceway of full ceramic ball bearing after the test. (**a**) Measuring point 1, (**b**) Measuring point 2.

The preliminary analysis of the reasons for the above phenomenon was that, when the full ceramic ball bearing ran in oil lubrication conditions, the debris on the outer ring of the cage and surface of the pockets were involved in the bearing contact micro area under the action of bearing rotation.

5.3. Chemical Composition and Qualitative Analysis of Surface Layer of Bearing Contact Area

The black film and cage on the outer racing surface of ceramic ball bearing were analyzed by Raman spectroscopy. In order to ensure the accuracy of the test, the components of the cage and raceway surface film were measured and analyzed twice, and the test results are shown in Figure 15.

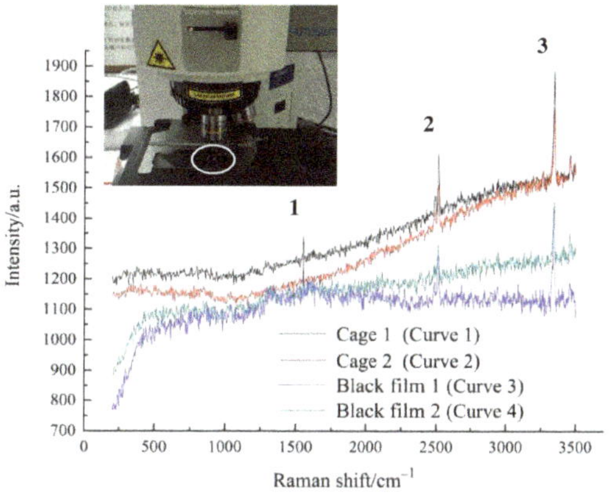

Figure 15. Raman spectra peaks of black film between cage and raceway after the test. 1–3. correspond to 3 characteristic peaks, respectively.

By comparing the four groups of measured curves in the figure, it can be seen that the spectral distribution trend of the film on the surface of the bearing cage and raceway were different after the test, that is, the 1/2 distribution trend of the curve is different from the 3/4 distribution trend of the curve. However, by comparing the three characteristic peaks of the four curves, it was found that the Raman wavelengths of the characteristic scattering peaks were almost the same. This indicates that the crystal interface between the cage and raceway surface film material was basically the same; we can preliminarily judge that the chemical compositions of the two groups of substances were similar.

XRD analysis was carried out on the powder of the outer ring raceway of the bearing and film layer on the surface of the ball and cage powder. The test results were shown in Figure 16.

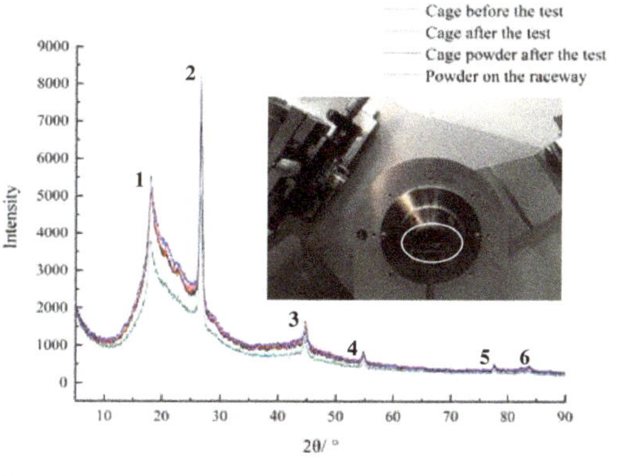

Figure 16. XRD spectrogram of cage and raceway black film. 1–6. correspond to 6 characteristic peaks, respectively.

The XRD results show that the phase diffraction peaks of the bearing cage and raceway film were basically the same before and after the test. Combined with the above analysis, it was preliminarily shown that PVX did not generate new substances by chemical reaction with the medium in the environment under the service conditions of extreme low variable temperature conditions. In Figure 15, the reason the distribution trend of curve 1/2 was different from that of curve 3/4 was that the film components formed by the retener powder under rolling action contain a very small amount of worn silicon nitride and sintering agent materials.

By comparing the above analysis results, it can be inferred that the powder formed by the cage under the action of physical conditions became the lubricating medium of the full ceramic ball bearing oil, which played a promoting role in the service of full ceramic ball bearings.

6. Conclusions

(1) In the service process of full ceramic ball bearing oil lubrication, there is an optimal oil supply. Under the action of the optimal oil supply lubrication, full-film lubrication can be achieved, and the bearing exhibits the optimal characteristics of friction, vibration, temperature rise, and so on. Compared with the load, the speed of the bearing has a decisive influence on the optimal oil supply. In this paper, it was found that the optimal oil supply of 6208CE silicon nitride full ceramic ball bearings was about 1.2 mL/min when the rotational speed was 5000 rpm. When the speed was 10,000 rpm, the optimal oil supply was about 1.6 mL/min.

(2) In the service process of full ceramic ball bearings, when the oil supply was less than the optimal oil supply, the full ceramic ball bearings were in the state of lacking oil lubrication, and the oil–solid mixed lubrication in the contact micro-zone led to the friction, vibration, and temperature rise of the bearings. When the oil supply was greater than the optimal oil supply, the viscous resistance generated by too much lubricating oil increased the friction and vibration of the bearing, but the friction and vibration were smaller than those in the state of poor oil lubrication. When the oil supply was too much, a large amount of lubricating oil reduced the temperature rise generated by the bearing, thus playing the role of lubrication and cooling, so the temperature rise of the bearing outer ring continued to decrease.

(3) When the rotational speed and load of the full ceramic ball bearing were constant, the extrusion effect and Hertz contact pressure increased gradually with the increase of the angle ψ from the minimum stress point of the rolling body, and the minimum oil film thickness and oil film pressure in the contact area of the rolling body decreased. When the speed and load of the bearing increased, the minimum oil film thickness and oil film pressure of the rolling body at the same position increased with the increase of the speed and load of the bearing, due to the influence of the dynamic pressure effect of the lubricating oil film.

(4) The full ceramic ball bearing can run under the above oil lubrication, as well as high speed and heavy load conditions. Key components, including ceramic ball, ceramic ring, and retainer, did not fail after detection. However, under the action of high contact stress, a film will be formed in the outer ring raceway of the bearing, which does not affect the service performance of the bearing. If the cage material is kept within the lubrication characteristics, the film has a certain promotion effect on the service of the bearing.

Author Contributions: Data curation, Z.B.; resources, Y.W.; software, Z.X., J.T. and L.G.; writing—original draft, J.Y. (Jinmei Yao); writing—review and editing, J.Y. (Jiaxing Yang) and J.S. All authors have read and agreed to the published version of the manuscript.

Funding: The authors acknowledge the collective support granted by the National Natural Science Foundation of China (grant No. 52105196), Department of Science and Technology of Liaoning Province (grant No. 2020-BS-159), and Young and Middle-aged Innovation Team of Shenyang (grant No. RC210343).

Data Availability Statement: Not applicable.

Conflicts of Interest: The authors declare no conflict of interest.

References

1. Yao, J.; Wu, Y.; Sun, J.; Xu, Y.; Wang, H.; Zhou, P. Research on the metamorphic layer of silicon nitride ceramic under high temperature based on molecular dynamics. *Int. J. Adv. Manuf. Technol.* **2020**, *109*, 1249–1260. [CrossRef]
2. Sun, J.; Wu, Y.; Zhou, P.; Li, S.; Zhang, L.; Zhang, K. Simulation and experimental research on Si3N4 ceramic grinding based on different diamond grains. *Adv. Mech. Eng.* **2017**, *9*, 9–14. [CrossRef]
3. Han, X.X.; Xu, C.H.; Jin, H.; Xie, W.H.; Meng, S.H. An experimental study of ultra-high temperature ceramics under tension subject to an environment with elevated temperature, mechanical stress and oxygen. *Sci. China Technol. Sci.* **2019**, *62*, 1349–1356. [CrossRef]
4. Sun, J.; Yang, J.X.; Yao, J.M.; Tian, J.X.; Xia, Z.X.; Yan, H.P.; Gao, L.F.; Li, S.H.; Bao, Z.G. The effect of lubricant viscosity on the performance of full ceramic ball bearings. *Mater. Res. Express* **2022**, *9*, 015201. [CrossRef]
5. Meyer, L.D.; Ahlgren, F.F.; Weichbrodt, B. An Analytic Model for Ball Bearing Vibrations to Predict Vibration Response to Distributed Defects. *J. Mech. Des.* **1980**, *102*, 205–210. [CrossRef]
6. Lynagh, N.; Rahnejat, H.; Ebrahimi, M.; Aini, R. Bearing induced vibration in precision high speed routing spindles. *Int. J. Mach. Tools Manuf.* **2000**, *40*, 561–577. [CrossRef]
7. Alfares, M.; Saleem, O.; Majeed, M. Analytical study of thermal variation impact on dynamics of a spindle bearing system. *Proc. Inst. Mech. Eng. Part K J. Multi-Body Dyn.* **2019**, *233*, 871–898. [CrossRef]
8. Jalali-Vahid, D.; Gohar, R.; Jin, Z.M.; Rahnejat, H. Comparison between experiments and numerical solutions for isothermal elastohydrodynamic point contacts. *J. Phys. D Appl. Phys. A Europhys. J.* **1998**, *31*, 2725. [CrossRef]
9. Hamrock, B.J.; Dowson, D. Isothermal Elastohydrodynamic Lubrication of Point Contacts: Part 1—Theoretical Formulation. *J. Lubr. Technol.* **1976**, *98*, 223–228. [CrossRef]
10. Ioannides, E.; Harris, T.A. A New Fatigue Life Model for Rolling Bearings. *J. Tribol.* **1985**, *107*, 367–377. [CrossRef]
11. Zhu, W.B.; Zhang, X.B.; Lu, Y.S.; Wang, D.L. Numerical Study on lubrication Performance of ball bearings. *J. Propuls. Technol.* **2019**, *4*, 892–901.
12. Nagare, P.N.; Kudal, H.N. Tribological Failure Analysis and Suitability of Grease Lubrication for Sugarcane Crushing Mill Journal Bearings. *J. Fail. Anal. Prev.* **2018**, *18*, 1311–1319. [CrossRef]
13. Lu, L.M.; Lu, J.F.; Li, Z.H. Lubrication Characteristics of roller bearing under Cyclic Impact Load. *Lubr. Eng.* **2021**, *12*, 78–85.
14. Kang, J.; Lu, Y.; Zhang, Y.; Liu, C.; Li, S.; Mueller, N. Investigation on the skidding dynamic response of rolling bearing with local defect under elastohydrodynamic lubrication. *Mech. Ind.* **2019**, *20*, 615. [CrossRef]
15. Antonio, G.; Linares, F.; Arias, M. Numerical Investigations of the Lubrication Conditions in Hydrodynamic Bearings with Shaft Misalignment Effect. *Ing. Investig. Y Tecnol.* **2013**, *14*, 89–98.
16. Cho, I. The influence of boundary conditions on the lubrication characteristics of a journal bearing for reciprocating compressors. *J. Mech. Sci. Technol.* **2015**, *29*, 751–758. [CrossRef]
17. Biswas, N.; Chakraborti, P.; Belkar, S. An analytical and experimental approach for pressure distribution analysis of a particular lobe and plain bearing performance keeping in view of all impeding varying parameters associating with fixed lubrication SAE20W40. *J. Mech. Sci. Technol.* **2016**, *30*, 2187–2193. [CrossRef]
18. Dmitrichenko, N.F.; Milanenko, A.A.; Hluhonets, A.A.; Minyaylo, K.N. Method of Forecasting Durability of Bearings Rolling and Optimal Choice of Lubricants under Conditions of Flood Lubrication and Oil Starvation. *Treniei Iznos* **2017**, *38*, 114–120.
19. Brizmer, A.; Gabelli, C.; Vieillard, G.E.; Morales, E. An Experimental and Theoretical Study of Hybrid Bearing Micropitting Performance under Reduced Lubrication. *Tribol. Trans.* **2015**, *58*, 829–835. [CrossRef]
20. Li, K.; Chen, Z.X.; Liu, P.P.; Li, G.; Ding, M.; Li, Z.X. Characterization and performance analysis of 3D reconstruction of oil-lubricated Si_3N_4–GCr15/GCr15–GCr15 friction and wear surface. *J. Therm. Anal. Calorim.* **2020**, *144*, 2127–2143. [CrossRef]
21. Xiang, G.; Wang, C.; Wang, Y.J.; Han, Y.F.; Wang, J.X.; Lv, Z. Dynamic Mixed Lubrication Investigation of Water-Lubricated Bearing With Unbalanced Rotor During Start-Up. *Tribol. Trans.* **2021**, *64*, 764–776. [CrossRef]
22. Ma, S.; Zhang, X.; Yan, K.; Zhu, Y.; Hong, J. A Study on Bearing Dynamic Features under the Condition of Multiball–Cage Collision. *Lubricants* **2022**, *10*, 9. [CrossRef]
23. Sun, X.; Zhang, W.H.; Tian, H. Theoretical analysis of cylindrical roller bearing with flexible rings mounted in groove elastic support:Papers. *J. Adv. Mech. Des. Syst. Manuf.* **2020**, *14*, JAMDSM0102. [CrossRef]
24. Zheng, Y.; Zhao, S. Research on the Performance of radial sliding Bearing with sliding Surface considering cavitation pressure. *Lubr. Eng.* **2017**, *7*, 54–59.
25. Du, F.M.; Chen, C.D.; Zhang, K.G. Fluid Characteristics Analysis of the Lubricating Oil Film and the Wear Experiment Investigation of the Sliding Bearing. *Coatings* **2022**, *12*, 67. [CrossRef]

26. Shetty, P.; Meijer, R.J.; Lugt, P.M. An Evaporation Model for Base Oil from Grease-Lubricated Rolling Bearings including Breathing. *Tribol. Trans.* **2021**, *64*, 891–902. [CrossRef]
27. Zhang, Y.; Luo, Y.; Ni, S.; Wu, R.; Zhang, Z.Q. High speed characteristics of lubricating oil film of heavy hydrostatic bearing with micro inclined plane. *J. Jilin Univ.* **2021**, *2*, 450–457.
28. Xu, G.; Geng, H.P.; Zhou, J.; Yu, L.; Yang, L.; Lu, J.M. Research on the Static and Dynamic Characteristics of Misaligned Journal Bearing Considering the Turbulent and Thermohydrodynamic Effects. *J. Tribol.* **2015**, *137*, 024504. [CrossRef]

lubricants

MDPI

Article

Dynamic Performance Analysis of Cage in Four-Point Contact Ball Bearing

Yeteng Li [1], Wenchao Li [1,2], Yongsheng Zhu [1,*], Gaobo He [1], Shuaijun Ma [1] and Jun Hong [1]

[1] Key Laboratory of Education Ministry for Modern Design & Rotor-Bearing System, Xi'an Jiaotong University, Xi'an 710049, China; liyeteng@stu.xjtu.edu.cn (Y.L.); zys_liwenchao@163.com (W.L.); gaobohe@stu.xjtu.edu.cn (G.H.); msj821@stu.xjtu.edu.cn (S.M.); jhong@mail.xjtu.edu.cn (J.H.)

[2] Luoyang Bearing Research Institute Co., Ltd., Luoyang 471039, China

* Correspondence: yszhu@mail.xjtu.edu.cn; Tel.: +86-13991149360

Abstract: Due to the special structure of double-half inner rings, four-point contact ball bearings are prone to uneven forces in the inner raceway during movement, which affects the dynamic performance of the rolling element and cage, and even leads to cage sliding. Dynamic performance of the cage is an important factor affecting the working stability of bearings. In this paper, in order to grasp the operation law of the cage so as to guide the application of four-point contact ball bearings, the dynamic model of four-point contact ball bearings is established by the secondary development of Automatic Dynamic Analysis of Mechanical Systems (ADAMS). The dynamic performance of the cage is analyzed and evaluated with the indexes of vortex radius ratio and vortex velocity deviation ratio of the cage centroid trajectory. The results show the following: the cage stability increases and then decreases to a certain degree with rotating speed-rise; it increases and then decreases with the increase in the pure axial load; under a combination of axial and radial load, the cage moves more smoothly with smaller radial force. Rotating speed has little effect on cage stability, while radial force has a great influence on cage stability, followed by axial load. In order to verify the simulation results, a test bench for rolling bearing cages is developed, and the accuracy of the simulation results is verified by the test results.

Keywords: load distribution; four-point contact ball bearing; bearing dynamics; cage motions experiment

Citation: Li, Y.; Li, W.; Zhu, Y.; He, G.; Ma, S.; Hong, J. Dynamic Performance Analysis of Cage in Four-Point Contact Ball Bearing. *Lubricants* **2022**, *10*, 149. https://doi.org/10.3390/lubricants10070149

Received: 14 April 2022
Accepted: 29 June 2022
Published: 11 July 2022

1. Introduction

As an important part of the high-speed railways, the bogie plays a role in connecting the body and the track, thus its motion stability affects the safety and comfort of the train [1,2]. Figure 1 shows a bogie. The four-point contact ball bearing (FPCBB) is one of the important components of the bogie, which could support axial loads in both directions due to its double-half structure. It is usually used in conjunction with cylindrical roller bearings for locating and supporting. Compared to a traditional paired configuration angular contact ball bearing, FCPBB requires less coaxiality and space between bore and shaft. However, in the motion process, the force between the rolling elements and the double-half raceway will change, inevitably making a four-point contact, three-point contact or two-point contact inside the bearing, which may lead to cage instability. In case of serious instability, it will even cause abnormal heat generation and the early failure of the bearing. Therefore, investigation of its dynamic characteristics is desired.

Research on FPCBB focuses more on wind turbines, aero-engines and robots joints, and most of them are static analysis. For example, Zhang [3] has analyzed the influence of radial load and overturning moment on the load distribution of FPCBB; Li [4] has analyzed the influence of positive and negative clearance on the load distribution of FPCBB; Li and Tang [5] have studied the influence of different parameters on the load–displacement relationship theoretically by establishing the geometric coordination equation of FPCBB.

In the research on the dynamic characteristics of FPCBB, researchers mostly pay more attention to the wind turbine bearings that mainly bear radial load and overturning torque, as well as the thin-walled FPCBB of industrial robots that prefer structural optimization and lightweight design. Wu [6] has analyzed the influence of different speed and preload on the dynamic characteristics of FPCBB of wind turbine generator; Shi [7] has studied the influence of different structural parameters on the dynamic characteristics of four point contact ball bearing under the combined action of axial load and overturning moment; Yang [8] has studied the precise elliptical contact area shape and contact state of thin-walled FPCBB based on the finite element method; Liu [9] has explored the optimization of structural parameters of thin-walled FPCBB from the aspects of contact force between ball and cage and its influence on axial vibration intensity of cage centroid by multi-body dynamics and Hertz contact theory. These studies provide great help for improving the service performance and the efficiency of bearing design.

Figure 1. The bogie of high-speed railway train.

In terms of dynamic analysis of bearings, the establishment of motion differential equations was started by Walters [10]. His model with a 4 degrees of freedom (4-DOF) ball and a 6-DOF cage laid the foundation for dynamic analysis of ball bearing. Based on the study of Walters, Gupta [11–13] studied the complex motion and contact state between the rollers and raceway, and comprehensively analyzed the steady and transient dynamic characteristics of the bearing. Hagiu [14] programmed the calculation of the interaction between bearing components, but the model did not take into account the lubrication. Wijnant [15] constructed the ball bearing dynamic model of considering lubrication film, but the calculation accuracy is dependent on the step size and is difficult to guarantee. Weinzapel [16] established a flexible cage and obtained the rigid–flexible coupling model in order to be closer to the actual working conditions, which makes the calculation more complex. In addition, some scholars [17–20] applied the finite element method to analyze contact characteristics of FPCBB, but due to the complex meshing, these studies are mostly limited to the analysis of large bearings under steady low speed conditions.

Furthermore, some researchers used the commercial software of multi-body dynamics to numerically analyze dynamic characteristics of bearings. For example, Ji [21] used ADAMS to analyze the dynamic characteristics of the instantaneous response of the cage, and studied cage centroid trajectory characteristics under different rotating speeds, loads, radial clearances and the number of steel balls. Hong [22] and Chan [23] also used ADAMS to establish the bearing dynamic model, so as to carry out corresponding researches. Hou [24] built a rigid–flexible hybrid dynamic model of the bearing based on ANSYS and ADAMS, and obtained the motion law of rigid components and the influence of vibration. However, there are relatively few analyses on the dynamic performance for bogie four-point contact ball bearings, a load condition in which axial load is dominant and radial load is supplementary. Wu [6] analyzed the bearing dynamic feature under main radial load. However, it did not put forward the corresponding evaluation index, and only analyzed the motion characteristics of the bearings under different working conditions. Zhang [25] analyzed the influence of cage structure shape on bearing dynamic characteristics under combined loads by ADAMS. Based on Hertz contact theory, Zhao [26] used ADAMS to establish a bearing dynamic model considering gear meshing, axial load, speed and overturning torque on bearing vibration and contact force.

As mentioned above, no clear evaluation indicators were given for the motion stability analysis of four-point contact ball bearings. The common method for bearing dynamic analysis is based on the multi-body dynamics commercial software, which is greatly affected by the limitation of the software itself and can only consider limited factors. In order to discuss the operation stability of bearings, a general methodology for dynamic performance analysis of cage will be proposed in this work. A dynamic model of FPCBB will be established based on ADAMS secondary development, and the cage centroid trajectory will be discussed under the conditions of radial load and axial load in the following sections. The variation of cage motion stability is described by vortex velocity deviation ratio and vortex radius ratio, the latter referring to the ratio of the difference between the maximum and minimum vortex radius to the average vortex radius. Finally, a test bench will be developed to collect the real speed and cage motion trajectory at different rotating speeds and load conditions to validate the simulation results.

2. Bearing Dynamic Model

2.1. Solid Model Creation

The bearing model is QJ215, which is generally used in the high-speed railway bogie. The structure is shown in Figure 2. The bearing inner rings, outer ring and rolling elements are made of bearing steel, and the cage is made of brass. Specific size and material parameters are shown in Table 1. In this paper, the parametric design method is used to model the FPCBB in ADAMS/View by macro commands and CMD program [27,28]. The 3D model, as shown in Figure 3, can be obtained by inputting corresponding structural parameters.

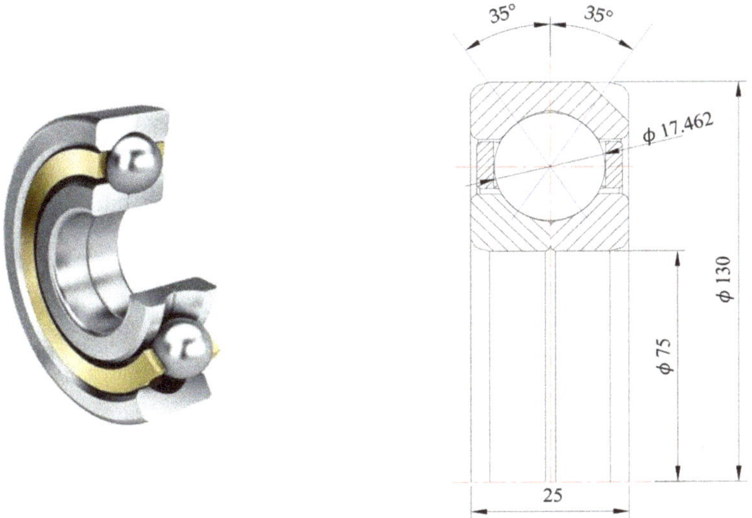

(a) (b)

Figure 2. The structure of QJ215. (**a**) A physical picture of QJ215; (**b**) A dimension drawing of QJ215.

Table 1. Dimensional parameters and material parameters of the four-point contact ball bearing.

Characters	Parameter Name	Numerical Value
d	Diameter of inner ring	75 mm
D	Diameter of outer ring	130 mm
D_w	Diameter of ball	17.462 mm
Z	Number of balls	15
B	Width of bearing	25 mm
α	Contact angle	35°
Pd	Radial clearance	27 μm
ΔD_c	Lead gap	20
E_1	Elastic modulus of bearing steel	2.1×10^{11} N/mm^2
ν_1	Poisson ratio of bearing steel	0.29
ρ_1	Density of bearing steel	7.81×10^{-6} kg/mm^3
E_2	Elastic modulus of brass	1.06×10^{11} N/mm^2
ν_2	Poisson ratio of brass	0.324
ρ_2	Density of brass	9.5×10^{-7} kg/mm^3

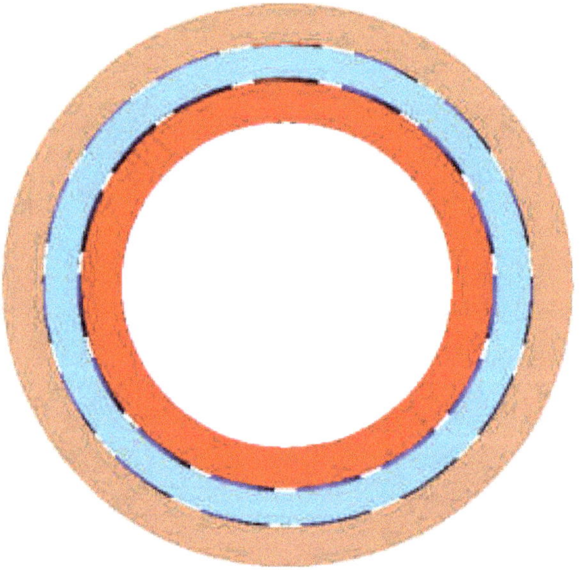

Figure 3. 3D model of QJ215.

2.2. Bearing Dynamics Model Construction Based on GFOSUB

The contact force between bearing components is greatly affected by the recognition accuracy of ADAMS solver on the solid model, and the results are prone to problems such as large values and burrs [29]. For more realistic results, the bearing contact force can be solved directly by GFOSUB subroutine, which can bypass the contact stiffness and ADAMS solver, reducing the negative effects of contact stiffness and other parameters. The process of building a bearing dynamics model based on Gforce Subroutine (GFOSUB) is shown in Figure 4.

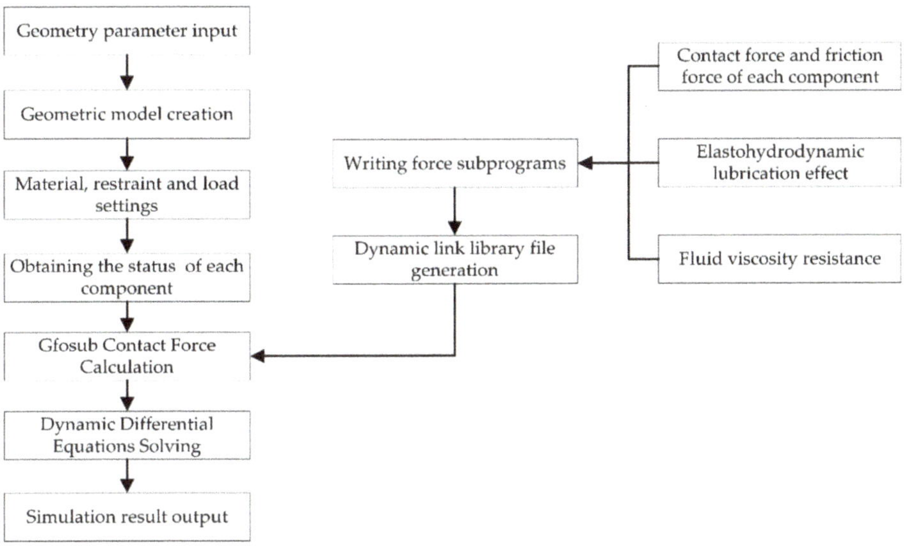

Figure 4. The process of building the bearing dynamics model based on GFOSUB.

Firstly, the geometric model of FPCBB is established by structural parameters, and boundary conditions such as materials, constraints and loads are set to obtain the state information of bearing components, including displacement, speed and acceleration, etc. Meanwhile, the GFOSUB bypasses the Adams contact solver to complete the calculation of force, finally the output results are solved by the differential equation solver. The command box of GFOSUB is shown in Figure 5. The parameters in the red box are ball diameter, groove curvature radius of outer raceway, outer raceway groove bottom diameter, inner raceway groove bottom diameter, dynamic friction coefficient, static friction coefficient, outer ring outer diameter, mass marker point of the inner ring center and mass marker point of the outer ring center. The yellow box is the name of the .dll file where GFOSUB is located. GFOSUB can be called by entering the corresponding parameters.

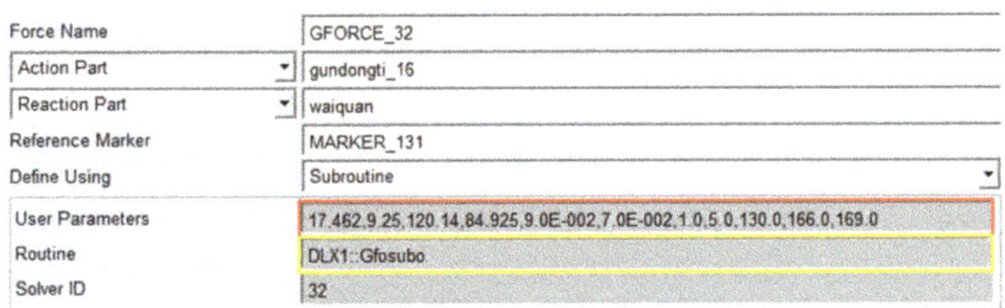

Figure 5. The command box of GFOSUB.

2.3. Boundary Condition Settings

Boundary condition settings mainly include adding freedom constraint, load, driving speed, contact force, etc.

1. Addition of freedom constraints;

In the unconstrained state, the bearing has six degrees of freedom for translation and rotation in X, Y and Z directions. In the actual working process, the outer ring is in a fixed state with 0 degrees of freedom, so it is necessary to add the corresponding restraint sub to restrain its motion. The inner ring has 4 degrees of freedom, i.e., translational movement in X, Y and Z directions and axial rotation: δ_x, δ_y, δ_z and θ_y; the rolling elements and the cage retain 6 degrees of freedom: δ_x, δ_y, δ_z, θ_x, θ_y and θ_z. The above freedom relationship is shown in Figure 6.

2. Addition of load and driving speed;

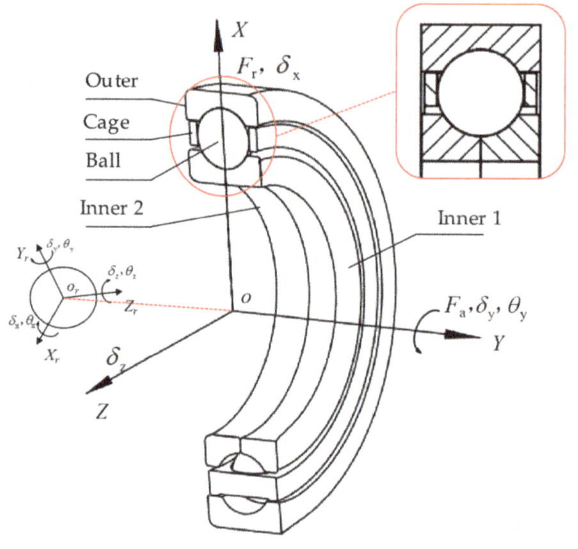

Figure 6. The dynamic model of FPCBB.

For the FPCBB model, it is only subjected to axial and radial loads in the working process, and the load acts on the center of the inner ring. In Figure 6, *Fa* is the axial load, *Fr* is the radial load. In addition, the driving speed is set in the direction of the axial rotation of the inner ring, and according to the design requirements, the rated speed of the bearing is 5988 r/min.

3. Addition of contact force.

The interaction between the components of the bearing is mainly transmitted according to the contact force. FPCBB has a wide range of contact elements; according to the contact elements, they can be divided into rolling elements and inner rings, rolling elements and outer rings, rolling elements and cages, cages and guide rings, etc. When using ADAMS for dynamic simulation, the contact parameters between the two elements mainly include stiffness, damping and related friction coefficient.

For the FPCBB, the contact model between the rolling elements and the inner and outer rings is point contact, and the bearing contact stiffness K_j (j = i,o) can be calculated by empirical formula [30], such as Equation (1):

$$K_j = 2.15 \times 10^5 \left(\sum \rho\right)^{-1/2} (n_\delta)^{-3/2} \tag{1}$$

where $\sum \rho$ is the curvature of the contact point, and n_δ is the contact deformation coefficient of elastomer.

The calculation of $\sum \rho_i$ and $\sum \rho_o$ is as follows:

$$\sum \rho_i = \frac{4}{D_w} + \frac{2}{d_m - D_w} - \frac{1}{r_i} \tag{2}$$

$$\sum \rho_o = \frac{4}{D_w} - \frac{2}{d_m - D_w} - \frac{1}{r_o} \tag{3}$$

where d_m is the equivalent diameter of bearing, which can be calculated by Equation (4):

$$d_m = (d + D)/2 \tag{4}$$

Replace the bearing parameters to obtain the value of K_j: $K_i = 1.25 \times 10^6$, $K_o = 1.1 \times 10^6$.

In ADAMS, the bearing contact damping is taken as 0.1~0.01% times the contact stiffness and the friction coefficient is about 0.1 [31].

At this point, the dynamics model of FPCBB is completed in ADAMS.

3. Verification of FPCBB Model

For the created bearing model, it is necessary to verify its accuracy to improve the reliability of the simulation results. In this paper, the correctness of the dynamics model is judged by the theoretical calculation results as well as the bearing load distribution curve under the static simulation and dynamics simulation.

Based on the above-mentioned parametric model, an axial load of 2000 N, a radial load of 500 N and a rotating angular speed of 1000 rpm are applied to the center of inner rings. The dynamic simulation of the bearing is implemented in ADAMS software, the velocities and dynamic contact forces of the bearing are shown in Figure 7.

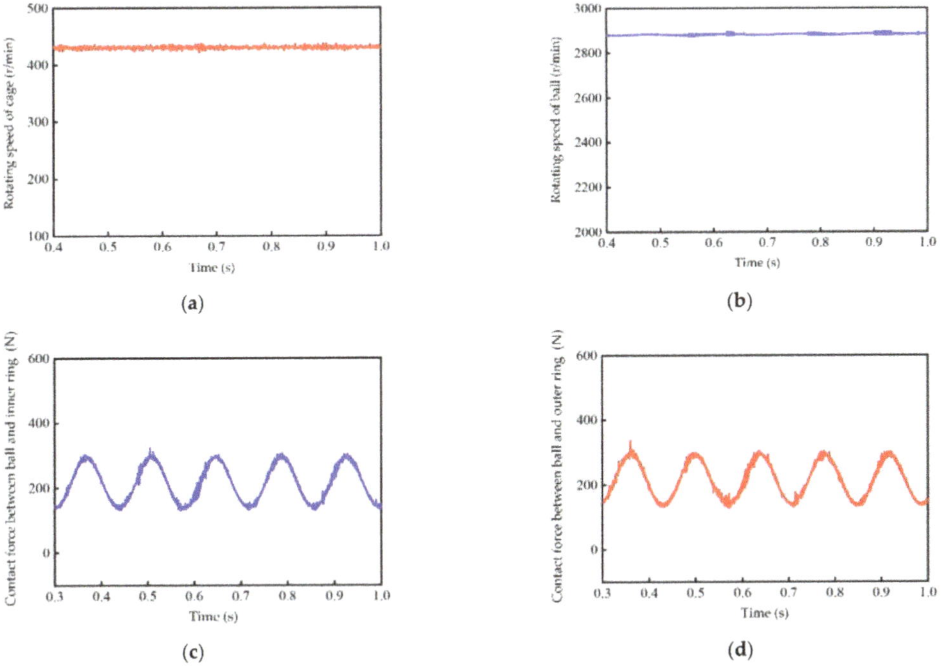

Figure 7. Velocities and dynamic contact forces of QJ215. (**a**) Rotating speed of the cage; (**b**) Rotating speed of the ball; (**c**) Contact force between ball and inner ring; (**d**) Contact force between ball and outer ring.

According to empirical formula, the bearing cage speed n_c is

$$n_c = \frac{1}{2}[n_i(1-\gamma) + n_o(1+\gamma)] \tag{5}$$

where n_i is the rotating speed of inner rings and n_o is outer rings speed.

$$\gamma = \frac{D_w \cos\alpha}{d_m} \tag{6}$$

where α is ball connect angle.

It is obtained that the cage speed is 430.21 r/min while the simulation value is 432.92 r/min, and the difference is 0.63%, so the simulation results can be considered reasonable.

For ball bearings subjected to combined axial and radial loads, due to the influence of centrifugal force and gyro torque, the indirect contact angle between the rolling element and raceway at different phase angles is different, and the contact load between the rolling element and the inner and outer rings is also different. If the applied radial load is recorded as Fr, axial load is Fa, and the included angle between adjacent rolling elements is ϕ, the rolling element at $\phi = 0$ will bear the maximum load $Q_{\max}$ [32].

The maximum contact load of ball bearings under combined load is:

$$Q_{\max} = \frac{Fr}{ZJ_r(\varepsilon)\cos\alpha} \tag{7}$$

where $J_r(\varepsilon)$ can be obtained by referring to Table 7.4 of the reference [33].

The load distribution function at different phase angles is:

$$Q_{\phi i} = Q_{\max}\left[1 - \frac{1}{2\varepsilon}(1-\cos\phi)\right]^{1.5} \tag{8}$$

Applying radial and axial forces of 500 N to bearing, the result calculated by empirical formulas is shown in Figure 2.

The bearing static load distribution curve reflects the bearing load distribution in the stationary or ultra-low speed conditions of the bearings. For bearings running at medium and high speeds, collisions are generated between the components during the motion, resulting in instantaneous increase in load values between the rolling element and raceway or cage. Considering that the load curve in dynamic analysis is more confusing, the obtained dynamic load curve needs to be filtered to eliminate the burr of the signal. Static simulation and dynamic simulation were carried out by applying the same load conditions to the bearing model in ADAMS, and the load magnitudes of the rolling element under different phase angles were obtained. Accordingly, the static load distribution curve and dynamic load distribution curve are plotted and the results are shown in Figure 8.

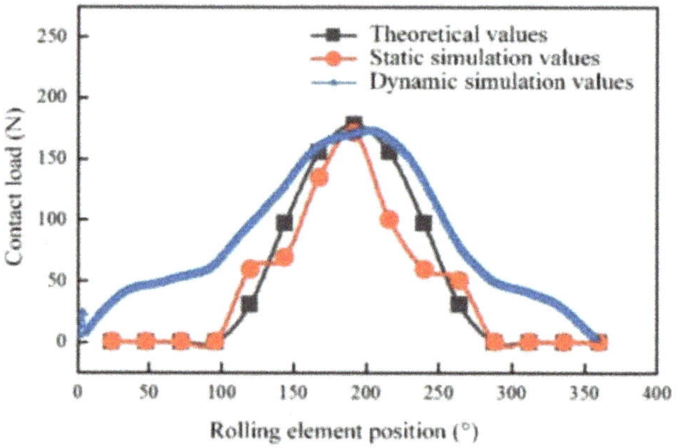

Figure 8. Comparison of dynamic simulated, static simulated and theoretical values of combined load.

According to Figure 8, the dynamic load curve is roughly similar to the static simulation results and the load curve profile obtained from theoretical calculation, which proves that the bearing dynamic model in Section 2 is correct. In the real service process, the dynamic load profile is obviously not as stable as the other two static load profiles, due to the random collision between the rolling element and inner and outer rings, rolling element and cage, cage and inner and outer rings during the bearing motion. Additionally, due to the existence of inertia moment and centrifugal force, the non-load-bearing area where the contact load between rolling body and raceway is 0 under theoretical or static simulation conditions, and the same contact load exists under dynamic simulation conditions. Therefore, it is reasonable that the dynamic load curves are partially different from the static simulation curves in the validation analysis of the dynamics model.

4. Analysis of Dynamic Characteristics of the Cage

The dynamic characteristics of the cage have different performances with the change in working conditions. In previous studies [30,33,34], it is usually judged according to the motion state of the centroid: when the trajectory of the cage centroid is one point, it can be considered as a completely stable state; when the cage centroid trajectory is not a point and a vortex occurs, the motion stability of the cage is judged according to the centroid vortex trajectory and the vortex velocity deviation ratio. When the cage vortex trajectory is almost circular, it indicates that the cage centroid is in a stable vortex state. When the value of the cage vortex velocity deviation ratio is small, it indicates that its motion is relatively stable.

The deviation ratio of cage vortex velocity is expressed as σ_{V}, and the calculation formula is as follows:

$$\sigma_{\mathrm{V}} = \frac{\sqrt{\sum_{i=1}^{n} (v_i - \overline{v}_{\mathrm{m}})^2/(n-1)}}{\overline{v}_{\mathrm{m}}} \tag{9}$$

where v_i is the speed of the cage centroid at any moment, $\overline{v}_{\mathrm{m}}$ is the average speed of the cage centroid, and n is the number of sampling points of the cage speed time domain curve.

4.1. Stability Variation of Cage with Rotating Speed

Apply 1000 N axial load to the bearing, observe the simulation results at driving speeds of 1000, 2000, 3000, 4000, 5000, 6000, 7000 and 8000 r/min, and analyze the influence of speed on the stability of cage.

The trajectory of the cage centroid after stabilization obtained by simulation analysis at the above 8 different rotating speeds is shown in Figure 9.

From Figure 9, it can be seen that the vortex radius is close to the guiding gap after stabilization. When the rotating speed is less than 3000 r/min, the cage centroid trajectory tends to change steadily and regularly with the increase in rotating speed; after exceeding 3000 r/min, the centroid trajectory tends to become cluttered with the increase in rotating speed; after exceeding 6000 r/min, it changes relatively invisibly.

Figure 10 shows the vortex radius ratio and vortex velocity deviation ratio of the cage centroid trajectory. The vortex radius ratio is the ratio of the difference between the maximum and minimum vortex radius to the average vortex radius, and it reflects the divergence degree of the cage centroid vortex trajectory. It can be seen from the figure that the ratio value decreases gradually with the increase in speed when the speed is less than 3000 r/min, when the speed exceeds 3000 r/min, the deviation ratio of centroid vortex velocity increases gradually with the increase in speed, and finally tends to be level off. It is noteworthy that both ratios almost have the same trend, but the vortex radius ratio has a greater magnitude of change. Even so, its range of variation with rotating speed is only 0.15, so the effect of speed on cage stability can be considered small.

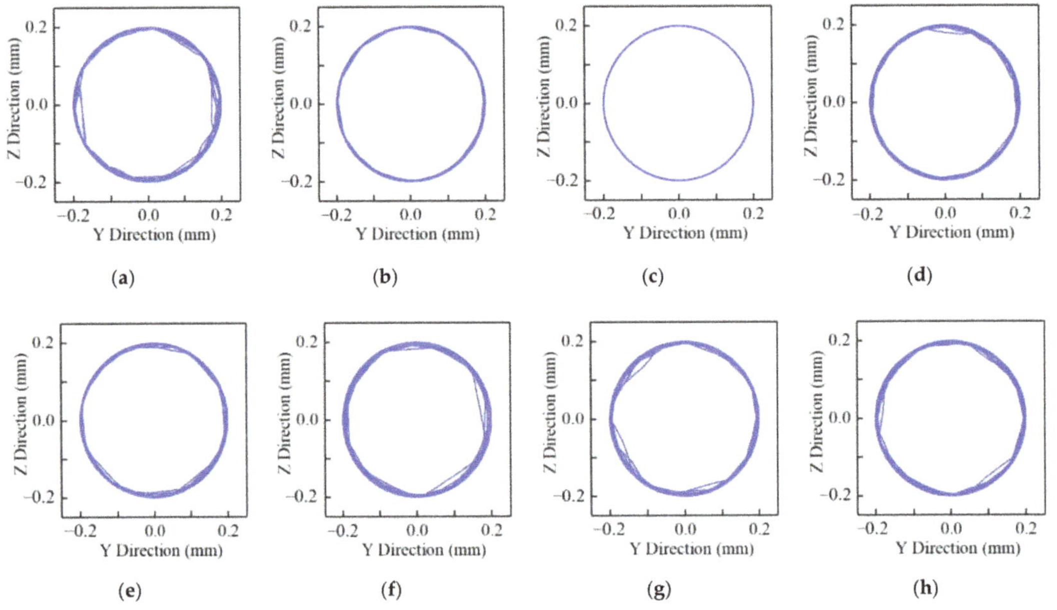

Figure 9. Cage centroid trajectory at different speeds. (**a**) 1000 r/min; (**b**) 2000 r/min; (**c**) 3000 r/min; (**d**) 4000 r/min; (**e**) 5000 r/min; (**f**) 6000 r/min; (**g**) 7000 r/min; (**h**) 8000 r/min.

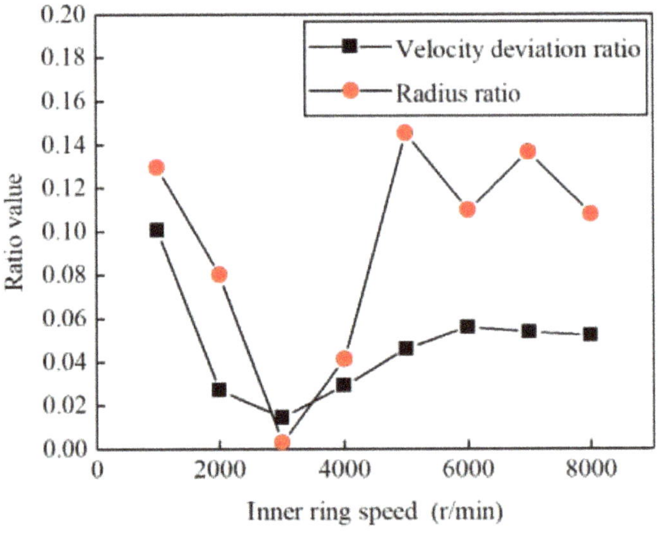

Figure 10. Effect of rotating speed on the velocity deviation ratio and radius ratio of cage centroid vortex velocity.

As can be seen from Figures 9 and 10, the motion stability of the cage gradually stabilizes with an increase in speed and then gradually deteriorates to a certain level. This is due to the fact that with the increase in rotating speed, the unbalanced force of the cage itself and the interaction with the steel ball gradually increase, but the increasing amplitude of the cage unbalanced force is large and plays a major role, prompting the cage to push towards the guide surface of the guiding ring and produce a stable vortex. When the rotating speed exceeds 3000 r/min, the interaction between the cage pocket and the steel ball increases, gradually changes the motion state of the cage, presents a trend of gradual confusion, and the motion stability becomes worse. Compared with the simulation results of Wen [34], the stability trend of the bearing cage in this paper is in good agreement with the simulation results of the angular contact ball bearing 7103AC, which verifies the correctness of the simulation results.

4.2. Stability Variation of Cage with Pure Axial Load

Apply the driving speed of 1000 r/min to the bearing and select the simulation analysis results of five different working conditions of 1000, 2000, 3000, 4000 and 5000 N to analyze the influence of axial load on the stability of cage. The trajectory of the cage centroid obtained under the action of the above five different axial forces is shown in Figure 11.

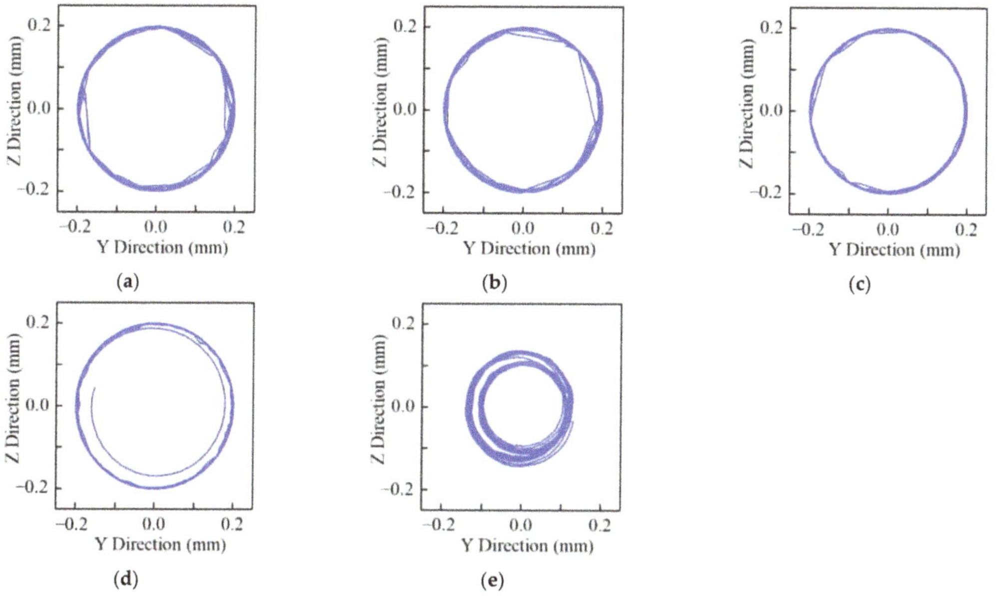

Figure 11. Cage centroid trajectory at different axial load. (**a**) Cage centroid trajectory at 1000 N; (**b**) Cage centroid trajectory at 2000 N; (**c**) Cage centroid trajectory at 3000 N; (**d**) Cage centroid trajectory at 4000 N; (**e**) Cage centroid trajectory at 5000 N.

It can be seen from Figure 11 that when the axial force is less than 3000 N, the trajectory of the cage centroid tends to be stable and regular with the increase in the axial force; after more than 3000 N, the trajectory of the centroid tends to be more complex with the increase in axial load.

The vortex radius ratio and vortex velocity deviation ratio are shown in Figure 12. It can be seen that when the axial force is less than 3000 N, the ratios gradually decrease with the increase in axial force, and when the axial force exceeds 3000 N, they enhance with the increase in axial force. The vortex radius ratio has a variation of 0.45, so the effect of axial force on cage stability is significant.

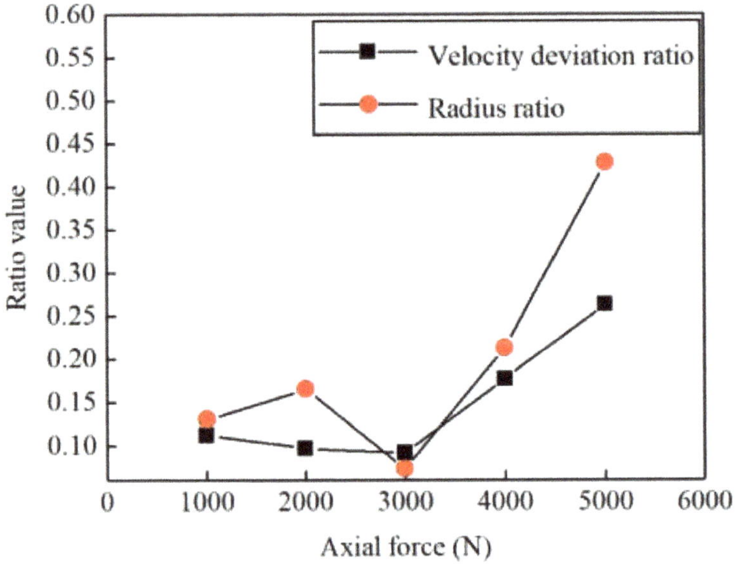

Figure 12. Effect of pure axial load on the velocity deviation ratio and radius ratio of the cage centroid trajectory.

It can be seen from Figures 11 and 12 that within a certain axial force range, the motion stability of the cage tends to be stable with the increase in the axial force. After exceeding the critical value, the stability of the cage gradually deteriorates. This is because within a certain range of axial force, with the increase in axial load, the load borne by each steel ball is gradually uniform. Therefore, the difference in the guiding ring drag force leads to the reduction in the fluctuation of rotating speed when the steel ball rotates to different azimuth angles, which reduces the collision between the steel ball and the cage pocket. The unbalanced force of the cage is dominant, which makes the cage show a stable vortex, which is conducive to the stability of the movement of the bearing cage. When the axial force is too large, the collision force fluctuation between the steel ball and the cage pocket hole increases, making the stability of the cage worse.

4.3. Stability Variation of Cage with Radial Load

FPCBB cannot bear pure radial force, the size of radial force it can bear should be less than 0.7 times of axial force. The simulation analysis results of 6 different radial force working conditions of 100, 200, 300, 400, 500 and 600 N are selected to analyze the influence of radial force on the stability of the cage under the combined load condition. The trajectory of the cage centroid obtained under the above six different working conditions is shown in Figure 13, from which it can be seen that the cage centroid trajectory is gradually complicated with the increase in radial force.

The vortex radius ratio and vortex velocity deviation ratio under the above six different load conditions are shown in Figure 14. It can be seen that the ratios have an upward trend and the trend tend to expand. The amplitude of the radius ratio exceeds 1.0 when the radial force is 600 N. It can be considered that the radial force has the greatest influence on the stability of the cage, and the influence enhances with the increase in the radial force.

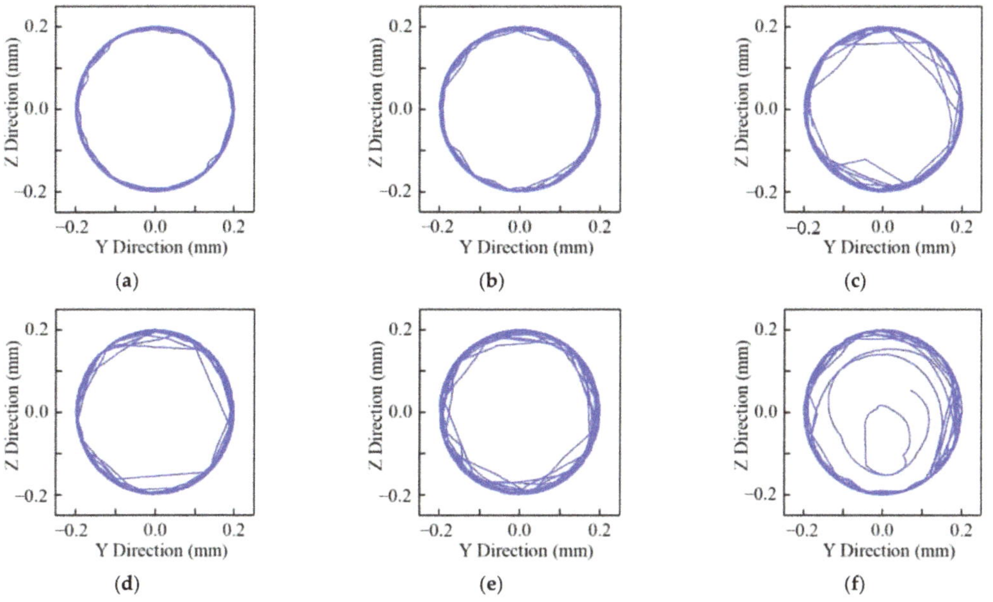

Figure 13. Cage centroid trajectory at different radial load under combined load. (**a**) 100 N; (**b**) 200 N; (**c**) 300 N; (**d**) 400 N; (**e**) 500 N; (**f**) 600 N.

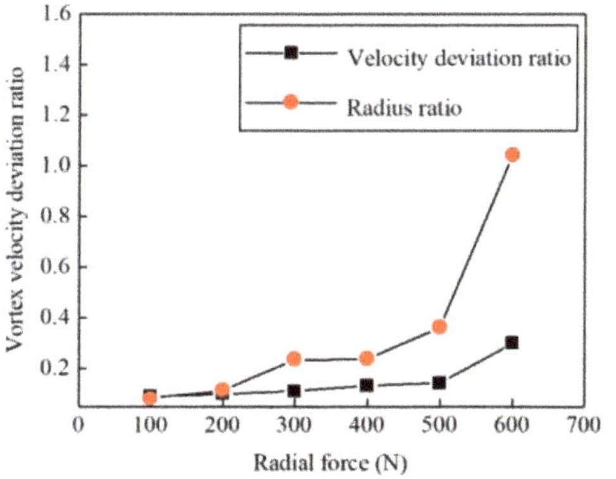

Figure 14. Effect of radial load on the speed deviation ratio and radius ratio of the cage centroid trajectory under combined load.

From Figures 13 and 14, it can be seen that the increase in radial force will prompt the cage centroid motion trajectory tends to be chaotic, and the stability becomes worse. This is because with the increase in radial force, the load of each steel ball is more and more uneven, which increases the frequency and amplitude of the collision between the steel ball and the cage pocket hole, thus changing the cage trajectory, making the trajectory irregular and reducing the stability of the cage.

4.4. Experimental Verification

In order to verify the accuracy of the analysis results of the bearing dynamic model established according to the parametric method in this paper, a rolling bearing cage performance analysis test bed is built. The test bed is composed of motorized spindle, coupling, support shaft, test shaft, loading device and test system. The main structure is shown in Figure 15. The test bearing is QJ215, its structure is shown in Figure 2. The structural parameters are consistent with the bearing used for simulation.

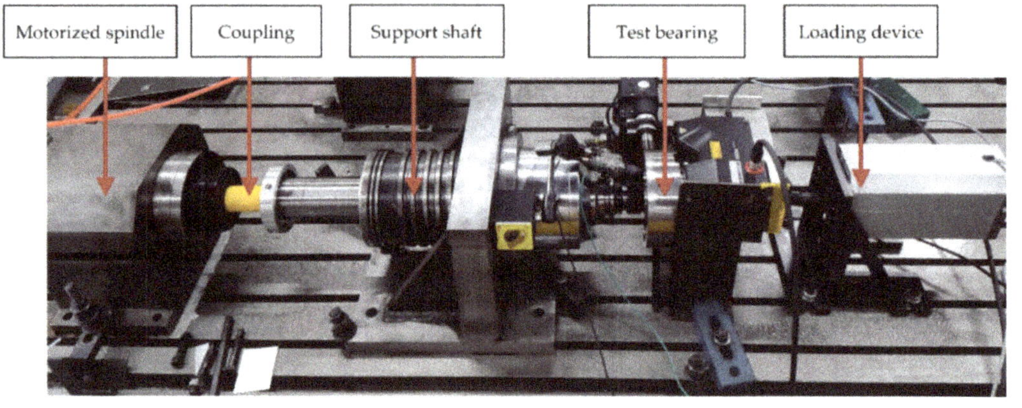

Figure 15. Test bed of rolling bearing cage performance analysis.

The test system includes a rotating speed measurement system and a motion displacement measurement system, as shown in Figure 16. The rotating speed measurement system includes laser speed sensor, reflector, signal acquisition system and computer, etc., and the motion displacement measurement system includes 2D laser profilers and processor.

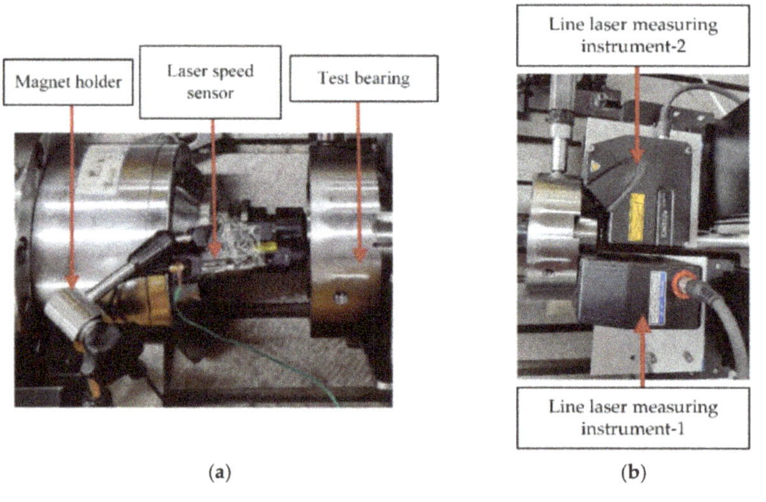

Figure 16. Cage speed and displacement measuring device. (**a**) Cage speed measuring device; (**b**) Cage displacement measuring device.

The principle of cage speed measurement is to put a reflector on the end surface of the cage, and align the laser to the reflector. When the reflector returns to the laser with one revolution of the cage, the laser sensor returns a pulse signal, which is converted into a rotational speed signal by the acquisition system, and finally, the curve of rotating speed of the cage is obtained. The laser sensor is Laser Tacho Probe-MM0360 from Denmark, and the signal acquisition device is PAK Mobile MK II data acquisition system from Millerbem, Germany.

The principle of cage trajectory measurement is shown in Figure 17. The vibration displacement information of the cage is monitored in real time by 2D laser profiler, and the positions of the cage profile edge points at different moments are subtracted to obtain the vibration displacement signal of the cage. Two sensors are used in the test: sensor 1 is placed in the direction parallel to the test bench, for measuring the displacement in the *X*-axis direction; sensor 2 is placed in the direction perpendicular to the test bench, to measure the displacement in the *Z*-axis direction. The two sensors are synchronous and the signal is transmitted to the monitor through a switch. The 2D laser profiler is the LJ-X8020 model from Keyence.

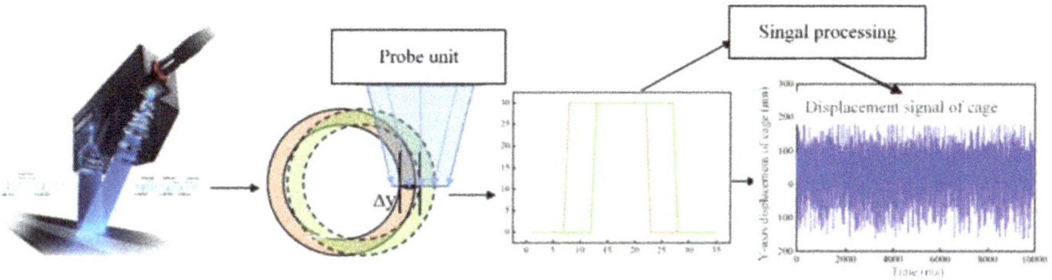

Figure 17. Measurement principle of cage trajectory.

The rotating speed of the cage was measured using the rolling bearing cage motion test bench. The test rotating speed of the electric spindle changes from 600 r/min to 2000 r/min under constant load. The radial force loading device kept 0 N, and the output values of the axial force loading device were from 1000 N to 2000 N. The comparison results between simulation and experiment are shown in Figure 18.

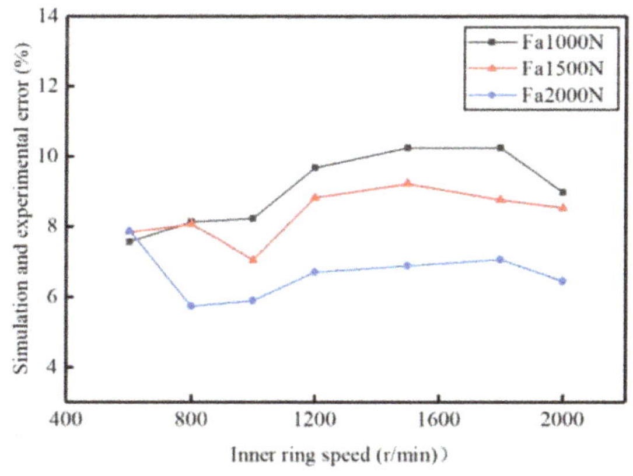

Figure 18. Simulation and experimental error curve.

The experimental speed is higher than the simulated speed, and the maximum error between them is 10.4%. It shows that the rotating speed has little influence on the error, and the error decreases with the increase in the axial load. The main reason for the error is that the test bearing adopts SKF's LGMT 3/1 grease, whose base oil viscosity is 125 mm^2/s. Its typical application is bearings with > 100mm inner diameter. It is larger than the viscosity recommended in the manual of QJ215. Thus, the drag force between the rolling body and the inner and outer rings is larger, which causes the cage speed to increase. Therefore, the simulation results can be considered to be consistent with the experimental results, and the accuracy of the simulation is verified.

The radial vibration signal of the cage is collected at the driving speeds of 1000, 1500 and 1800 r/min, as well as 1000 N pure axial load condition. The collected vibration displacement signals are filtered to eliminate the rotating frequency and external interference, and the centroid trajectory is obtained by synthesizing the two vibration displacement signals. The synthetic centroid trajectory of the cage is shown in Figure 19.

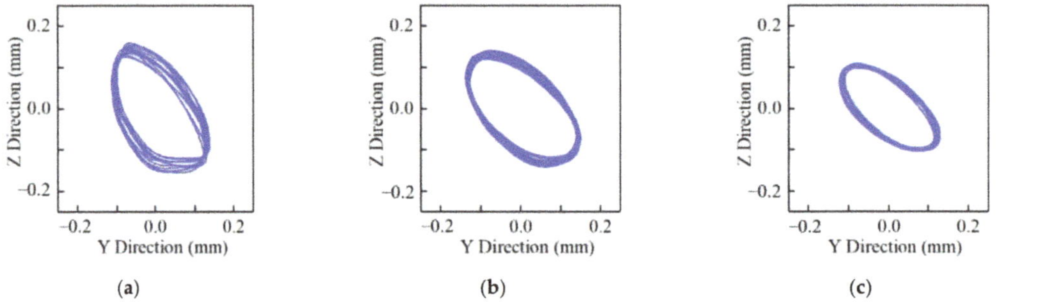

Figure 19. Cage centroid trajectory at different driving speeds. (**a**) 1000 r/min; (**b**) 1500 r/min; (**c**) 1800 r/min.

From Figure 19, it can be found that the centroid trajectory is elliptical, rather than the circle shown in the simulation result of Figure 9, this is caused by the difference in the stiffness of the bearing in all directions due to the actual bearing installation error or the deflection of the loading force. As shown in the figure, in the speed range of 1000–1800 r/min, the centroid trajectory of the cage shows vortex motion and tends to change steadily and regularly with increasing speed. It is consistent with the trend of simulation results of Figure 9.

Similarly, with regard to a change in the operating conditions, the centroid trajectory was analyzed at the speed of 1000 r/min and pure axial loads of 1000, 1500 and 1800 N, respectively, and the results are shown in Figure 15.

It can be seen from the figure that the centroid trajectory with axial force of 1500N and 1800N is better concentrated than that at 1000N, and the cage vortex situation is more stable, which is consistent with the trend of simulation results of Figure 11. From Figures 19 and 20, it can be seen that the change rules of the cage centroid trajectory measured by the experiment is consistent with the change in the simulation results, indicating that the change rules of the centroid trajectory summarized before may be correct, which also verifies the correctness of the simulation. There are still many shortcomings of the experiment. For example, only pure axial force loading is realized. Due to the limitation of conditions, there is no way to stably load radial force and axial force at the same time. High speed makes the test bench vibrate seriously, so for safety and accuracy, only the operation at low speed is simulated.

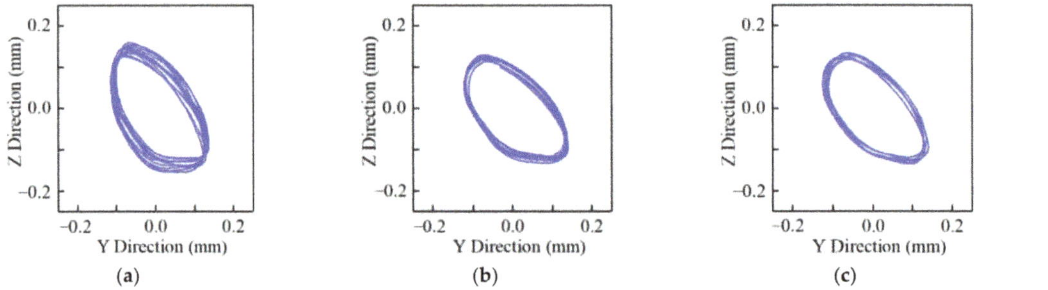

Figure 20. Cage centroid trajectory at different axial forces. (**a**) 1000 N; (**b**) 1500 N; (**c**) 1800 N.

5. Conclusions

In this paper, the dynamic model of FPCBB is established according to the multi-body dynamics simulation software ADAMS and its subroutine GFOSUB, which makes full use of the advantages of ADAMS in solving dynamic equations with high stability. The correctness of the dynamic model is verified by comparing the theoretical calculation results with the static simulation results of the cage speed and bearing load distribution. The effects of rotating speed and load on the stability of the cage are evaluated with the indexes of vortex radius ratio and vortex speed deviation ratio of the cage centroid trajectory. A test bench is established, and measurement methods of the cage speed and centroid vortex trajectory are proposed. The motion stability of cage reflects the stability of bearings; therefore, it is of practical significance to master the operation law of cage and guide the application of four contact point ball bearing. Some conclusions have been given as follows:

The dynamic bearing load curve is similar to the static load curve in overall trend, but the dynamic load curve is less stable than the static load curve, due to the existence of collision force between components.

The cage stability is affected by rotating speed and load conditions. The vortex radius ratio and the vortex velocity decrease and then increase as the inner ring getting faster. When the axial load is small, these two indicators are relatively stable, but after the axial force exceeds 3000 N, the cage becomes more and more unstable. Under both axial load and radial load, the increased radial force reduces cage stability.

The experimental results verify the rationality of the simulation results to a certain extent. It also verifies the feasibility of the speed and centroid vortex trajectory measurement method proposed in this paper. For different four-point contact ball bearings, further simulation and experiments are needed to verify the conclusions of this paper.

Author Contributions: Conceptualization, Y.Z. and J.H.; methodology, Y.L., S.M. and Y.Z.; software, S.M. and G.H.; validation, W.L.; formal analysis, G.H.; funding acquisition, J.H.; data curation, W.L., Y.L. and G.H.; writing—original draft preparation, Y.L. and Y.Z.; writing—review and editing, Y.L. All authors have read and agreed to the published version of the manuscript.

Funding: This research received was funded by the National Science Foundation of China (52175250).

Institutional Review Board Statement: Not applicable.

Informed Consent Statement: Not applicable.

Data Availability Statement: Not applicable.

Acknowledgments: The authors would like to thank the National Science Foundation of China (52175250) for the financial support.

Conflicts of Interest: The authors declare no conflict of interest.

References

1. Tang, W.C.; Chen, G.D.; Sun, Y.C.; Xu, L. Dynamic research and fault simulation on high-speed railway axle box bearings based on ANSYS/LS-DYNA. *Mod. Mach.* **2015**, *5*, 5–10.
2. Lundberg, G.; Palmgren, A. *Dynamic Capacity of Rolling Bearing*; Acta Polytechnica Mechanical Engineering Series; Royal Swedish Academy of Engineering Sciences: Stockholm, Sweden, 1947.
3. Zhang, L.Y.; Wang, Y.S.; Yuan, Q.Q.; Ji, Y.G. Influence of load and geometric parameters on load distribution of single-row four-point contact ball slewing bearings. *J. Mech. Transm.* **2013**, *37*, 27–29.
4. Li, Y.F.; Wu, Z.Y.; Lu, B.H.; Zhao, G.Y.; Sun, L.M. Influence of clearance on load distribution of single row four-point contact ball slewing bearing. *J. Mech. Transm.* **2010**, *34*, 56–58.
5. Li, K.Y.; Tang, W.C. Load-displacement relationship model and measurement of deep groove ball bearing and 4-point contact ball bearing. *J. Mech. Sci. Technol.* **2021**, *35*, 3045–3058. [CrossRef]
6. Wu, L. *Multi-Body Dynamics Analysis of Rigid-Flexible Coupling of Thin-Walled Four-Point Contact Ball Bearings*; Kunming University of Science and Technology: Kunming, China, 2017.
7. Shi, P.F. *Performance Analysis of Thin-Wall Four-Point Contact Ball Turntable Bearings*; Henan University of Science and Technology: Henan, China, 2013.

8. Yang, B.; Yao, T.Q. Finite element analysis of contact characteristics of thin-walled four-point contact ball bearing. *Softw. Guide* **2019**, *19*, 70–73.
9. Liu, Z.M.; Yao, T.Q.; Zhang, Y.W.; Cheng, X. Influence of the structural parameters of thin-wall four-point contact ball bearings on the mechanical. *J. Mach. Des.* **2021**, *38*, 7–13.
10. Walters, C.T. The Dynamics of Ball Bearings. *J. Lubr. Technol.* **1971**, *93*, 1–10. [CrossRef]
11. Gupta, P.K. *Advanced Dynamics of Rolling Elements*; Springer: New York, NY, USA, 1984.
12. Gupta, P. Modeling of Instabilities Induced by Cage Clearances in Cylindrical Roller Bearings. *ASLE Trans.* **1991**, *34*, 1–8. [CrossRef]
13. Gupta, P. Some Dynamic Effects in High-Speed Solid-Lubricated Ball Bearings. *ASLE Trans.* **1983**, *23*, 393–400. [CrossRef]
14. Hagiu, G.D.; Gafitanu, M.D. Dynamic characteristics of high speed angular contact ball bearings. *Wear* **1997**, *211*, 22–29. [CrossRef]
15. Wijnant, Y.H.; Vensing, J.A.; Nijen, G.C. The influence of lubrication on the dynamic behavior of ball bearings. *J. Sound Vib.* **1999**, *222*, 579–596. [CrossRef]
16. Weinzapel, N.; Sadeghi, F. A discrete element approach for modeling cage flexibility in ball bearing dynamics simulations. *J. Tribol.* **2009**, *131*, 110–119.
17. Daidié, A.; Chaib, Z.; Ghosn, A. 3D Simplified Finite Elements Analysis of Load and Contact Angle in a Slewing Ball Bearing. *Mech. Des.* **2008**, *130*, 082601. [CrossRef]
18. Iker, H.; Josu, A.; Mikel, A.; Ibai, C.; Iñigo, E. Load distribution and friction torque in four-point contact slewing bearings considering manufacturing errors and ring flexibility. *Mech. Mach. Theory* **2019**, *137*, 23–36.
19. Josu, A.; Mikel, A.; Rafael, A.; Igor, F.B. General static load-carrying capacity for the design and selection of four contact point slewing bearings: Finite element calculations and theoretical model validation. *Finite Elem. Anal. Des.* **2012**, *55*, 23–30.
20. Ammarullah, M.I.; Afif, I.Y.; Maula, M.I.; Winarni, T.I.; Tauviqirrahman, M.; Akbar, I.; Basri, H.; Heide, E.; Jamari, J. Tresca Stress Simulation of Metal-on-Metal Total Hip Arthroplasty during Normal Walking Activity. *Materials* **2021**, *14*, 7554. [CrossRef]
21. Ji, B.W.; Jing, M.Q.; Liu, Q. Dynamic Analysis and Simulation of Ball Bearing Cage Based on ADAMS. *Mach. Build. Autom.* **2014**, *43*, 113–116.
22. Hong, J.C.; Zhang, T.Z.; Cui, B.R.; Fu, C.X. *Parametric Modeling and Dynamic Simulation of Rolling Bearing Based on ADAM*; Qingdao University: Qingdao, China, 2014.
23. Chan, W.B.; Yao, T.Q. Multi-body Dynamics Analysis of Angular Contact Ball Bearings. *New Technol. New Process* **2014**, *2*, 46–49.
24. Hou, Y.; Xiong, X.Y.; Wang, X.; Niu, X.M. Dynamic co-simulation based on ANSYS and ADAMS and its application. *Min. Process. Equip.* **2014**, *42*, 111–115.
25. Zhang, Y.W.; Yao, T.Q.; Liu, Z.M.; Cheng, X. Stability Study of Thin-Walled Four-Point Contact Ball Bearing Cage. *Light Ind. Mach.* **2021**, *39*, 1–6.
26. Zhao, J.; Yao, T.Q.; Chen, R.B. Dynamics Simulation Analysis of Single Row Four Point Contact Ball Slewing Bearing. *J. Mech. Transm.* **2020**, *44*, 130–136.
27. Li, C. *ADAMS/View Parametric Design Technology and the Application in Mechanical Engineering*, 1st ed.; Science Press: Beijing, China, 2018; pp. 111–116.
28. Ma, S.J.; Zhang, X.; Yan, K.; Zhu, Y.S.; Hong, J. A Study on Bearing Dynamic Features under the Condition of Multiball–Cage Collision. *Lubricants* **2022**, *10*, 9. [CrossRef]
29. Deng, S.E.; Jia, Q.Y. *Roller Bearing Design Principle*, 2nd ed.; Standards Press of China: Beijing, China, 2008; pp. 85–87.
30. Ye, Z.H. *Research on Dynamic Behavior of High Speed Rolling Bearing in Aeroengines*; Harbin Institute of Technology: Harbin, China, 2013.
31. Yao, T.Q.; Wang, L.H.; Liu, X.B.; Huang, Y.Y. Multibody dynamics simulation of thin-walled four-point contact ball bearing with interactions of balls, ring raceways and crown-type cage. *Multibody Syst. Dyn.* **2020**, *48*, 337–372. [CrossRef]
32. Lu, Q. *Dynamic Analysis and Structural Optimization of the Angular Contact Ball Bearing on ADAMS*; Guangdong University of Technology: Guangdong, China, 2016.
33. Liu, X.H. *High-Speed Rolling Bearing Dynamics Analysis Model and Cage Dynamic Performance Research*; Dalian University of Technology: Dalian, China, 2011.
34. Wen, B.G. *Dynamic Characteristics of Angular Contact Ball Bearing Cage and Experimental Study*; Dalian University of Technology: Dalian, China, 2017.

lubricants

Article

Thermal Analysis Based on Dynamic Performance of Rocker Arm Full-Type Needle Bearings

Guangtao Zhang, Bing Su *, Fengbo Liu, Wenhu Zhang and Haisheng Yang

School of Mechatronics Engineering, Henan University of Science and Technology, Luoyang 471003, China; hkd_zhanggt@163.com (G.Z.); hnly_lfb@163.com (F.L.); 9905721@haust.edu.cn (W.Z.); yanghaisheng1975@163.com (H.Y.)
* Correspondence: subing@haust.edu.cn; Tel.: +86-0379-64231479

Abstract: Based on a dynamic analysis of rolling bearings, the equations for rocker arm full-type needle bearings were established by considering the traction coefficients of FVA-M reference lubricating oil, and then they were solved by the GSTIFF (Gear Stiff) integer algorithm with variable steps. The influence of working conditions on friction power consumption and the lubricant's convective coefficients were investigated. Then, on the basis of the heat generation and heat transfer mechanisms, the frictional power consumption was used as the boundary condition of the bearings' simulation model. Finally, temperature fields were calculated by the finite element method. The results showed that the overall value of frictional power consumption increased gradually with the increase in either the radial load or the rotation speed. The presence or absence of lubricating oil film in the contact area affected the heat conduction of the bearing, resulting in a temperature difference. Compared with the temperature of the radial load exerted on the bearing, the maximum temperature was more sensitive to the variations in the rotation speed. When running under the conditions of a fatigue life test, the steady-state temperature value of the bearing gradually decreased from the outer raceway to the needle roller and the outer ring surface, and then to the central shaft. The maximum temperature rise was 25.9 °C relative to the ambient temperature.

Keywords: full-type needle bearing; dynamics performance; temperature fields; lubricating oil

Citation: Zhang, G.; Su, B.; Liu, F.; Zhang, W.; Yang, H. Thermal Analysis Based on Dynamic Performance of Rocker Arm Full-Type Needle Bearings. *Lubricants* **2022**, *10*, 104. https://doi.org/10.3390/lubricants10050104

Received: 4 April 2022
Accepted: 12 May 2022
Published: 23 May 2022

1. Introduction

A rocker arm full-type needle bearing is used between a cam and a rocker to control the reciprocating movement of a valve lever in completing the intake and exhaust functions [1]. Due to the complex structure of the valve train and the flexible and changeable positions of the components, it is necessary to analyze the dynamics and temperature field of the valve train under different loads and speeds [2,3]. Dynamic performance analysis can determine the stress and friction torque of the main contact parts of the bearing, thereby determining the heat source value that can affect the bearing's performance. The changes in bearing temperature caused by frictional heat will cause the thermal deformation of components, greatly affecting the internal structural dimensions of the bearing [4]. A reduction in the working clearance can lead to damage on the surface of the rolling bearing and, in severe cases, shorten the life of the bearing. Those phenomena will affect the running accuracy of the bearing and cause varying degrees of harm to the engine.

Thermal analysis of rolling bearings is the study of the heat generated by friction between bearing components and lubrication. The friction properties of the lubricants affect the tribological behavior of the rolling bearings [5,6]. Palmgren [7] carried out friction tests on bearings of various types and sizes, and regressed the test data to obtain an empirical formula for calculating bearing friction torque. Ma et al. [8] classified and discussed the existing experimental research regarding heat generation and heat transfer of bearing components, and briefly described the calculation method of combining theoretical models to predict the thermal problems of high-speed bearings. Hatazawa et al. [9] established a mathematical model of the friction torque of thrust needle roller bearings and studied the influence of working conditions and structural parameters on the dynamic performance of thrust needle roller bearings, but that study did not mention the thermal analysis of the bearings. Li et al. [10] used ADAMS software to analyze the dynamics of cylindrical roller bearings and determined the friction and positive pressure between the rolling elements and the inner raceway at different positions. Then, they analyzed the contact stress between rollers and the inner raceway with ANSYS Workbench. Based on the dynamic method, the friction heat inside the bearing was obtained. On the basis of this information, the temperature field of the bearing was analyzed. Wang et al. [11,12] established an analysis model based on the quasi-static method to calculate the heat source of high-speed cylindrical roller bearings and studied the effects of speed, load, lubricating oil, and working temperature on heat generation. Subramaniam et al. [13] used the Finite Element Method to analyze heat transfer in a ceramic conventional ball bearing and study the heat dissipation, temperature profile, and thermal stresses occurring in a bearing as a function of rotational speed. Liming et al. [14] used the finite element method to calculate the temperature field of the sliding block under different loads and speeds, and verified the feasibility of the finite element method in analyzing the temperature field.

For the finite element method to solve the bearing temperature field problem, the accurate determination of the boundary conditions determines the accuracy of the bearing thermal analysis. In the process of determining the boundary conditions, most of the current literature assumes that the bearing's rolling elements have the same heat transfer form during the rotation of the bearing. The above review of the relevant literature indicated that the internal temperature of the bearing was generally evenly distributed, and the calculation of the bearing was based on the heat in the stable operation stage. The research results were obtained by using the finite element method to deal with such boundary conditions and were verified by experiments; the results were comparable to previous experimental data, such as the data provided in [15,16].

Many previous studies focused on the calculation of the friction torque based on the lubrication and the load effects to determine the heat source of the needle roller bearing. However, the variations in the elasto-hydrodynamic traction coefficient affect the interaction force of the bearing's elements. Based on dynamic analysis, the programming language FORTRAN was used in this paper to compile custom functions to realize the analysis of the force balance of various components in the bearing. The heat of bearings lubricated by FVA-M lubricating oil was calculated. Considering the heat transfer coefficient of the lubricating oil film in the contact area, steady-state thermal analysis was used to determine the internal equilibrium temperature field of the bearing.

2. Dynamic Model of a Rocker Needle Roller Bearing

The research object was a certain type of rocker arm full-type needle bearing, which is used for the rocker arm valve mechanism and mainly includes an outer ring in rolling contact with the engine camshaft. A central shaft is fixed on the rocker arm, and multiple rolling elements are arranged between the outer ring and the central shaft. Together, these components form the rocker bearing.

When the bearing is working, the central shaft is fixed. Therefore, it is assumed that the central shaft is connected with the ground in the dynamic analysis, and only the outer ring rotates together with the action of the cam. The axis of rotation is along the X direction. The center of mass of each component coincides with the center of gravity, and the surface of the component is an ideal surface. In order to establish a dynamic model of the bearing, the following three coordinate systems, shown in in Figure 1, were defined: $\{O; X, Y, Z\}$ is a fixed coordinate system; $\{o_o; x_o, y_o, z_o\}$ is a coordinate system of the outer ring; and $\{o_{rj}; x_{rj}, y_{rj}, z_{rj}\}$ is the center of the mass coordinates of the needles.

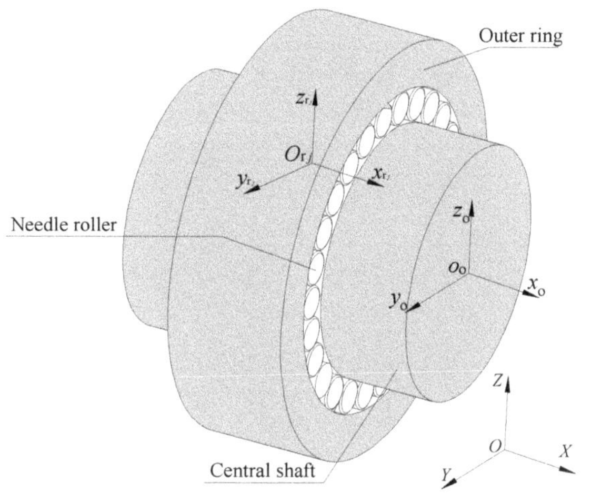

Figure 1. Coordinate systems.

Based on the following assumptions, Newton's law of motion and classical Euler dynamic equations can be used to derive the force and motion of the components of the rocker needle roller bearing, as follows:

1. The contact deformation between the needle roller and the raceway conforms to Hertzian elliptical contact theory, and the contact state is full line contact.
2. The influences of the frictional moment caused by the skew and tilt of the needle in motion are not considered.
3. The needle rollers are evenly distributed in the initial state of simulation, with a certain circumferential clearance.

2.1. Motion Analysis of Needle Rollers

The movement relationship among the components is shown in Figure 2. The central shaft is fixed. Suppose the rotation angular velocity of the outer ring is Ω_o, and the needle rollers' revolution angular velocity is Ω_m, rotating at a uniform speed. If the needle rollers are fixed, the movement of the bearing assembly can be equivalent to the angular velocity of the outer ring $(\Omega_o - \Omega_m)$, and the angular velocity of the central shaft relative to the needle roller is Ω_m. Therefore, the velocities of the contact area can be obtained, which are $\vec{V}_{rj}$, $\vec{V}_i$, $\vec{V}_o$, respectively.

$$\vec{V}_{rj} = \frac{1}{2} D_w \vec{\omega}_{rj} \tag{1}$$

$$\vec{V}_i = \frac{1}{2}(D_{wp} - D_w)\vec{\Omega}_m \tag{2}$$

$$\vec{V}_o = \frac{1}{2}(D_{wp} + D_w)\left(\vec{\Omega}_o - \vec{\Omega}_m\right) \tag{3}$$

The relative sliding speeds of the contact points between the central shaft, the outer ring raceway, and the needle rollers and adjacent rollers are shown as Equations (4)–(6), respectively.

$$\vec{V}_{ij} = \vec{V}_i - \vec{V}_{rj} \tag{4}$$

$$\vec{V}_{oj} = \vec{V}_o - \vec{V}_{rj} \tag{5}$$

$$\Delta\vec{V}_{rj} = \vec{V}_{rj} - \vec{V}_{r(j+1)} \tag{6}$$

where the subscript 'i' represents the central shaft; 'o' represents the outer ring; 'r' represents the needle roller; '*j*' represents the needle number; ω is the angular velocity of the needle roller; D_{wp} is the center diameter of the needle roller group; and D_{w} is the needle diameter.

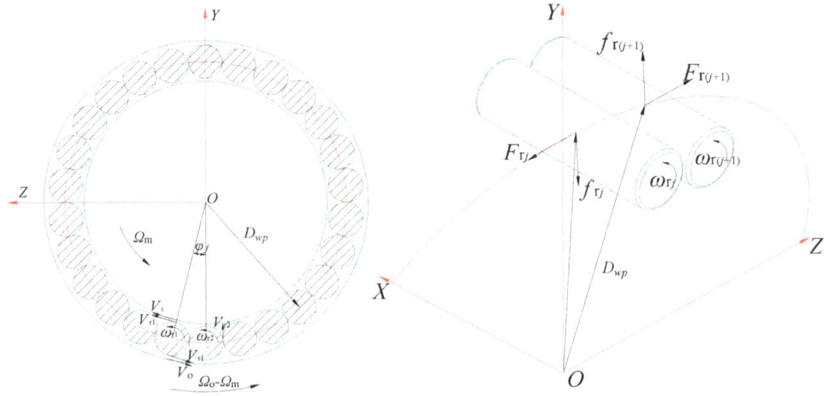

(**a**) (**b**)

Figure 2. Motion and load status of full-type needle bearing: (**a**)Motion status; (**b**) Load status.

2.2. Force Analysis of Needle Roller

As shown in Figure 3, N_j^{i} and N_j^{o} are the normal contact forces between the *j*th needle and the raceway. T_j^{i} and T_j^{o} are the traction forces between the *j*th needle and the raceway. $M_{\mathrm{N}j}^{\mathrm{i}}$ and $M_{\mathrm{N}j}^{\mathrm{o}}$ are the moments caused by the normal contact force. $M_{\mathrm{T}j}^{\mathrm{i}}$ and $M_{\mathrm{T}j}^{\mathrm{o}}$ are the moments caused by traction forces between the *j*th needle and the raceway. $F_{\mathrm{r}j}$ is the centrifugal force of the *j*th needle roller; q_{jm}^{i}, q_{jm}^{o} are the normal contact forces between the *m*th needle roller and the raceway. T_{jm}^{i}, T_{jm}^{o} are the traction forces between the *m*th needle roller and the raceway. *n* is the number of slices of the needle roller. The calculation of the forces and moments can then be carried out [17].

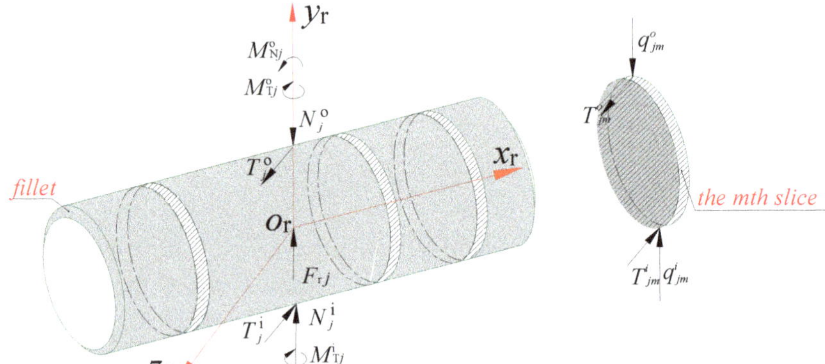

Figure 3. Schematic diagram of needle forces.

On the basis of the dynamic relationship, the nonlinear equation of the *j*th needle roller can be written as follows:

$$\left\{\begin{array}{c} m_r\ddot{y}_{rj} = N_j^o\cos\varphi_j - N_j^i\cos\varphi_j - F_{rj}\cos\varphi_j \\ +T_j^i\sin\varphi_j - T_j^o\sin\varphi_j \\ m_r\ddot{z}_{rj} = -N_j^o\sin\varphi_j + N_j^i\sin\varphi_j + F_{rj}\sin\varphi_j \\ +T_j^i\cos\varphi_j - T_j^o\cos\varphi_j \\ J_{rx}\dot{\omega}_{rx} = T_j^i\frac{D_w}{2} + T_j^o\frac{D_w}{2} \\ J_{ry}\dot{\omega}_{ry} = M_{Tj}^i\cos\varphi_j + M_{Tj}^o\cos\varphi_j - M_{Nj}^i\sin\varphi_j \\ -M_{Nj}^o\sin\varphi_j \\ J_{rz}\dot{\omega}_{rz} = -M_{Tj}^i\sin\varphi_j - M_{Tj}^o\sin\varphi_j - M_{Nj}^i\cos\varphi_j \\ -M_{Nj}^o\cos\varphi_j \end{array}\right. \quad (7)$$

where m_r is the mass of the needle; $\ddot{y}_{rj}$ and $\ddot{z}_{rj}$ are the displacement accelerations of mass of the *j*th needle mass center in {O; X, Y, Z} ; J_{rx}, J_{ry}, and J_{rz} are the moments of inertia under the needle in {O; X, Y, Z}; and $\dot{\omega}_{rx}$, $\dot{\omega}_{ry}$, and $\dot{\omega}_{rz}$ are the angular accelerations of the needle in {O; X, Y, Z}.

The interaction model between two adjacent rollers is shown in Figure 2. According to the theory of contact and collision between rigid bodies, the expression of the mathematical model of the force of adjacent rollers is obtained as follows:

$$F_{rj(j+1)} = \left\{\begin{array}{cc} 0 & (q \le q_1) \\ K_n(q_1 - q) - c_{max}\dot{q}step(q, q_1 - d, 1, q_1, 0) & (q > q_1) \end{array}\right\} \quad (8)$$

where q is the distance variable; q_1 is the relative speed; $\dot{q}$ is the derivative of the relative speed; e is the exponent of the force; K_n is the stiffness coefficient; c_{max} is the maximum damping coefficient; d is the cut-in amount when the damping reaches the maximum value; and *step* (*) is the step function.

Therefore, the friction force f between adjacent needle rollers can be obtained as follows:

$$f_{rj} = \mu F_{rj} \quad (9)$$

$$f_{r(j+1)} = -\mu F_{r(j+1)} \quad (10)$$

The traction coefficient μ, relating to the sliding speed between the needle rollers, can be determined according to the given traction model.

2.3. Nonlinear Equation of the Outer Ring

Similarly, the dynamic equation of the outer ring can be obtained via the following equations from the relationship between motion and force:

$$\left\{\begin{array}{c} m_o\ddot{y}_o = \sum_{j=1}^{RN}\left(-N_j^o\cos\varphi_j + T_j^o\sin\varphi_j\right) - F_r \\ m_o\ddot{z}_o = \sum_{j=1}^{RN}\left(N_j^o\sin\varphi_j + T_j^o\cos\varphi_j\right) \\ J_{ox}\dot{\omega}_{ox} = \sum_{j=1}^{RN} T_j^o\frac{D_w}{2} \\ J_{oy}\dot{\omega}_{oy} = \sum_{j=1}^{RN}\left(M_{Tj}^o\cos\varphi_j - M_{Nj}^o\sin\varphi_j\right) \\ J_{oz}\dot{\omega}_{oz} = \sum_{j=1}^{RN}\left(-M_{Tj}^o\sin\varphi_j - M_{Nj}^o\cos\varphi_j\right) \end{array}\right. \quad (11)$$

where m_o is the mass of the outer ring; $\ddot{y}_o$ and $\ddot{z}_o$ are the displacement accelerations of the center of mass of the outer ring in inertial coordinate system {O;X,Y,Z} ; J_{ox}, J_{oy}, and J_{oz} are the moments of the outer ring in inertial coordinate system {O;X,Y,Z}, $\dot{\Omega}_{ox}$, $\dot{\Omega}_{oy}$ and $\dot{\Omega}_{oz}$ are the angular accelerations of the outer ring in inertial coordinate system {O;X,Y,Z}; F_r is the radial load; and RN is the number of needle rollers.

In the above equations, the traction coefficient μ_{jm} of the oil is a variable value, which is obtained through experiments. The calculation formula of the oil traction coefficient is fitted from the test data according to the Gupta's four-parameter model. The FVA-M lubricant was provided by German Schaeffler Company; its traction coefficient was tested using the lubricant traction characteristic test ring, and can be expressed in the following equation [18]:

$$\mu_{jm} = (\mathrm{A} + \mathrm{BS})\mathrm{e}^{-\mathrm{C}S_{jm}} + \mathrm{D} \tag{12}$$

$$\begin{cases} \mathrm{A} = -0.5164\overline{W}^{0.7102}\overline{U}^{0.5353}\overline{T}^{-0.4159} \\ \mathrm{B} = 4.6747 \times 10^{4}\overline{W}^{-0.0725}\overline{U}^{0.2831}\overline{T}^{0.079635} \\ \mathrm{C} = 3.4129 \times 10^{-6}\overline{W}^{-0.021}\overline{U}^{0.274}\overline{T}^{-0.4796} \\ \mathrm{D} = 0.5164\overline{W}^{0.7102}\overline{U}^{0.5353}\overline{T}^{-0.4159} \end{cases} \tag{13}$$

where $\overline{W}$ is the dimensionless load parameter; $\overline{U}$ is the dimensionless speed parameter; $\overline{T}$ is the dimensionless temperature parameter; and S_{jm} is the sliding-rolling ratio (the ratio of the sliding velocity to the rolling velocity) at the contact point between the mth slice of the jth needle and the inner or outer raceways.

3. Heat Generation and Heat Transfer Calculation Model for Rocker Needle Roller Bearings

3.1. Calculation Model of Heat Generation for Rocker Needle Roller Bearings

Due to a certain friction phenomenon among the rolling elements, the raceways, and the lubricating film, the bearing friction is not constant. According to Palmgren, the total frictional power consumption of a bearing should be determined by the frictional power between the rolling element and the ring, caused by the load and the lubricant's viscous frictional power consumption when the bearing is at a medium speed and subjected to a medium external load. Harris [19] modified the overall calculation of bearing friction power consumption through experiments based on Palmgren's work. He proposed a calculation method for local heat generation that takes into account the six major factors of rolling bearing friction. The dynamic parameters of the bearing can be analyzed by using a computer program, and then the local frictional heat calculation formula of the bearing element can be solved simultaneously.

Based on the local heat generation model proposed by Harris, the total friction power consumption of a rocker arm full-type needle bearing is comprised of the friction power consumption caused by the elastic hysteresis between the needle rollers and the raceways, the friction power consumption caused by the sliding friction between the needle rollers and the raceways, the friction power consumption caused by the viscous resistance between the oil and the bearing components, and the friction power consumption caused by the sliding between adjacent needle rollers.

3.1.1. Frictional Power Consumption Caused by the Elastic Hysteresis between the Needle Rollers and the Raceways

When the needle roller rolls on the central shaft and the outer ring raceway, the frictional power consumption due to the elastic hysteresis of the material can be calculated as follows:

$$M_{\mathrm{E}} = \sum_{j=1}^{RN}\sum_{m=1}^{n}\left(\xi\sqrt{\frac{\pi q_{jm}^{\mathrm{i(o)}}}{2\eta D_{\mathrm{w}}}}\delta_{jm}^{\mathrm{i(o)}}\right) \tag{14}$$

$$E_{\mathrm{R}} = M_{\mathrm{E}}\left|\omega_{\mathrm{i(o)}} - \omega_m\right| \tag{15}$$

where ξ is the elastic hysteresis coefficient of the material, which is 0.01 preferable for steel; η is the comprehensive elastic constant of the two contact bodies; and $\delta_{jm}^{\mathrm{i(o)}}$ is the elastic deformation between the mth slice of the jth needle roller and rings.

3.1.2. Friction Power Consumption Caused by Sliding Friction between the Needle Rollers and the Rings

When the needle roller rolls on the central shaft and the outer ring raceway, the frictional power consumption caused by relative sliding can be calculated as follows:

$$E_{\mathrm{D}} = \sum_{j=1}^{RN} \sum_{m=1}^{n} \left(\mu_{jm}^{\mathrm{i(o)}} q_{jm}^{\mathrm{i(o)}} \Delta v_{jm}^{\mathrm{i(o)}}\right) \tag{16}$$

where μ_{jm}^{i} and μ_{jm}^{o} are, respectively, the traction coefficients of the oil film between the mth slice of the jth needle and the inner and outer raceways, which are obtained by interpolation from the relationship between the slip-roll ratio and the elasto-hydrodynamic tration coefficient; and $\Delta v_{jm}^{\mathrm{i}}$ and $\Delta v_{jm}^{\mathrm{o}}$ are, respectively, the relative sliding speeds between the mth slice of the jth needle and the inner and outer raceways.

3.1.3. Friction Power Consumption Caused by Viscous Resistance

Due to the stirring effect of the lubricating oil during the rotation of the needle, the frictional power consumption caused by the lubricating oil film formed at the contact between the rolling elements and the inner and outer raceways is calculated as follows [17]:

$$E_{\mathrm{oil}} = \int_{0}^{2\pi} \frac{1}{8} C_{\mathrm{d}} \rho_m D_{\mathrm{w}} l (D_{\mathrm{wp}} \omega_{\mathrm{r}})^2 \mathrm{d}\varphi \tag{17}$$

where C_{d} is the flow resistance coefficient; ρ_m is the density of oil and gas mixture; l is the length of needle roller; and φ is the position angle of the needle roller.

3.1.4. Sliding Friction Power Consumption between Adjacent Needle Rollers

The frictional power consumption due to the contact and collision between adjacent rollers can be expressed as follows:

$$E_{\mathrm{r}} = \sum_{j=1}^{RN} \mu F_{\mathrm{r}j} \left|\Delta \vec{V_{\mathrm{r}}}\right| \tag{18}$$

The total frictional power consumption can be expressed as follows:

$$E = E_{\mathrm{R}} + E_{\mathrm{D}} + E_{\mathrm{oil}} + E_{\mathrm{r}} \tag{19}$$

3.2. Heat Transfer Calculation Model for Rocker Needle Roller Bearings

For a bearing system, heat convection mainly includes the convection between the lubricant and the surface of the inner or outer raceway and the convection between the lubricant and the surface of the needle rollers. When the lubricant is oil, the performance of the lubricating oil film has a greater impact on improving the working conditions and the service life of the bearing; the convection of the lubricating oil film in the contact area should also be considered. The convection coefficient can be approximately calculated by Equations (20)–(21).

3.2.1. Convective Coefficient between Lubricant and Bearing Element Surface

$$\alpha_1 = 0.332 \frac{k}{D_{\mathrm{x}}} P_{\mathrm{r}}^{\frac{1}{3}} \left(\frac{V_{\mathrm{x}}}{\upsilon_0}\right)^{\frac{1}{2}} \tag{20}$$

where k is the thermal conductivity of the lubricant; P_r is the Prandtl number of the lubricant; ν_0 is the kinematic viscosity of the lubricating oil; D_{x} is the radius of rotation, the outer ring taking the groove bottom diameter D_{o}, the central axis taking the groove bottom diameter D_i, and the roller needle taking D_{wp}; V_x is the velocity of bearing component, $V_x = \frac{\pi}{60} n_r D_x$; $n_r = \frac{1}{2}(n_i(1 - \frac{D_w}{D_{wp}}) + n_o(1 + \frac{D_w}{D_{wp}}))$; n_r is the revolution speed of the needle roller; and n_i and n_o are the speeds of the central shaft and the outer ring, respectively.

3.2.2. Thermal Conductivity of the Lubricating Oil Film in the Contact Area

$$\alpha_2 = k / H_{i(o)0} \tag{21}$$

where $H_{i(e)0}$ is the average center oil film thickness.

According to the analysis of Hertz, the contact state of the needle roller and the raceway under the load state is line contact. Higginson and Dowson provided a numerical solution to the linear contact elasto-hydrodynamic lubrication problem, and a method for calculating the thickness of the linear contact elasto-hydrodynamic oil film, as detailed in [20]. The formula for the dimensionless minimum oil film thickness is calculated as follows:

$$H_{min} = 2.65 \frac{G^{0.54} U^{0.7}}{W^{0.13}} \tag{22}$$

$$h_{min} = [2.65 \alpha^{0.54} (\eta_0 U)^{0.7} R^{0.43} L^{0.13}] / (E'^{0.03} W^{0.13}) \tag{23}$$

The relationship between the average center oil film thickness and the minimum oil film thickness is shown as follows:

$$H_0 = 4/3 \times h_{min} \tag{24}$$

where U and W are the introduced intermediate variables related to speed and load; $G = \alpha_{oil} E_0$; E_0 is the equivalent elastic modulus; and α_{oil} is the lubricating oil viscosity coefficient.

3.3. Friction Power Consumption Solve Process

The Fortran language was used to program the bearing element interaction force and bearing friction power consumption calculation custom subroutines, and to link the rocker arm full-type needle bearing to complete the simulation analysis modules. The solution procedure of the dynamics differential equations is shown in Figure 4. The initial conditions of dynamics differential equations—namely, the relative position and the motion vector of various components—are obtained based on the initial estimated values of the bearing component's position and motion constraint. Then, the dynamic differential equation of the rocker arm full-type needle bearing is established, and the Gear stiff (GSTIFF) integer algorithm with variable steps is used to solve it. Equations (6)–(13) can be solved by the GSTIFF variable-step integration algorithm to obtain the dynamic characteristics of the bearing. To verify whether the error meets the convergence error set, it was set to 1×10^{-3}. If "yes" (see Figure 4), the process continues to the next solution after obtaining the outputs of the motion parameters. If "no" (see Figure 4), the process is to choose a smaller step value and repeat the solving until the solving error meets the convergence error. Finally, the frictional power consumption of the rocker arm full-type needle bearing is calculated by Equations (14)–(19).

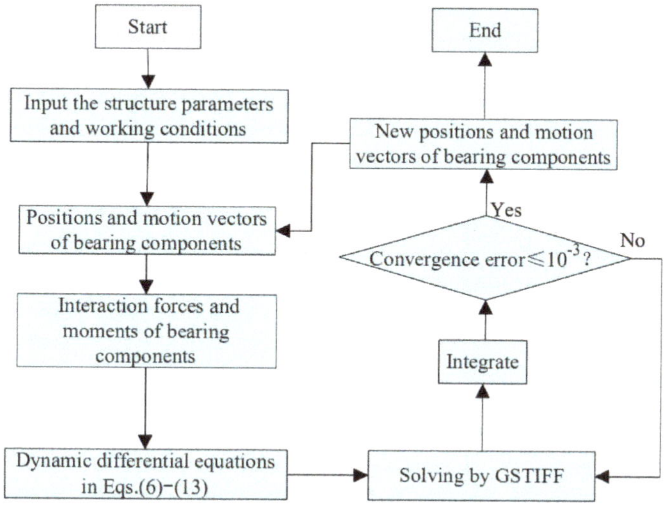

Figure 4. Solution procedure of dynamic of differential equation.

4. Simulation Analysis and Results

The bearing was selected as the simulation example, as is typical of a rock arm application, and its design parameters are shown in Table 1. The ambient temperature was 40 °C, FVA-M lubricating oil was used for lubricant, and the bearing material was GCr15.

Table 1. Basic parameters of the bearing.

Parameters	Value
Outer diameter/mm	24
Outer ring width/mm	6.9
Central shaft width/mm	17.3
Pitch diameter/mm	16.4
Needle diameter/mm	2
Needle length/mm	6.8
Number of rollers	26

4.1. The Influence of Working Conditions on Bearing Frictional Heat Value

In order to analyze the value of frictional heat, the outer ring initial speed was set as 7000 rpm, and the radial force was 2000 N~4500 N. The initial radial load was set as 3000 N, and the outer ring speed was 4000 rpm~9000 rpm. By solving the dynamics model, the minimum oil film thickness, the force of adjacent rollers, the half width of the contact area, and the contact force of the bottom roller can be calculated. The results obtained from the dynamic simulations are shown in Table 2. Furthermore, the partial and overall frictional heat can be extracted during the stable working of the full-type needle bearing. Figures 5 and 6 depict the variations, in average values with working conditions.

From the perspective of the primary and secondary heat sources, the order of internal heat generation is as follows: heat generated by needle rollers stirring oil, heat generated by sliding between needle rollers, and heat generated by elastic hysteresis or differential sliding. It has the same distribution law as other common heat generation models. The heat generated by the mixing of rolling elements and lubricating oil is the main part. The full-type needle bearing that only bears a radial load has load-bearing and non-loading areas, which affect the operation of the bearing and change the speed of the rolling element at this place of transition. The phenomenon generates contact collision force and friction between needle rollers, so the sliding heat generation between the rolling elements of the full-type needle bearing follows. The heat value caused by the elastic hysteresis and differential sliding in the dynamic process is very small.

Table 2. The results obtained from dynamic simulations.

Speed/r/min	Radial load/N	The Minimum Inner Oil Film Thickness/mm	The Minimum Outer Oil Film Thickness/mm	Contact Width/mm	The Contact Force/N	The Force of Adjacent Rollers/N
4000	3000	2.30×10^{-4}	2.70×10^{-4}	0.0297	479	0.64
5000		2.80×10^{-4}	3.40×10^{-4}	0.0297	479	0.67
6000		3.10×10^{-4}	3.50×10^{-4}	0.0297	479	1.17
7000		3.40×10^{-4}	3.90×10^{-4}	0.0297	479	1.42
8000		3.70×10^{-4}	4.20×10^{-4}	0.0297	479	1.72
9000		4.00×10^{-4}	4.60×10^{-4}	0.0297	479	2.54
7000	2000	3.70×10^{-4}	4.20×10^{-4}	0.0242	320	1.18
	2500	3.50×10^{-4}	4.00×10^{-4}	0.0271	400	1.47
	3000	3.40×10^{-4}	3.90×10^{-4}	0.0297	479	1.42
	3500	3.30×10^{-4}	3.80×10^{-4}	0.0322	638	1.38
	4000	3.30×10^{-4}	3.80×10^{-4}	0.0343	728	1.36
	4500	3.20×10^{-4}	3.70×10^{-4}	0.0365	818	1.52

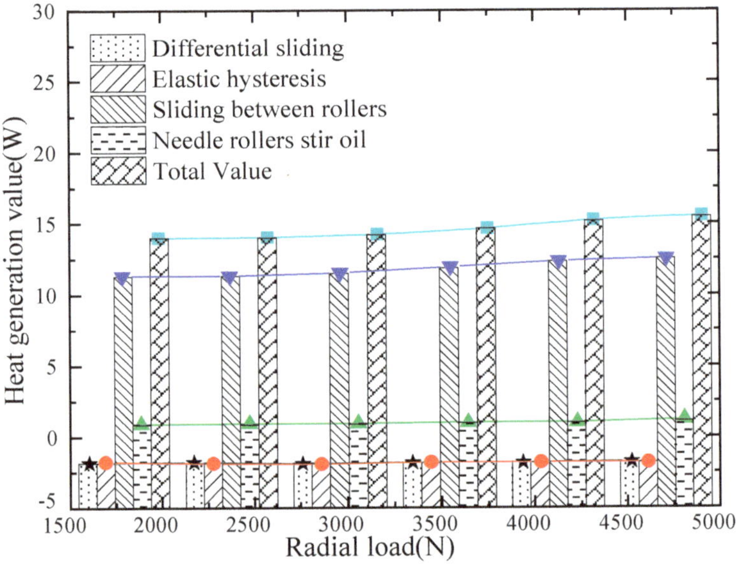

Figure 5. The influence of radial load on heat generation.

From the perspective of the change in heat generation with working conditions, the change in speed has a braking effect on the change in heat generation. Due to the gradual increase of the lubricating oil film thickness with the increase of the speed and the centrifugal force of the rolling elements, the value of the heat generated also gradually rises. However, the change in radial load barely affects the change in frictional heat generation. The frictional heat generation rises slightly with an increase in radial load.

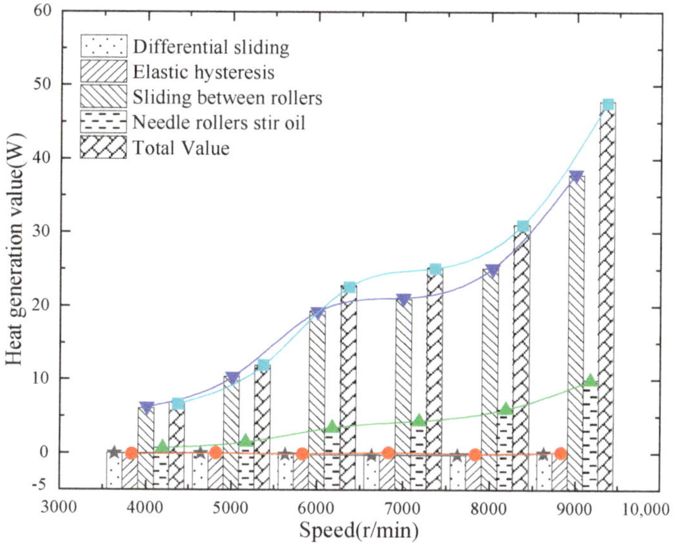

Figure 6. The influence of speed on heat generation.

4.2. Calculation of Convective Coefficient

4.2.1. Convective Coefficient between the Bearing's Components and the Heat Transfer Medium

In the working of open bearings, it is not only necessary to consider the convection of the lubricating oil on the surface of the bearing component, but also to consider the convection between air and the surface of the component. he convective coefficient between the bearing components and the heat transfer medium at different speeds may be calculated. The results are shown in Figures 7 and 8.

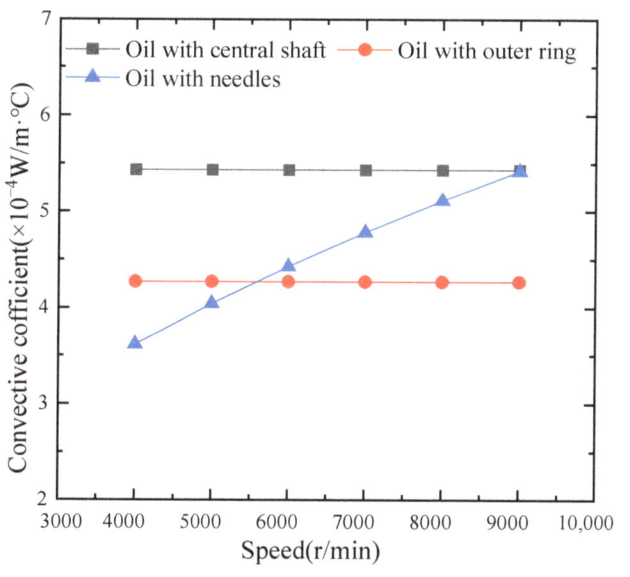

Figure 7. The influence of speed on the convective coefficient of oil.

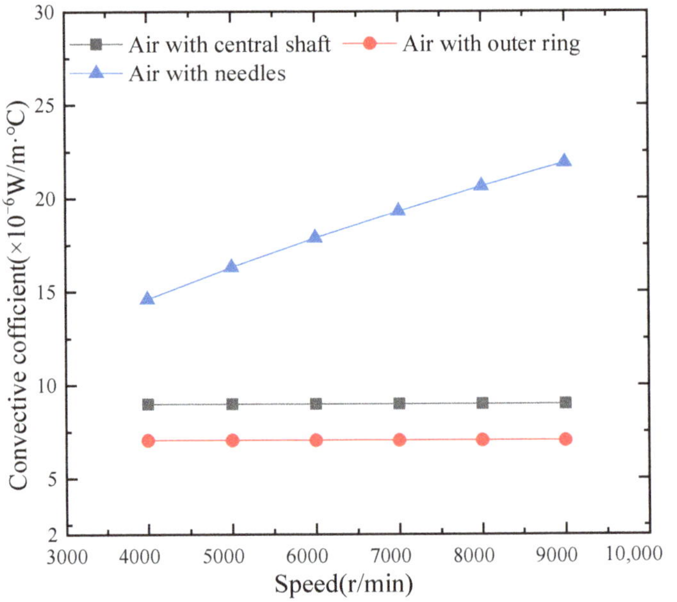

Figure 8. The influence of speed on the convective coefficient of air.

4.2.2. Thermal Conductivity Coefficient of Oil Film

During the rotation of oil-lubricated bearings, the two friction pairs between the contact areas are often separated by a layer of oil film to reduce friction and improve the bearing capacity of the bearing. Therefore, it is necessary to consider the thermal conductivity coefficient of the contact oil film between the needle roller and rings of the bearing. The changes of the thermal conductivity coefficient with the radial load and the outer ring speed are shown, respectively, in Figures 9 and 10.

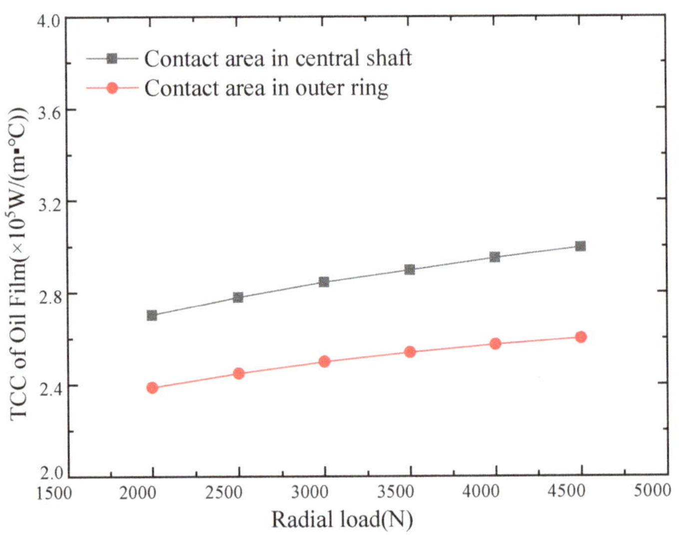

Figure 9. The influence of radial load on the heat transfer coefficient of oil film.

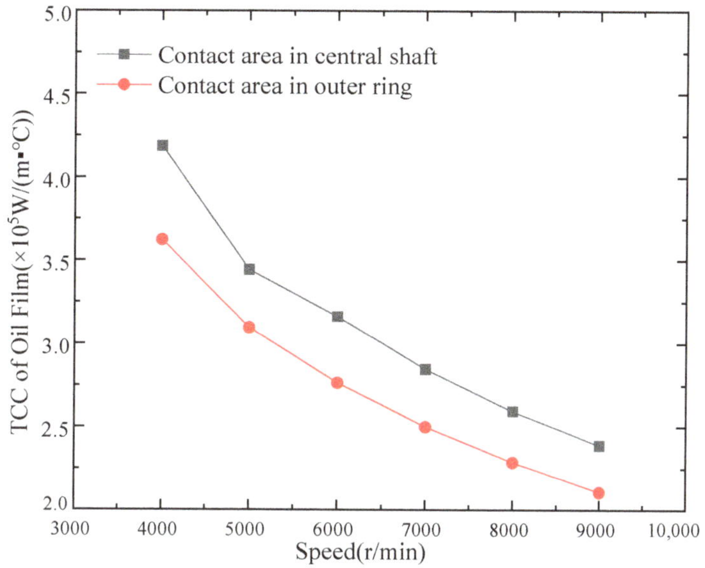

Figure 10. The influence of speed on the heat transfer coefficient of oil film.

From Figure 9, it can be seen that as the load increases, the convective coefficient in the contact area gradually increases. This is because when the shear force and pressure inside the oil film increase, it is difficult to form a lubricating oil film in the contact area. Therefore, it causes the thickness to gradually decrease. From Figure 10, it can be seen that as the rotation speed increases, the convective coefficient of the contact area decreases. This phenomenon is mainly caused by the lubricating oil film on the contact areas, which forms easily and gradually increases.

4.3. *Thermal Analysis Simulation Process*

The finite element method can be used to obtain the temperature field of each component of the bearing by solving the discretization area. Because the calculation model is consistent with the actual structure, the finite element method can clearly calculate the balance of each component in the cross-section and in the adjacent area. At the same time, the temperature distribution can better characterize the heat loss. The above descriptions are of two basic thermal field solving problems—namely, steady-state temperature field analysis and transient temperature field analysis. In the process of bearing service, the equilibrium temperature reached by the bearing neither generates heat nor loses heat. It is the operating temperature at which the bearing is running smoothly.

Because the bearing is a symmetrical rotating assembly, each needle roller has the same form of heat generation and heat transfer during high-speed rotation. It is assumed that the axial thermal characteristics of the system are consistent, and that the heat transfer only occurs in the radial direction [21]. In the working process, forced convection occurs between the lubricating oil and the needle rollers, as well as in the inner and outer raceway surfaces of the bearing. Natural convection occurs between the contact area of the end of the central shaft and the outside part of the shaft with the outside air. The friction heat generated by the bearing is evenly distributed to the needle roller and the ferrule, in a ratio of 1:1 [22].

4.3.1. Thermal Boundary Conditions

In ANSYS Workbench, the bearing is modeled and meshed according to known parameters. The finite element model of the bearing is meshed by a mapping method. The meshing produces a total of 114,578 elements and 525,468 nodes. The divided model is shown in Figure 11.

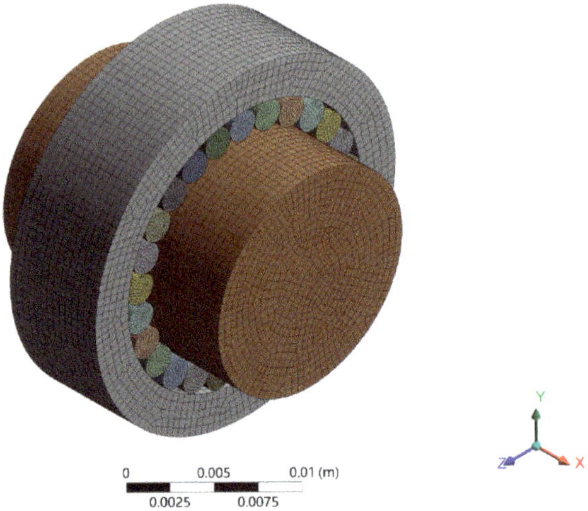

Figure 11. Meshed model.

In the software, we entered the material properties, such as Young's modulus, the Poisson ratio, the density, and the thermal conductivity. We set the bearing material properties in ANSYS and added the thermal conductivity of the bearing material to 40.1 W/(m·K). Figure 12 shows the boundary conditions for the steady-state temperature field analysis of the bearing. The main settings in the steady-state temperature analysis module included three types of boundary conditions. The first type of boundary condition was setting the ambient temperature to 40 °C, i.e., the temperature of the bearing components and the external environment. The second type of boundary condition was applying a thermal load to the bearing and loading a calorific value in the form of heat flow to the inner and outer surfaces of the rolling element and the raceway in the contact area. The heat generation of the bearing was mainly on the raceway surface and the needle roller surface in the contact area. Therefore, according to the calculation results in Figures 5 and 6, the heat generation of the inner ring, the outer ring, and the needle rollers were distributed, respectively, to the three contact surfaces. The third type of boundary condition was applying the oil film convective coefficient to the contact pair and loading the convective coefficient of the air or lubricating oil on the inner and outer rings and the outer surfaces of the rolling elements. Among them, the convective heat transfer coefficients of the outer ring, the central shaft, and the surface of the needle rollers were set according to the calculation results in Figures 7 and 8, respectively. The properties of the oil film in the thermal analysis were considered as the value of the oil film conductivity coefficient in the contact pair setup. The value was related to the oil film thickness and the thermal conductivity of the lubricating oil. This was regarded as the thermal resistance between the two contact pairs. The above boundary conditions included the heat transfer and the heat dissipation boundary conditions of the bearing.

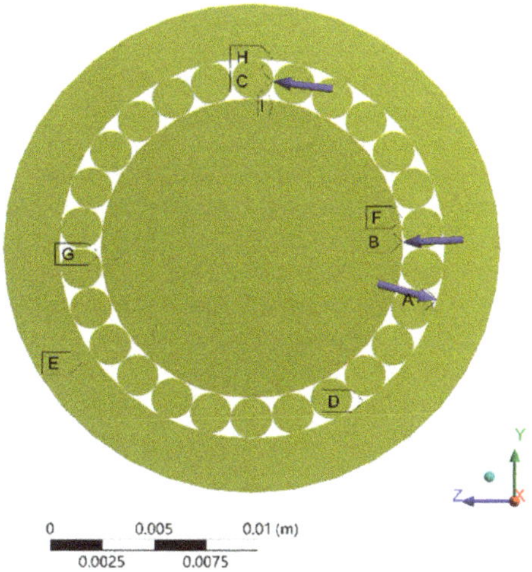

Figure 12. Boundary conditions for steady-state temperature field analysis.

4.3.2. Analysis of Steady-State Temperature Field

1. The influence of oil film thermal conductivity coefficient on bearing temperature.

In order to analyze whether the convective coefficient of the lubricating oil film affects the heat dissipation inside and outside the bearing, a "path" from the center of the central axis along the positive direction of the Y axis to the outer surface of the outer ring could be set. In addition, the temperature on the "path" could be studied under the two conditions of the oil film convective coefficient and no-oil film convective coefficient. The law of change is shown in Figure 13. It can be seen from the temperature distribution that, due to the compact structure of the needle roller bearing, the temperature difference between the internal parts of the bearing was small. The temperature range of the bearing had the same distribution in each needle and contact area, and only changed in temperature along the radial direction.

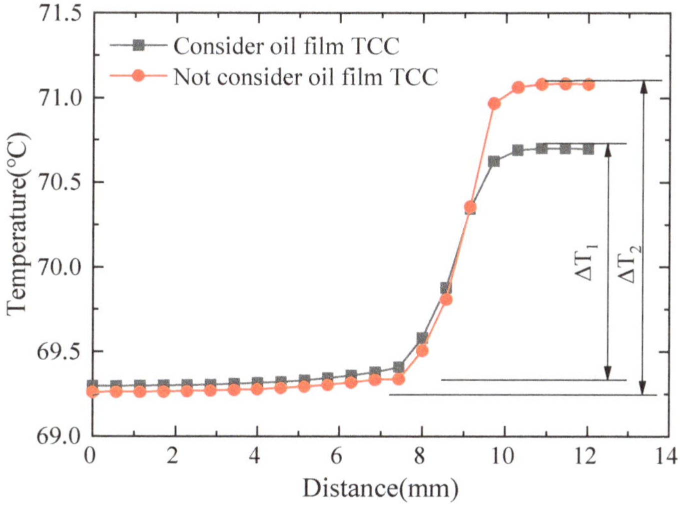

Figure 13. Extraction of bearing internal temperature value.

Because the outer ring rotates and the load mainly acts on the outer ring, the heat generated in the contact area between the needle roller and the outer ring is greater than the heat generated between the needle roller and the central shaft in the dynamic simulation. Accordingly, in the temperature field analysis, it can be seen that the temperature from the central shaft core to the outer ring rose and then decreased gradually, and the temperature was highest at the contact point between the needle rollers and the outer ring. In addition, when there was no heat transfer coefficient of the lubricating oil film, it meant that the bearing needles were in direct contact with the outer ring and the central shaft. As a result, the heat conduction effect was better, and the bearing temperature difference was smaller. However, when there was a lubricating oil film heat transfer coefficient, the temperature difference between the inside and outside of the bearing was large. The above phenomenon indicated that the lubricating oil films affected the heat conduction in the bearing contact area.

2. The influence of operating parameters on bearing temperature.

Using the local heat generation model, the frictional heat generation, and the convective coefficient obtained in Section 4.1 as the boundary conditions for the temperature field calculation, the bearing temperature under different loads and different speeds could be calculated. The maximum temperature of the outer ring raceway, the needle roller, and the central shaft surface could be extracted.

Figure 14 shows the effect of different radial loads on the maximum temperature when the outer ring rotated at a speed of 7000 rpm and an initial temperature of 40 °C. It can be seen that the maximum temperature rise was in the outer ring raceway. The maximum temperature value gradually decreased from the outer ring raceway, the outer surface of the outer ring, and the needle roller surface to the central shaft surface. The radial load had no effect on the heat generation of the needle roller bearing. Therefore, under the radial load of 2000 N~4500 N, the temperature of the bearing gradually increased, and the maximum temperature value was between 65 °C and 70 °C.

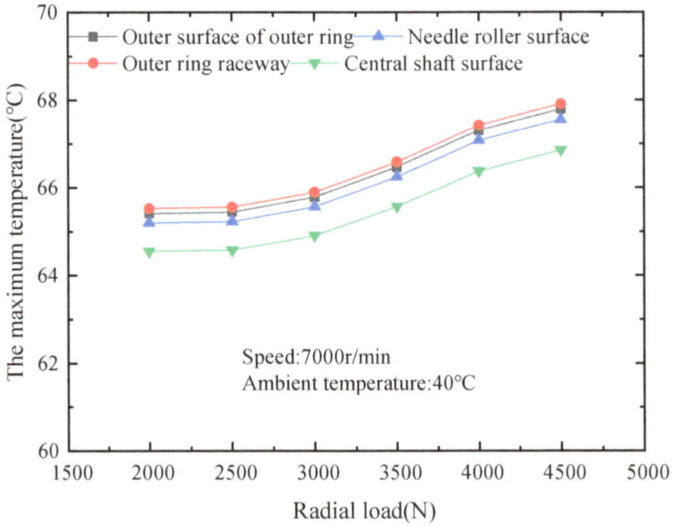

Figure 14. The influence of radial load on the maximum temperature of the bearing.

Figure 15 shows the influence of different speeds on the maximum temperature at a radial load of 3000 N and an initial temperature of 40 °C. It can be seen that the maximum temperature rise was also in the outer ring raceway. The maximum temperature value gradually decreased from the outer ring raceway, the outer ring outer surface, and the needle roller surface to the central shaft surface temperature. When the speed was 9000 rpm, the maximum bearing temperature was 85 °C.

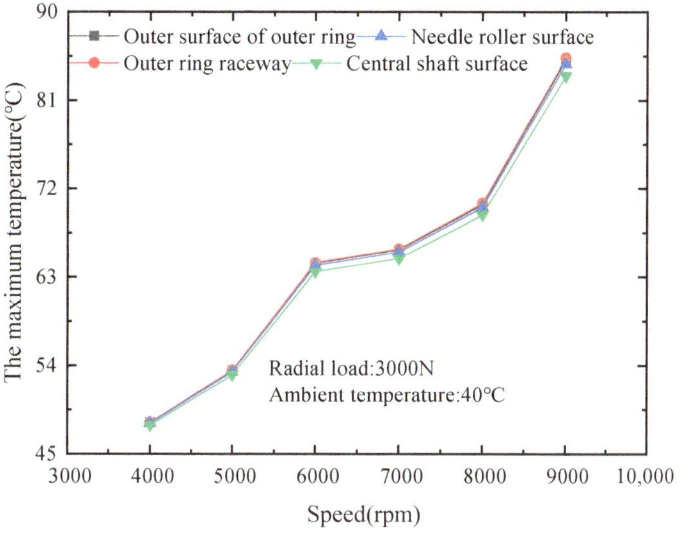

Figure 15. The influence of speed on the maximum temperature of the bearing.

5. Discussion of Simulation Results

In order to verify the correctness of the simulation method in the study, its results were calculated and compared based on the reliability test conditions of a full-type needle bearing for rocker arms of the same specifications. The rated dynamic load of the bearing was calculated as 8.6 KN. The limit speed was referred to in the SKF bearing catalog (sample) for the needle roller bearings of the same specification. The test lubricating oil was engine oil 10W-30, and its performance was similar to that of the FVA-M lubricating oil. The test conditions that were determined are shown in Table 3 [23].

Table 3. Test condition for bearing.

Parameters	Value
Load/N	2580
Rotational speed of outer ring/rpm	7000
Lubricant	engine oil 10W-30
Oil temperature/°C	40
Ambient temperature/°C	40

The following points can be observed from Figure 16. The highest temperature distribution was on the raceway. Similarly, the raceway surface was the most prevalent failure site due to excessive temperature, which was consistent with the results of [10]. Considering the influence of load and speed, the maximum temperature of this bearing under the selected FVA-M lubricating oil was 65.9 °C, and the maximum temperature rise was 25.9 °C. That proved the rationality of the bearing conforming to the tribological design. On the one hand, the working load used in the simulation analysis was 0.2~0.3 times the rated dynamic load, and the working speed was 0.2~0.6 times the limit speed within the rated thermal speed range of the bearing specified in the fatigue life test. As a result, the temperature variation range was in line with the national standards of bearing life and the reliability test and evaluation [24]. In other words, the bearing temperature rose generally and did not exceed 55 °C above the ambient temperature, and the maximum temperature of the outer ring of the bearing did not exceed 95 °C when lubricated by circulating oil. On the other hand, the same thermal boundary conditions as those in [15] were used in the thermal analysis of the example bearing. It is believed that when the bearing rotates at a high speed, the temperature of all parts of the raceway is basically uniform. When the bearing was modeled, the temperature coupling of the inner and outer ring surfaces, respectively was performed. It was proved by the related temperature field experiment that the simulation results were basically in agreement with the experimental data.

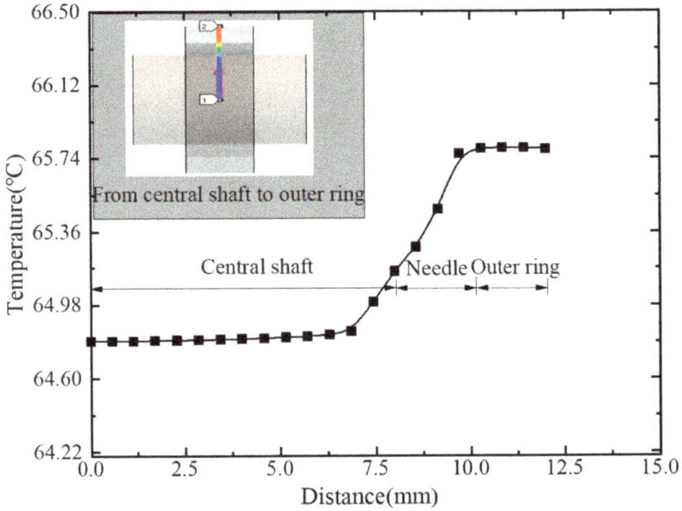

Figure 16. Temperature distribution under test conditions.

Thermal analysis using the finite element method can significantly enhance calculation efficiency and repeatability. In future work, we can design a thermal study of the ANSYS and Adams co-simulation based on the bearing model acquired in the secondary-built Adams commercial program. MATLAB was used to process the steady friction power consumption and convection coefficient, and the ANSYS parameterized command flow was invoked to complete the thermal analysis of the bearing. This allowed for the optimization of bearing structure characteristics, as well as for the analysis of working conditions. The method developed for analyzing bearing thermal behavior can be applied to the examination of comparable bearings. This research is extremely important for tribological design and bearing application choices.

6. Conclusions

Combined with the dynamic method of rolling bearings, the friction power consumption of full-type needle bearings for rocker arms under FVA-M oil lubrication is clarified. This work established a dynamic model, and the influence of the working conditions on the bearing temperature field, from the perspective of theoretical analysis and finite element calculation, was analyzed. The main conclusions are as follows:

1. The multi-body dynamics method was used to obtain the dynamic performance of bearing, and the influence of friction power consumption on the bearing temperature was studied. The results showed that as the speed increases, the frictional power consumption of the bearing increases, and the maximum temperature of the bearing gradually increases. With the increase in the load, the frictional power consumption gradually increased, and the maximum temperature of the bearing also gradually increased. Compared with the speed, the load change had a relatively small effect on the temperature of the bearing.

2. We selected appropriate boundary conditions in the bearing thermal model to study the steady-state temperature field. The results showed that the temperature of the outer raceway is the maximum, and the temperatures of the needle roller surface, the outer ring surface, the inner raceway, and the central shaft core decreased, in order. The temperature difference between the inside and outside of a bearing was affected by the lubricating oil film in the contact area of the bearing. The presence or absence of the oil film can affect the temperature difference within and between the bearing and its external environment.
3. According to an analysis of the heat generation characteristics of the bearing under the influence of multiple factors, the heat value created by the collision and sliding in a full-type needle bearing accounted for a considerable share. As a result, effective lubrication and timely lubricant replenishment should be employed in bearing applications. Under a high-speed condition, the influence of bearing thermal effects must also be considered. The temperature distribution of the bearing showed regularity. In addition, the highest temperature was distributed at the raceway. The high temperature rise reduced the hardness of the material, resulting in a contact fatigue failure between the needle roller and the raceway. To avoid that, the characteristics of the bearing material should be improved by controlling the heat treatment.

Author Contributions: Conceptualization, G.Z., B.S. and F.L.; data curation, G.Z. and F.L.; formal analysis, G.Z. and B.S.; funding acquisition, W.Z.; investigation, G.Z., F.L. and B.S.; methodology, B.S. and W.Z.; project administration, B.S. and H.Y.; resources, H.Y.; software, G.Z.; supervision, W.Z. and B.S.; validation, W.Z. and F.L.; writing—original draft, G.Z. and B.S.; writing—review & editing, B.S., W.Z., and H.Y. All authors have read and agreed to the published version of the manuscript.

Funding: This research was funded by Youth Program of the National Natural Science Foundation of China (51905152).

Institutional Review Board Statement: Not applicable.

Informed Consent Statement: Not applicable.

Data Availability Statement: The data used to support the findings of this study are available from the corresponding author upon request.

Acknowledgments: The authors would like to thank Youth Program of the National Natural Science Foundation of China (51905152) for the financial support.

Conflicts of Interest: The authors declare no conflict of interest.

References

1. Liu, Z.; Yu, X. Research on Cam Contact Force of OHC. *J. Intern. Combust. Engines* **2014**, *6*, 40–43. (In Chinese)
2. Zhang, K.; Yuan, Z.; Sun, P. Dynamic Analysis on Rocker Arm Bearings Based on ANSYS/LS-DYNA. *J. Bear.* **2018**, *3*, 6–9. (In Chinese)
3. Zhang, K.; Yuan, Z.; Sun, P. Analysis on Temperature Field of Rocker Arm Bearings Based on ANSYS/LS-DYNA. *J. Bear.* **2018**, *11*, 9–13. (In Chinese)
4. Yanshuang, L.Z.W. Current situation and expectation of research on temperature field for rolling bearing. *J. Mech. Transm.* **2010**, *34*, 88–91. (In Chinese)
5. Nassef, M.G.A.; Soliman, M.; Nassef, B.G.; Daha, M.A.; Nassef, G.A. Impact of Graphene Nano-Additives to Lithium Grease on the Dynamic and Tribological Behavior of Rolling Bearings. *Lubricants* **2022**, *10*, 29. [CrossRef]
6. Wang, H.; Huang, H.; Yu, S.; Gu, W. The Effect of Surface Morphology of Tapered Rolling Bearings in High-Speed Train on Grease Lubrication. *Lubricants* **2020**, *8*, 76. [CrossRef]
7. Palmgren, A. *Ball and Roller Bearing Engineering*, 3rd ed.; Burbank: Philadelphia, PA, USA, 1959; pp. 30–40.
8. Ma, Z.; Li, J.; He, D.; Liu, B.; Zhang, J. Analysis on Thermal Characteristics of High-Speed Cylindrical Roller Bearings Based on Finite Element Method. *J. Bear.* **2018**, *1*, 19–22, 51. (In Chinese)
9. Hatazawa, T.; Kawaguchi, T. Frictional torque of needle roller thrust bearings. *J. Tribol. Trans.* **2008**, *52*, 127–132. [CrossRef]
10. Li, B.; Wang, H.; Bi, L. Analysis of temperature field of cylindrical roller bearing based on friction effect. *J. Lubr. Eng.* **2016**, *41*, 5–9. (In Chinese)
11. Wang, L.Q.; Chen, G.C.; Gu, L.; Zheng, D.Z. Study on heat generation of high-speed cylindrical roller bearings. *J. Lubr. Eng.* **2007**, *32*, 8–11. (In Chinese)

12. Wang, L.Q.; Chen, G.C.; Gu, L.; Zheng, D.Z. Study on operating temperature of high-speed cylindrical roller bearings. *J. Aerosp. Power* **2008**, *23*, 179–183. (In Chinese)
13. Subramaniam, S.B.; Harsha, R.D.; Vivek, M.; Prasad, V.D. Thermal Analysis of Ceramic Conventional Ball Bearings. *J. Indian J. Sci. Technol.* **2016**, *9*, 1–8. [CrossRef]
14. Liming, L.; Guowen, Z. FEM of Temperature Field of Rolling-sliding Compound Bearing. *J. Mecha. Trans.* **2016**, *40*, 139–142, 176. (In Chinese)
15. Lei, Z.; Xiao, S.; Wu, L. Analysis of the Oil Film Thermal Resistance and Its Influence on the Temperature Field of Angular Contact Ball Bearing. *J. Mach. Des. Manuf.* **2014**, *5*, 201–204. (In Chinese)
16. Yang, Y.; Zhu, R.; Miaomiao, L.I.; Qin, W.; Xie, P. Temperature Field Analysis of Angular Contact Ball Bearings Based on ANSYS Workbench. *J. Mach. Build. Autom.* **2019**, *48*, 67–69. (In Chinese)
17. Deng, S.E.; Jia, Q.Y. *Design Principle of Rolling Bearings*, 2nd ed.; China Standard Press: Beijing, China, 2014; pp. 245–260.
18. Ma, Y.; Su, B.; Liu, Y.Y. An Improvement on Traction Measurement System. In Proceedings of the 2018 International Conference on Computer, Communications and Mechatronics Engineering, Cuernavaca, Mexico, 26–29 November 2018; Destech Publications: Lancaster, PA, USA, 2018; pp. 678–681.
19. Harris, T.A.; Kotzalas, M.N. *Advanced Concepts of Bearing Technology: Rolling Bearing Analysis*; CRC Press: Boca Ratoon, FL, USA, 2006; pp. 148–153.
20. Wen, S.Z.; Huang, P. *Principle of Tribology*, 5th ed.; Tsinghua University Press: Beijing, China, 2018; pp. 53–60.
21. Pouly, F.; Changenet, C.; Ville, F.; Velex, P.; Damiens, B. Investigations on the power losses and thermal behavior of rolling element bearings. *JPI Mech. Eng. J J. Eng.* **2010**, *224*, 925–933.
22. Butron, R.A.; Staph, H.E. Thermally activated seizure of angular contact bearing. *J. ASLE Trans.* **1967**, *10*, 408–417.
23. Watanabe, Y.; Katayama, A. Rolling Bearing for Rocker Arm: US Patent. US7614374 B2, 10 November 2009.
24. GB/T 24607-2009. Rolling Bearings-Test and Assessment for Life and Reliability. China Standards Press: Beijing, China, 2009; pp. 1–15.

lubricants

Article

Research on the Effect of Spindle Speed on the Softening and Hardening Characteristics of the Axial Operating Stiffness of Machine Tool Spindle

Jiandong Li [1,2,*], Qiang Wang [1], Xurui Sun [3], Jue Qu [2], Ang Qiu [2], Wei Kang [4] and Shuaijun Ma [4]

1 Equipment Management and Unmanned Aerial Vehicle Engineering School, Air Force Engineering University, Xi'an 710051, China; caption_wang@21cn.com
2 Air and Missile Defend School, Air Force Engineering University, Xi'an 710051, China; qujue402@sina.com (J.Q.); qiuanng@163.com (A.Q.)
3 Baqiao District Economic Information Research Center, Baqiao District Development and Reform Commission, Xi'an 710051, China; sunxurui@126.com
4 Key Laboratory of Education Ministry for Modern Design and Rotor-Bearing System, Xi'an Jiaotong University, Xi'an 710049, China; kang_wei@stu.xjtu.edu.cn (W.K.); msj821@stu.xjtu.edu.cn (S.M.)
* Correspondence: zuiwojinxiao@stu.xjtu.edu.cn

Abstract: Spindle stiffness is one of the most critical indicators for evaluating and measuring the service performance of spindles. The traditional static stiffness indexes only involve static analysis and rarely focus on the study of spindle-carrying capacity under operating conditions. In this paper, the explicit solution approach is used to develop a mechanical model of the spindle's axial operating stiffness. This model was then used to explore the influence of rotational speed on the softening and hardening features of the spindle axial operating stiffness, and experimental verification was carried out. According to studies, the speed of a fixed-position preload spindle can lead its operating stiffness to exhibit a "stiffness-hardening" feature. However, when the axial displacement of the spindle is small, the operating stiffness curve of the spindle displays a noticeable "fluctuation" phenomenon for low-speed spindles. Furthermore, the speed-induced preload has a significant impact on the test results when testing spindle axial operating stiffness.

Keywords: axial operating stiffness; stiffness hardening; stiffness softening; machine tool

Citation: Li, J.; Wang, Q.; Sun, X.; Qu, J.; Qiu, A.; Kang, W.; Ma, S. Research on the Effect of Spindle Speed on the Softening and Hardening Characteristics of the Axial Operating Stiffness of Machine Tool Spindle. *Lubricants* **2022**, *10*, 132. https://doi.org/10.3390/lubricants10070132

Received: 5 June 2022
Accepted: 20 June 2022
Published: 22 June 2022

1. Introduction

Cutting forces are the most common external loads applied to machine tool spindles and are widely regarded as the best performance estimator for machining operations [1–3]. Spindle stiffness of the machine tool, which reflects its capacity to resist deformation when subjected to external loads, is one of the most essential indicators for evaluating the service performance of a spindle [4–6]. Low spindle stiffness causes chatter [1], unwanted back cutting, and excessive cutter tilt, all of which affect the cutting surface quality [7–9] and the machining accuracy [10,11], as well as causing the rolling elements of the spindle bearing to slip, aggravating spindle component wear [12] and weakening machine tool reliability [13–15]. As a result, high stiffness becomes one of the most important aims in precision machine tool spindle design [16].

Spindle static stiffness is a regularly used metric for measuring and evaluating spindle stiffness under constant or slowly variable quasi-static loads [17]. Static and dynamic analysis are two approaches to study spindle static stiffness [18]. The static analysis of spindle static stiffness has been investigated previously and is considered to be rather advanced. D. Olvera [18] proposed a static stiffness measurement method along the turning center kinematic chain, and analyzed the tool tip radial stiffness of turn-milling centers. With the help of a loaded double-ball bar and linear variable differential transformer, Laspas [19] proposed a new method to measure and identify the full translational stiffness matrices of the five-axis machining center by using the quasi-static circular trajectory, and realized the accurate identification of the quasi-static stiffness of the five-axis machine tool. As the accuracy of spindle machining improves, the spindle's service speed must also increase. When the spindle is rotating, however, the state characteristics are very different from when it is halted. Based on the proposed thermo-mechanical model of the spindle system, Li [20] discovered that while the spindle is operating at high speed, the clearance of its components is much different from that at standstill, and this has a considerable impact on spindle preload and component temperature. Subsequently, Li [21] then evaluated the transient preload of a fixed-position preload spindle in real-time and discovered that at 8000 rpm, the preload rose from 483 N at standstill to 720 N, while the bearing temperature increased from 24 °C to 36.5 °C. Large variations in the machine spindle's characteristics during operation are sure to impair the spindle stiffness performance, which, in turn, affects the machine tool's quality and efficiency. As a result, the dynamic analysis of spindle static stiffness (in this paper, referred to as operating stiffness to distinguish it from static and dynamic stiffness) is gaining attention.

In order to better investigate the effect of dynamic effects generated by spindle operation on its static stiffness, A. Matsubara [22] designed a magnetic loading device to measure the radial operating stiffness of the spindle and found that the velocity and thermal effects can lead to significant softening and hardening characteristics in the spindle radial stiffness. Wang [10] proposed a three-step identification algorithm for spindle radial stiffness based on stiffness theory modeling, which solves the difficult problem of the measured value being often coupled in the spindle-tool stiffness during spindle radial operating stiffness testing, and improves the accuracy of spindle radial operating stiffness testing. For drilling machines, vertical spindle surface grinders, and other axially loaded machine tools, the axial operating stiffness of the spindle system should be of increased importance throughout its entire life cycle. Tsuneyoshi [23] discovered that the spindle axial load–axial displacement curve showed a non-linear relationship in his investigation into spindle preload testing methodologies. Li [4] further explored the non-linear relationship of the spindle axial load–displacement curve and found that the machine tool spindle static stiffness exhibits hardening and softening characteristics depending on the preload. However, the speed-induced centrifugal effect generates a non-linear change in bearing stiffness, which will inevitably lead to a change in spindle stiffness, as the bearing stiffness determines the overall stiffness of the spindle bearing [24,25]. It is a pity that the influence of spindle speed on the softening and hardening characteristics of the axial operating stiffness of machine tool spindles has not been discovered in any of the preceding studies.

In this paper, the spindle axial operating stiffness of a widely configured fixed-position preload spindle for precision machine tools was investigated. The explicit solution approach is used to develop a mechanical model of the spindle's axial operating stiffness. This model was then used to explore the influence of rotational speed on the softening and hardening features of the spindle axial operating stiffness. A spindle operating stiffness test bench was also created to evaluate the model's validity and accuracy, with the impact of speed-induced preload on spindle operating stiffness being studied in particular.

2. Axial Operating Stiffness Model of Spindle

In this section, an explicit solution approach is used to build an angular contact ball-bearing mechanics model and analyze the relationship between speed, preload, and contact angle. Following that, an analytical model of the fixed-position preload spindle operating stiffness is constructed based on the spindle preload principle and the aforesaid bearing mechanics model. The influence of speed on the axial operating stiffness of the machine tool spindle can be investigated by adjusting the speed parameter in the spindle stiffness model.

2.1. Bearing Mechanics Model

Different service conditions, such as external load and speed, vary the bearing contact angle. In general, when a bearing is assembled, a certain preload is given to minimize bearing clearance to achieve the desired bearing stiffness, accuracy, and other characteristics [26,27]. The bearing contact angle will change from the initial contact angle α_f to α_p, and the outer raceway groove curvature center, ball center, and inner raceway groove curvature center will all be co-linear at this point (as shown in Figure 1). When the bearing rotates, the centrifugal force and gyroscopic moment effect generated by the speed act on the ball, causing the outer raceway groove curvature center, ball center, and inner raceway groove curvature center to lose co-linearity, and the ball-inner contact angle α_i and the ball-outer contact angle α_o to no longer be equal.

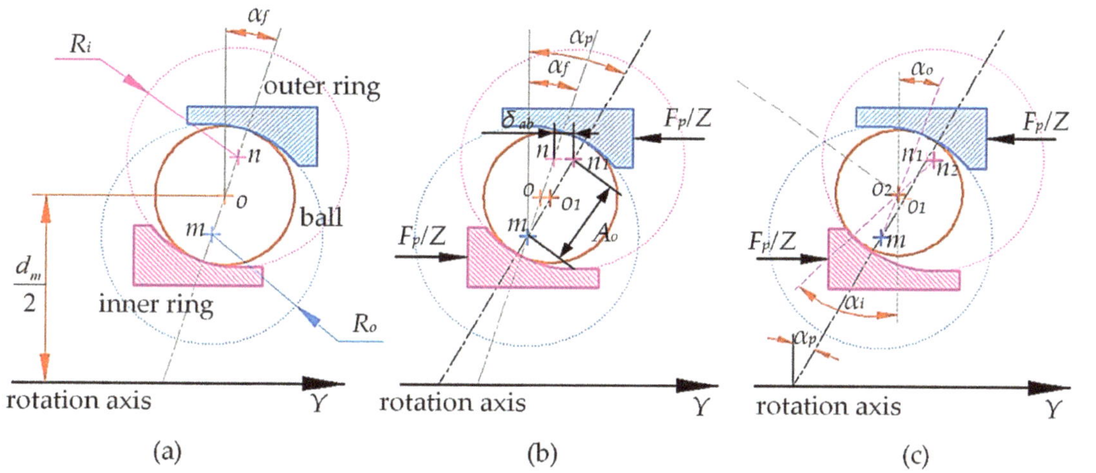

Figure 1. Bearing contact angle variation diagram: (**a**) free state; (**b**) withstand preload; (**c**) rotation status.

According to Ref. [4], the contact angle of the bearing with preload applied can be calculated as:

$$\alpha_p = \arcsin\left(\frac{A_o \sin\alpha_f + \delta_{ab}}{\sqrt{A_o^2\cos^2\alpha_f + \left(A_o\sin\alpha_f + \delta_{ab}\right)^2}}\right) \tag{1}$$

where α_p means the bearing contact angle with preload applied; α_f indicates the initial contact angle of the bearing, A_o is the distance between raceway groove curvature centers; and δ_{ab} means the axial displacement of the bearing.

When the bearing rotates, the ball is subjected to centrifugal force and gyroscopic moment, as shown in Figure 2. The centrifugal force can be split into two components: the component force F_u in the normal direction and the component force F_w parallel to the tangential direction of the contact point. The bearing load distribution is affected differently by the two components of centrifugal force. The component force F_u increases the normal load in the bearing outer ring on the ball, whereas the component force F_w compresses the ball farther against the bearing inner and outer rings. The ball center travels along the force's direction when component force F_w is applied. At this time, the ball-inner and the ball-outer contact angles are no longer equal, satisfying Equation (2).

$$\frac{1}{\tan\alpha_o} - \frac{1}{\tan\alpha_i} = \frac{7}{5}\frac{ZF_c}{F_p} \tag{2}$$

where Z indicates the number of balls in the bearing; F_p means preload applied to bearing; and F_c refers to centrifugal force, which can be calculated from Equation (3).

$$F_c = \frac{d_m}{2} m_b \omega_c^2 \tag{3}$$

where d_m denotes diameter of bearing pitch circle; m_b refers to ball mass; and ω_c is bearing cage speed.

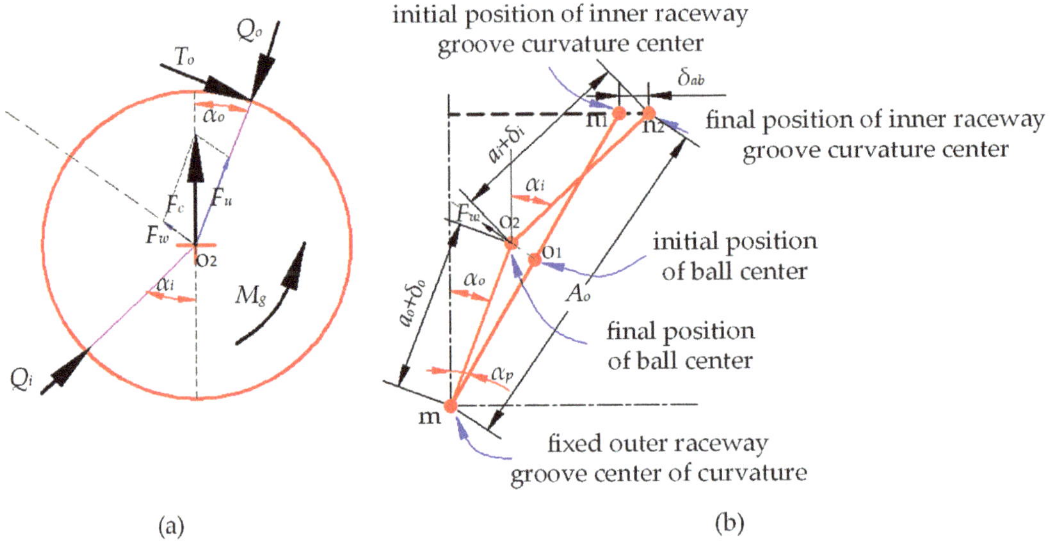

Figure 2. Parameter variation and load distribution on the ball after bearing rotation: (**a**) bearing ball load distribution; (**b**) variation in bearing groove curvature center position.

In order to construct a system of equations for calculating the contact angle α_o and α_i, another set of functional relationships is required in addition to Equation (2). According to the relationship between the curvature position of each bearing groove shown in Figure 2b, the contact angles α_o and α_i can also be derived from Equation (4).

$$(a_o + \delta_o)\cos\alpha_o + (\alpha_i + \delta_i)\cos\alpha_i = (a_o + a_i)\cos\alpha_f \tag{4}$$

where a indicates the distance between the raceway groove curvature center and the ball center and δ indicates the ball-race deformation. The subscripts o and i, respectively, concern outer ring and inner ring.

According to the explicit solution algorithm of Ref. [28], the ball-inner contact angle α_i and the ball-outer contact angle α_o can be obtained by combining the system of Equations (2) and (4).

The bearing axial displacement δ_{ab} under the preload force F_p can be obtained by substituting the results of the contact angles α_o and α_i into Equation (5).

$$\delta_{ab} = (a_o + \delta_o)\sin\alpha_o + (a_i + \delta_i)\sin\alpha_i - (a_o + a_i)\sin\alpha_f \tag{5}$$

2.2. Spindle Axial Operating Stiffness Model

With fixed-position preload, the combined bearing's axial relative position remains constant throughout use. Figure 3 depicts a schematic representation of the fixed-position preload spindle construction. When the spindle bearings are mounted back-to-back, the variation in width between the inner and outer spacers can modify the preload.

According to the analysis in Section 2.1, the relationship between the axial load F_{ab} and its corresponding axial displacement δ_{ab} can be obtained. Here, F_{ab} and δ_{ab} satisfy the following relationship for convenience:

$$F_{ab} = f(\delta_{ab}, n) \tag{6}$$

where $f(\cdot)$ represents the non-linear mapping relationship and n refers to the spindle bearing speed.

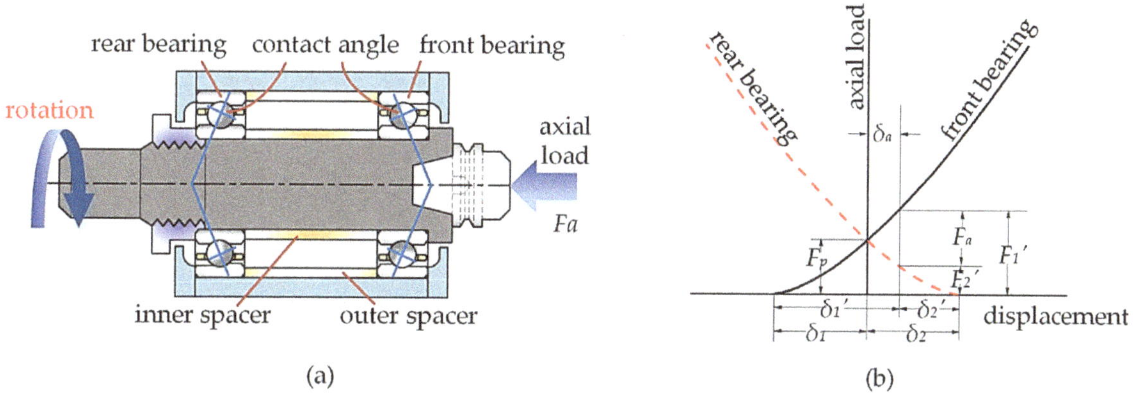

Figure 3. One side of the spindle is not completely unloaded during the pressure phase: (**a**) fixed position preload spindle construction; (**b**) spindle load distribution.

As shown in Figure 3b, the initial preload of the spindle is assumed to be F_p, and the load distribution of its front and rear bearings satisfies the following relationship:

$$F_p = f(\delta_1, n) = f(\delta_2, n) \tag{7}$$

where δ_1 donates the pre-deflection of the front bearing and δ_2 means pre-deflection of the rear bearing.

The spindle system is gradually subjected to axial pressure, which increases the stress on the front bearing of the spindle while lowering the force on the rear bearing. When the axial displacement of the spindle δ_a is less than the pre-deflection of the rear bearing δ_2, the force relationship of the spindle is given as:

$$f(\delta_1 + \delta_a, n) = F_a + f(\delta_2 - \delta_a, n) \tag{8}$$

Continue applying axial pressure until the spindle axial displacement δ_a equals or surpasses the rear bearing pre-deflection δ_2 (as shown in Figure 4), the rear bearing is completely unloaded, and the axial load is borne entirely by the spindle's front bearing. The force relationship for the spindle can now be expressed as:

$$f(\delta_1 + \delta_a, n) = F_a \tag{9}$$

The analytical process remains the same when the spindle is subjected to axial tension. The spindle system is gradually supplied axial tension, and as the force on the rear bearing grows, the force on the front bearing of the spindle diminishes. Before the front bearing of the spindle is entirely unloaded (i.e., the axial displacement of the spindle δ_a is less than the pre-deflection of the front bearing δ_1), so the force relationship of the spindle is as follows:

$$F_a + f(\delta_1 - \delta_a, n) = f(\delta_2 + \delta_a, n) \tag{10}$$

Continue to apply axial tension until the axial displacement δ_a of the spindle is greater than or equal to the pre-deflection of the front bearing δ_1, at which point the front bearing of the spindle is completely unloaded and the axial load is completely borne by the rear bearing of the spindle. The force balance equation of the spindle can be given as:

$$F_a = f(\delta_2 + \delta_a, n) \tag{11}$$

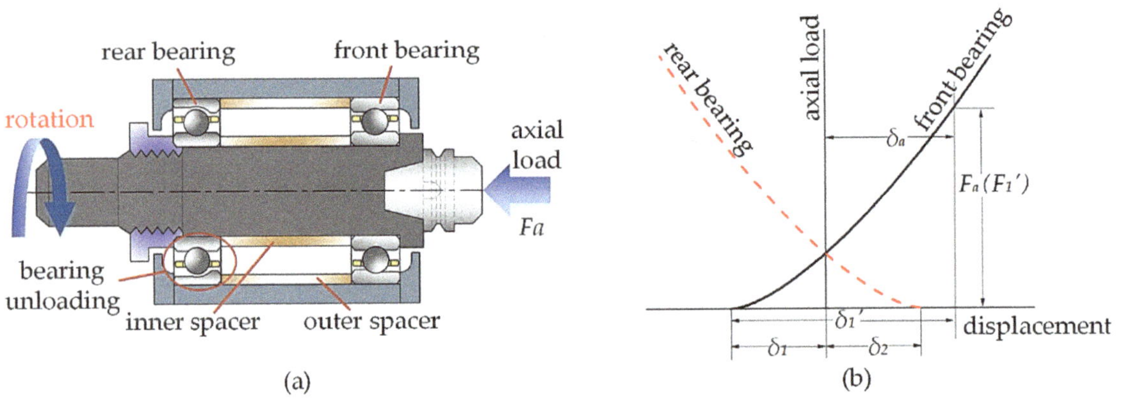

Figure 4. One side of the spindle is completely unloaded during the pressure phase: (**a**) fixed-position preload spindle construction; (**b**) spindle load distribution.

By solving the preceding Equations (7)–(11) together, the relationship between axial load applied to spindle F_a and its associated axial displacement δ_a can be obtained. Subsequently, the spindle axial operating stiffness K_a can be calculated, as given in Equation (12).

$$K_a = \frac{dF_a}{d\delta_a} \tag{12}$$

3. Effect of Spindle Speed on Spindle Axial Operating Stiffness

According to Ref. [4], the preload has a softening and hardening influence on the axial static stiffness of the spindle. Therefore, this paper explores the effect of speed on the axial operating stiffness of the fixed-position preload spindle by adjusting the spindle bearing speed parameters in the model under working circumstances with varying initial preload, with simulation results given in Figures 5 and 6. Among them, the bearing model utilized in the theoretical simulation is NSK®7014CTYNSULP4, the dimensional parameters of which are provided in Table 1.

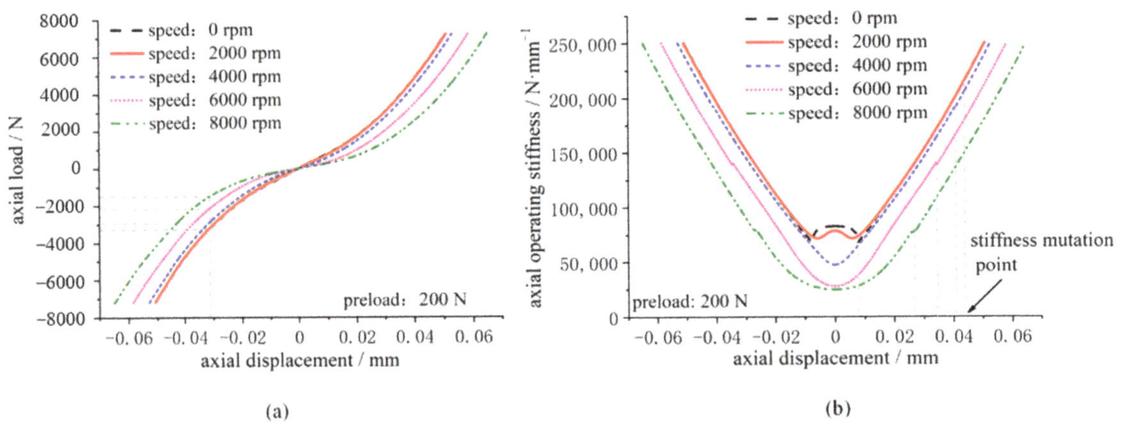

Figure 5. Effect of speed on axial operating stiffness of fixed-position preload spindle (preload: 200 N): (**a**) axial load–displacement curve; (**b**) axial operating stiffness curve.

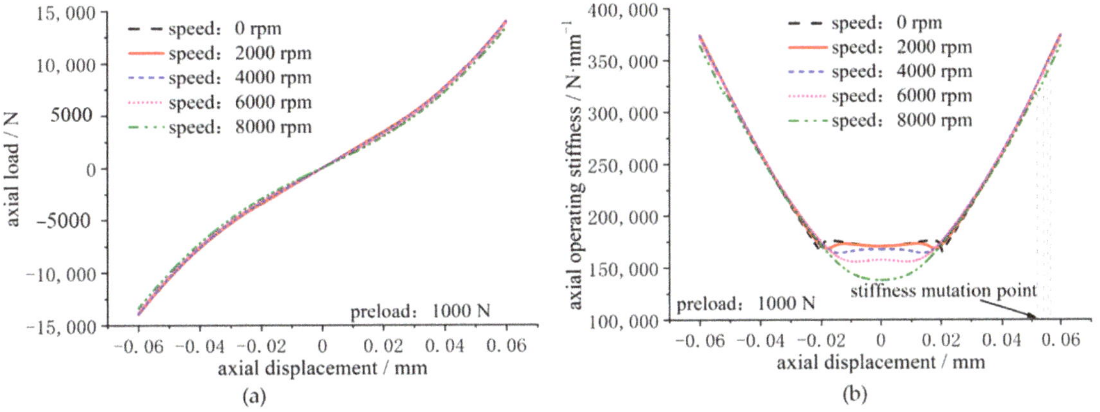

Figure 6. Effect of speed on axial operating stiffness of fixed-position preload spindle (preload: 1000 N): (**a**) axial load–displacement curve; (**b**) axial operating stiffness curve.

Table 1. The bearing parameters (NSK®7014CTYNSULP4).

Parameters	Values
Number of balls	20
Ball diameter	11.906 mm
Bearing pitch diameter	90 mm
Inner raceway groove curvature radius	6.19112 mm
Outer raceway groove curvature radius	6.19112 mm
Initial contact angle of bearing	15°

Figure 5 shows that for the fixed-position preload spindle with a smaller initial preload, the rotational speed has a greater influence on the load–displacement relationship and axial operating stiffness (200 N). As seen in Figure 5, as the speed increases, the load–displacement curve of the spindle becomes "smoother," meaning that the overall axial stiffness of the spindle reduces. In comparison to Figure 6, the spindle speed has less effect on the load–displacement relationship and axial operating stiffness when the spindle's initial preload is greater (1000 N). Because the bearing stiffness "softening" effect caused by rotational speed is more noticeable when the spindle preload is low, and the fixed-position preload spindle stiffness is a parallel relationship between the spindle front and rear bearing stiffness, the higher the spindle speed is, the lower the overall axial stiffness is. As shown in Figure 7, when the spindle preload is higher, the "softening" effect of bearing stiffness due to rotational speed is lessened, as is its effect on overall spindle stiffness.

Furthermore, as shown in Figure 5b, when the spindle is rotated, the axial displacement at the stiffness mutation point is greater than the axial displacement at zero speed. In the meantime, when the rotational speed increases, the axial displacement of the spindle's axial operating stiffness mutation point decreases.

The following are the reasons behind this: When the initial preload of the spindle remains constant, the initial pre-deflection of the bearing decreases as the spindle speed increases, but is greater than the pre-deflection at zero speed, which is the macroscopic manifestation of the bearing "stiffness softening" effect caused by the spindle speed (as shown in Figure 8). The spindle axial displacement must entirely balance the spindle's initial pre-deflection in order for the abrupt change in spindle axial operational stiffness to occur. As a result, the axial displacement at the stiffness mutation point is larger than the axial displacement at zero speed when the spindle is rotated. At the same time, when the speed of the spindle rises, the axial displacement at the stiffness mutation point decreases.

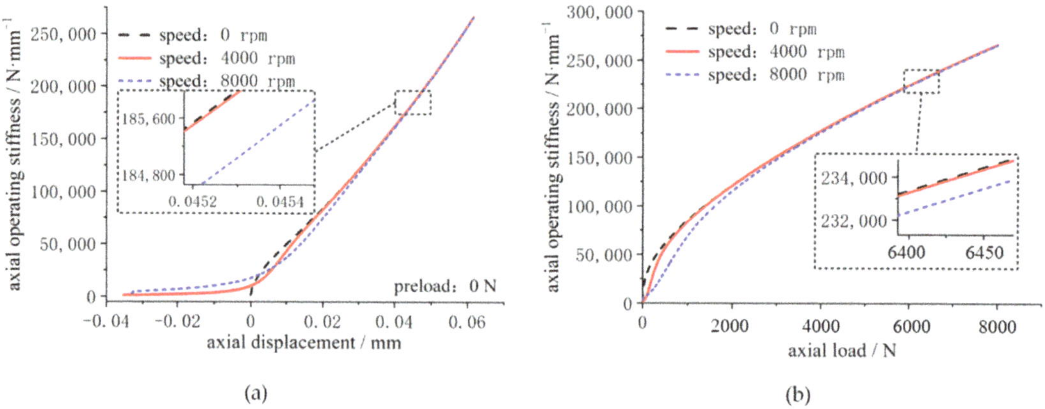

Figure 7. Single bearing axial operating stiffness at different speeds (bearing type: 7014C): (**a**) axial operating stiffness–axial displacement curve; (**b**) axial operating stiffness–axial load curve.

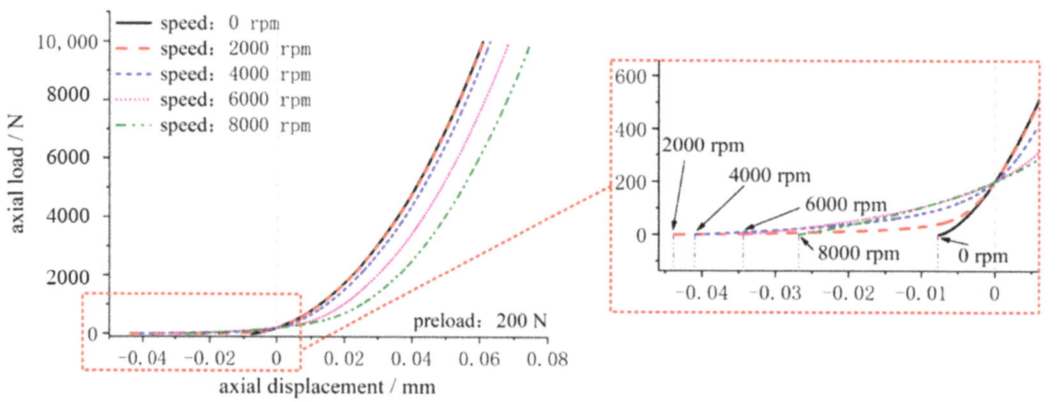

Figure 8. Single bearing load–displacement curve at different speeds (preload: 200 N).

It can also be observed in Figures 5 and 6 that when the spindle speed increases, the "sag" [1] in the axial stiffness curve of the spindle becomes more prominent (i.e., the spindle axial stiffness increases with the gradual increase in axial displacement before the abrupt change in spindle stiffness). When the spindle is rotating slowly, the axial stiffness of the spindle fluctuates (not exactly monotonically increasing or decreasing) in the phase with small displacement increments (e.g., −0.02 mm to 0.02 mm in Figure 6b). However, as the spindle speeds up, the stiffness fluctuation is gradually suppressed. When the spindle is operating at high speeds, the spindle axial stiffness displays a clear "sag" phenomenon, which shows the "stiffness hardening" characteristic.

This is because while the spindle is under axial load, one side bearing of the spindle progressively compresses while the other gradually unloads. Given that the fixed-position preload spindle stiffness is equal to the parallel connection of the front and rear bearing stiffness (superposition relationship) [4], the different change rates of single bearing stiffness with speed will result in different softening and hardening characteristics presented by the spindle axial operating stiffness.

When the spindle is at a lower speed, the local maximum and local minimum points in the bearing stiffness change rate curve are relatively obvious (as shown in the curve of the speed of 4000 rpm in Figure 9). The spindle pre-compression displacement value is typically positive because the spindle bearing in service requires a sufficient preload to keep the inner and outer rings and balls in contact. When the spindle is loaded, the relationship between the spindle pre-compression and the local maximum and local minimum values of the change rate in bearing stiffness differs, resulting in the increase rate in bearing stiffness on one side that differs from the decrease rate in bearing stiffness on the other. Before the spindle bearing is completely unloaded, the spindle stiffness depends on the superposition effect of the increase rate of the one-side bearing stiffness and the decrease rate of the other-side bearing stiffness. At this time, there is a certain fluctuation of the spindle axial operating stiffness (i.e., exhibiting the stiffness softening and hardening effect). When the spindle bearing is completely unloaded, the spindle stiffness is transformed into the single-side bearing stiffness, and the spindle axial stiffness increases with the gradual increase in axial displacement (i.e., the spindle shows stiffness hardening characteristics).

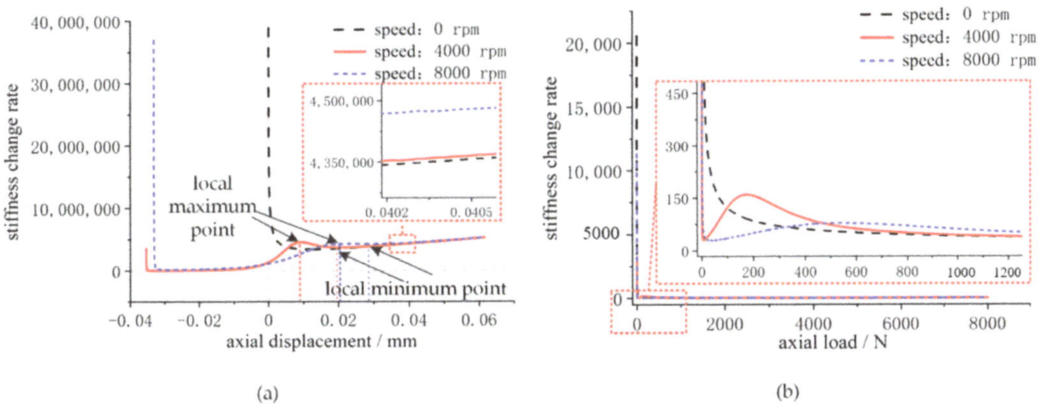

Figure 9. Single bearing stiffness change rate (corresponds to Figure 7): (**a**) relationship between stiffness change rate and axial displacement; (**b**) relationship between stiffness change rate and axial load.

When the spindle is at a higher speed, the local maximum and local minimum points in the bearing stiffness curve disappear gradually (as shown in Figure 9 for a speed of 8000 rpm), and the shape of the bearing stiffness curve resembles that of the bearing stiffness curve at zero speed (similar to a "V" shape that slopes severely to the right). As the spindle rotates, the axial displacement at the minimum point of the spindle stiffness change rate curve is very close to the axial displacement at the point of abrupt change in spindle stiffness, so the spindle pre-compression displacement is generally much larger than the axial displacement corresponding to the minimum value of the spindle stiffness change rate curve. Thus, at the stage where the front and rear bearings of the spindle are jointly loaded, the increase rate of one-side bearing stiffness is always greater than the decrease rate of the other-side bearing stiffness, and the axial stiffness of the spindle increases with the gradual increase in the axial displacement. When the spindle axial displacement is greater than the pre-deflection of single-side bearing, the spindle bearing is completely unloaded, and the spindle axial stiffness is transformed into the other-side bearing stiffness, and the spindle axial stiffness increases with the gradual increase in axial displacement.

4. Experimental Verification

As indicated in Figure 10, a test rig was built to investigate the spindle axial operating stiffness experimentally. The fixed-position preload experimental spindle is a mechanical spindle that is driven by an electric spindle and has a variable-frequency motor to change the speed. During the test, the system control and data acquisition box's motor control switch can be rotated to change the direction of motor rotation on the axial loading device. Adjust the motor to rotate forward first, then apply axial pressure to the spindle with the axial loading device. Adjust the motor reversal and the axial loading equipment to gradually release the axial pressure load to zero, and then provide a suitable amount of axial tension to the spindle. Finally, adjust the motor to rotate forward again, and the axial loading equipment will gradually reduce axial tension to zero. The force signal and related displacement signal are automatically captured in real-time by the NI®9215 data collection card during the test and sent to the computer for processing.

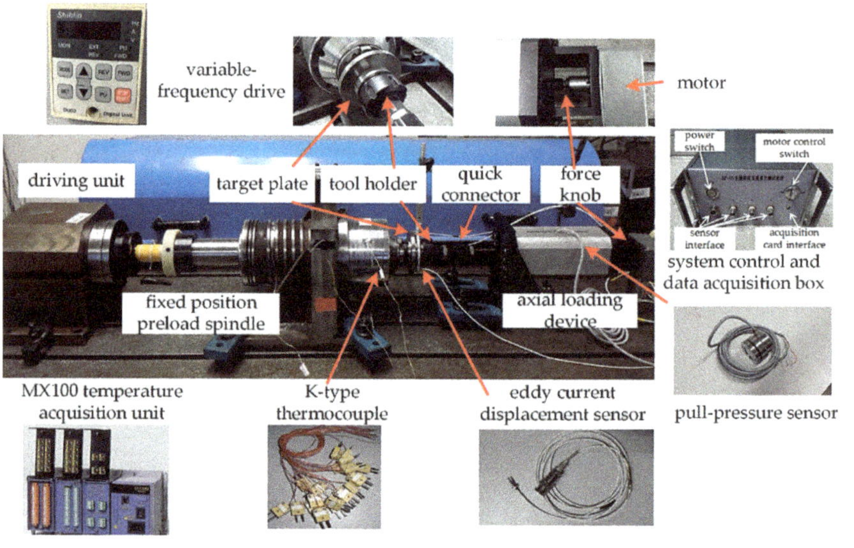

Figure 10. Spindle operating stiffness test rig.

The test spindle was set up in a back-to-back configuration, and the spindle's initial preload was set at 507 N by altering the width difference between the inner and outer spindle spacers. Before the test, the spindle components were cooled to room temperature. The influence of rotational speed on the spindle temperature field may be neglected here, since the test duration for each group is so brief. The spindle preload increases after the operation as a result of centrifugal load [29,30], and the change in spindle preload has a bigger impact on the spindle axial operating stiffness [4]. As a result, the real preload of the spindle at each speed of the experiment was computed during its execution. The test speeds were 0 rpm, 3000 rpm, and 6000 rpm, with the results displayed in Figure 11.

From Figure 11a–c, it can be seen that when the spindle axial displacement is small, the axial operating stiffness curve of the spindle with 0 rpm and 3000 rpm has certain fluctuations, but when the speed reaches 6000 rpm, the spindle axial operating stiffness curve fluctuation disappears, and the "sag" phenomenon appears, then the spindle exhibits a "stiffness hardening" characteristic. In addition, it can also be seen that the experimental results of the axial operating stiffness of the fixed-position preload spindle are consistent with the theoretical simulation results, and the two are in good agreement, where the maximum errors between the experimental and simulation results are 12.3%, 10.8%, and 10.5% for the spindle speeds of 0 rpm, 3000 rpm, and 6000 rpm, respectively. These errors can be attributed to three factors. To begin with, the filtered experimental data are distorted, and the filtering technique might be modified to lessen the error. Second, while bearing stiffness is the primary determinant of spindle stiffness, the experimental spindle stiffness also takes into account the stiffness of other spindle components, such as the draw bar mechanism. To lessen the error, the theoretical model can be enhanced further. Finally, there is a deviation between the real-time preload of the spindle and the true value of the preload in this experiment, which may be improved with more study into the transient preload measuring technique.

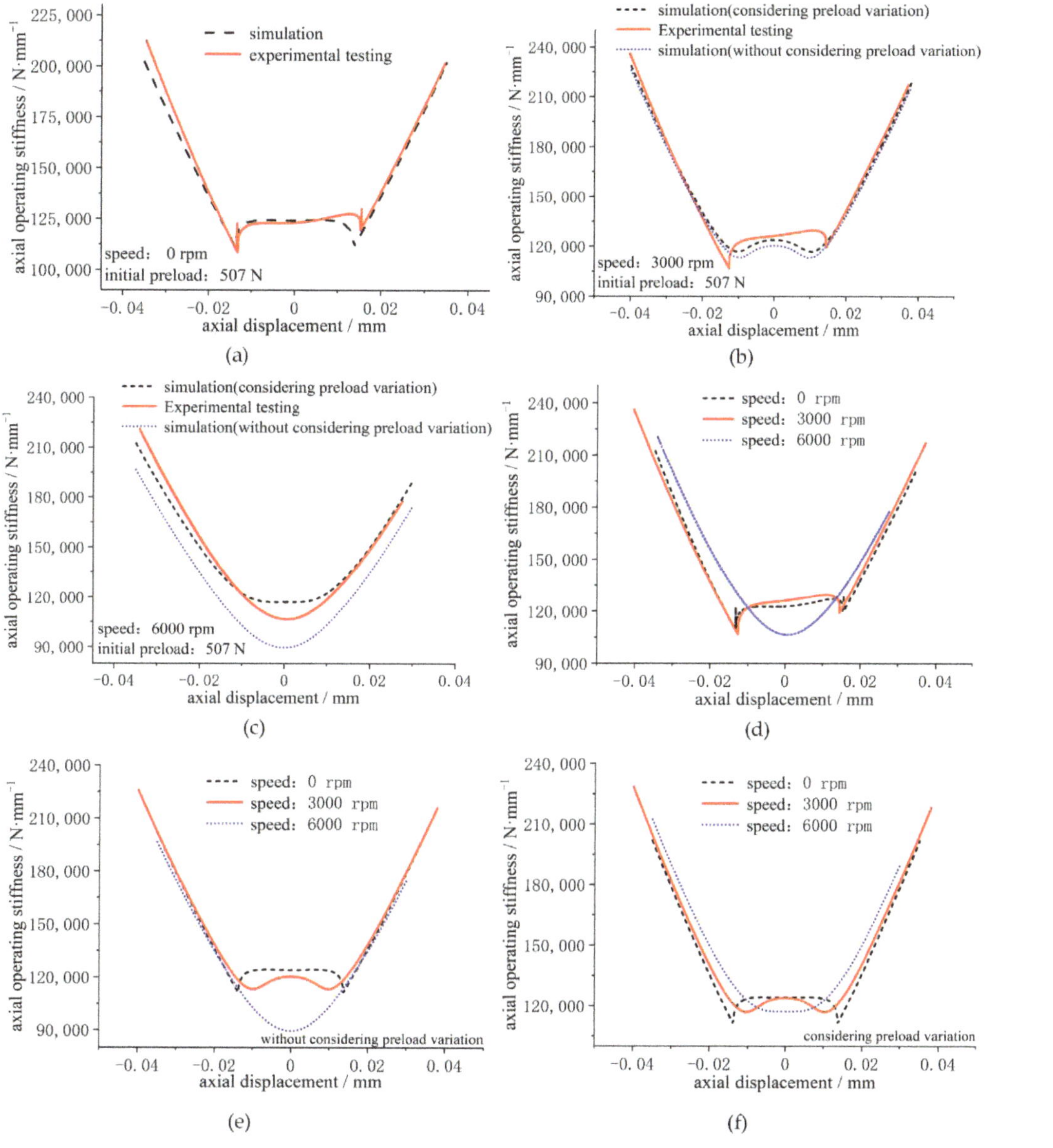

Figure 11. Experimental test on the operating stiffness of spindle at different speeds: (**a**) experimental testing (speed: 0 rpm); (**b**) experimental testing (speed: 3000 rpm); (**c**) experimental testing (speed: 6000 rpm); (**d**) experimental data comparison; (**e**) simulation data comparison (without considering preload variation); (**f**) simulation data comparison (considering preload variation).

As shown in Figure 11d, it can be seen that when the spindle speed is high, the spindle axial operating stiffness shows an obvious "sag" phenomenon, and the spindle exhibits the "stiffness hardening" characteristic. When comparing Figure 11d–f, it is clear that the experimental test findings in Figure 11d differ significantly from the simulation results, without considering preload variation in Figure 11e, but the trend is similar and less distinct from the simulation results when considering preload variation in Figure 11f, indicating that the change in spindle preload due to the speed factor has a large impact on this experimental test of spindle axial operating stiffness and is an important factor causing the change in spindle axial operating stiffness.

5. Conclusions

In this paper, the axial operating stiffness model of the fixed-position preload spindle is proposed, and the effect of the rotational speed on the softening and hardening characteristics of the spindle stiffness is studied based on the model. An experimental bench for testing the axial operating stiffness of the spindle was built to verify the accuracy and effectiveness of the model. In the analysis of the experimental results, the influence of speed-induced preload on the spindle axial operating stiffness is specially analyzed. The conclusions of the article are as follows:

1. For the fixed-position preload spindle with a smaller initial preload, the rotational speed has a greater influence on the load–displacement relationship and axial operating stiffness. However, the spindle speed has less effect on the load–displacement relationship and axial operating stiffness when the spindle's initial preload is greater.
2. When the spindle is rotated, the axial displacement at the stiffness mutation point is greater than the axial displacement at zero speed. In the meantime, when the rotational speed increases, the axial displacement of the spindle's axial operating stiffness mutation point decreases.
3. When the spindle is rotating slowly, the axial stiffness of the spindle fluctuates in the phase with small displacement increments. However, as the spindle speeds up, the stiffness fluctuation is gradually suppressed. When the spindle is operating at high speeds, the spindle axial stiffness displays a clear "sag" phenomenon, which shows the "stiffness hardening" characteristic.
4. During the experimental test of the axial operating stiffness of the spindle, the change in spindle preload due to the speed factor has a large impact on test results and is also an important factor causing the change in spindle axial operating stiffness.

Author Contributions: Conceptualization, J.L., Q.W. and X.S.; Data curation, J.L., W.K. and S.M.; Formal analysis, J.L., J.Q. and A.Q.; Funding acquisition, J.L., W.K. and S.M.; Investigation, J.L., Q.W. and X.S.; Methodology, J.L. and J.Q.; Project administration, J.L., W.K. and S.M.; Resources, J.L., A.Q. and W.K.; Software, J.L.; Supervision, J.Q.; Validation, J.L., J.Q. and A.Q.; Visualization, J.L., J.Q. and A.Q.; Writing—original draft, J.L.; Writing—review and editing, J.L. and X.S. All authors have read and agreed to the published version of the manuscript.

Funding: This research received no external funding.

Acknowledgments: The authors would like to thank Yongsheng Zhu and Ke Yan from Xi'an Jiaotong University for their help in the experimental testing.

Conflicts of Interest: The authors declare no conflict of interest.

Nomenclature

Capital Letter

F_a	Axial load applied to spindle
F_{ab}	Axial load applied to bearing
F_c	Centrifugal force
F_p	Preload applied to bearing
Z	Number of balls

Lowercase Letters

a	Distance between the raceway groove curvature center and the ball center
d_m	Diameter of bearing pitch circle
m_b	Ball mass
n	Spindle bearing speed

Greek Letters

α_f	Initial contact angle of bearing
α_p	Bearing contact angle with preload applied
α_i	The ball-inner contact angle
α_o	The ball-outer contact angle
δ	Ball-race deformation
δ_a	Spindle axial displacement
δ_{ab}	Bearing axial displacement
δ_1	Pre-deflection of front bearing
δ_2	Pre-deflection of rear bearing
ω_c	Bearing cage speed

Subscripts

o	Outer ring
i	Inner ring

References

1. De Lacalle, L.L.; Lamikiz, A.; Sánchez, J.; De Bustos, I.F. Recording of Real Cutting Forces along the Milling of Complex Parts. *Mechatronics* **2006**, *16*, 21–32. [CrossRef]
2. Van Luttervelt, C.; Childs, T.; Jawahir, I.; Klocke, F.; Venuvinod, P.; Altintas, Y.; Armarego, E.; Dornfeld, D.; Grabec, I.; Leopold, J.; et al. Present Situation and Future Trends in Modelling of Machining Operations Progress Report of the CIRP Working Group 'Modelling of Machining Operations'. *CIRP Ann.* **1998**, *47*, 587–626. [CrossRef]
3. Ehmann, K.; Kapoor, S.; DeVor, R.; Lazoglu, I. Machining Process Modeling: A Review. *J. Manuf. Sci. Eng. Trans. ASME* **1997**, *119*, 655–663. [CrossRef]
4. Li, J.; Zhu, Y.; Yan, K.; Yan, X.; Liu, Y.; Hong, J. Research on the Axial Stiffness Softening and Hardening Characteristics of Machine Tool Spindle System. *Int. J. Adv. Manuf. Technol.* **2018**, *99*, 951–963. [CrossRef]
5. Li, X.; Zhang, Y.; Hong, J.; Zhao, H.; Li, H. Experiment Analysis of Spindle Performance with Rolling Bearing under Non-Uniform Preload. *Proc. Inst. Mech. Eng. Part C J. Mech. Eng. Sci.* **2016**, *230*, 3135–3146. [CrossRef]
6. Hwang, Y.-K.; Lee, C.-M. A Review on the Preload Technology of the Rolling Bearing for the Spindle of Machine Tools. *Int. J. Precis. Eng. Manuf.* **2010**, *11*, 491–498. [CrossRef]
7. Uriarte, L.; Herrero, A.; Zatarain, M.; Santiso, G.; de Lacalle, L.L.; Lamikiz, A.; Albizuri, J. Error Budget and Stiffness Chain Assessment in a Micromilling Machine Equipped with Tools Less than 0.3 Mm in Diameter. *Precis. Eng.* **2007**, *31*, 1–12. [CrossRef]
8. Hsu, W.-Y.; Chen, J.-S. Error Analysis and Auto-Calibration for a Cartesian-Guided Tripod Machine Tool. *Int. J. Adv. Manuf. Technol.* **2004**, *24*, 899–909. [CrossRef]
9. Kim, B.-S.; Ro, S.-K.; Park, J.-K. Development of a 3-Axis Desktop Milling Machine and a CNC System Using Advanced Modern Control Algorithms. *Int. J. Precis. Eng. Manuf.* **2010**, *11*, 39–47. [CrossRef]
10. Wang, L.; Zhang, B.; Wu, J.; Zhao, Q.; Wang, J. Stiffness Modeling, Identification, and Measuring of a Rotating Spindle. *Proc. Inst. Mech. Eng. Part C J. Mech. Eng. Sci.* **2020**, *234*, 1239–1252. [CrossRef]
11. Yan, K.; Yan, B.; Wang, Y.; Hong, J.; Zhang, J. Study on Thermal Induced Preload of Ball Bearing with Temperature Compensation Based on State Observer Approach. *Int. J. Adv. Manuf. Technol.* **2018**, *94*, 3029–3040. [CrossRef]
12. Mahfoudi, F.; Boulanouar, L.; List, G. Experimental Study of the Influence of the Static Stiffness of Lathes on the Tool Wear Behaviour. *Int. Rev. Mech. Eng.* **2007**, *1*, 495–501.
13. Ma, S.; Zhang, X.; Yan, K.; Zhu, Y.; Hong, J. A Study on Bearing Dynamic Features under the Condition of Multiball—Cage Collision. *Lubricants* **2022**, *10*, 9. [CrossRef]
14. Cao, H.; Li, B.; Chen, X.; He, Z. Centrifugal Expansion of High-Speed Spindle and Its Influences on Bearing Dynamic Characteristics. *J. Mech. Eng.* **2012**, *48*, 59–64. [CrossRef]
15. Qianjian, G.; Jianguo, Y. Application of Projection Pursuit Regression to Thermal Error Modeling of a CNC Machine Tool. *Int. J. Adv. Manuf. Technol.* **2011**, *55*, 623–629. [CrossRef]
16. Matsubara, A.; Tsujimoto, S.; Kono, D. Evaluation of Dynamic Stiffness of Machine Tool Spindle by Non-Contact Excitation Tests. *CIRP Ann.* **2015**, *64*, 365–368. [CrossRef]
17. Slocum, A.H. *Precision Machine Design*; Society of Manufacturing Engineers: Michigan, MI, USA, 1992.
18. Olvera, D.; de Lacalle, L.; Compeán, F.; Fz-Valdivielso, A.; Lamikiz, A.; Campa, F. Analysis of the Tool Tip Radial Stiffness of Turn-Milling Centers. *Int. J. Adv. Manuf. Technol.* **2012**, *60*, 883–891. [CrossRef]
19. Laspas, T.; Theissen, N.; Archenti, A. Novel Methodology for the Measurement and Identification for Quasi-Static Stiffness of Five-Axis Machine Tools. *Precis. Eng.* **2020**, *65*, 164–170. [CrossRef]
20. Li, J.; Zhu, Y.; Yan, K.; Hong, J.; Yan, X. An Improved Thermo-Mechanical Model for Spindle Transient Preload Analysis. *Proc. Inst. Mech. Eng. Part J J. Eng. Tribol.* **2019**, *233*, 1698–1711. [CrossRef]
21. Li, J.; Zhu, Y.; Yan, K.; Yan, X.; Hong, J. A Novel Approach for Preload Measurement and Analysis of Fixed Position Preload Spindle. *Proc. Inst. Mech. Eng. Part C J. Mech. Eng. Sci.* **2019**, *233*, 3619–3632. [CrossRef]

22. Matsubara, A.; Yamazaki, T.; Ikenaga, S. Non-Contact Measurement of Spindle Stiffness by Using Magnetic Loading Device. *Int. J. Mach. Tools Manuf.* **2013**, *71*, 20–25. [CrossRef]
23. Tsuneyoshi, T. Spindle Preload Measurement and Analysis. In Proceedings of the ASPE Summer Topical Meeting, State College, PA, USA, 11–12 June 2007; pp. 35–38.
24. Zivkovic, A.; Zeljkovic, M.; Tabakovic, S.; Milojevic, Z. Mathematical Modeling and Experimental Testing of High-Speed Spindle Behavior. *Int. J. Adv. Manuf. Technol.* **2015**, *77*, 1071–1086. [CrossRef]
25. Rabréau, C.; Noël, D.; Loch, S.L.; Ritou, M.; Furet, B. Phenomenological Model of Preloaded Spindle Behavior at High Speed. *Int. J. Adv. Manuf. Technol.* **2017**, *90*, 3643–3654. [CrossRef]
26. Harris, T.A.; Kotzalas, M.N. *Advanced Concepts of Bearing Technology: Rolling Bearing Analysis*, 5th ed.; CRC Press: Boca Raton, FL, USA, 2006.
27. Harris, T.A.; Kotzalas, M.N. *Rolling Bearing Analysis: Essential Concepts of Bearing Technology*; Wiley: New York, NY, USA, 2007.
28. Antoine, J.-F.; Abba, G.; Molinari, A. A New Proposal for Explicit Angle Calculation in Angular Contact Ball Bearing. *J. Mech. Des.* **2006**, *128*, 468–478. [CrossRef]
29. Zhang, J.; Fang, B.; Hong, J.; Zhu, Y. Effect of Preload on Ball-Raceway Contact State and Fatigue Life of Angular Contact Ball Bearing. *Tribol. Int.* **2017**, *114*, 365–372. [CrossRef]
30. Zhang, J.; Fang, B.; Zhu, Y.; Hong, J. A Comparative Study and Stiffness Analysis of Angular Contact Ball Bearings under Different Preload Mechanisms. *Mech. Mach. Theory* **2017**, *115*, 1–17. [CrossRef]

lubricants

Article

Hard Negative Samples Contrastive Learning for Remaining Useful-Life Prediction of Bearings

Juan Xu [1], Lei Qian [1], Weiwei Chen [2] and Xu Ding [3,*]

[1] School of Computer and Information, Hefei University of Technology, Hefei 230601, China; xujuan@hfut.edu.cn (J.X.); 2020171144@mail.hfut.edu.cn (L.Q.)
[2] Shanghai Aerospace Control Technology Institute, Shanghai 201109, China; youthjiang@126.com
[3] The Institute of Industry and Equipment Technology, Hefei University of Technology, Hefei 230601, China
* Correspondence: dingxu@hfut.edu.cn

Abstract: In recent years, deep learning has become prevalent in Remaining Useful-Life (RUL) prediction of bearings. The current deep-learning-based RUL methods tend to extract high dimensional features from the original vibration data to construct the Health Indicators (HIs), and then use the HIs to predict the remaining life of the bearings. These approaches ignore the sequential relationship of the original vibration data and seriously affect the prediction accuracy. In order to tackle this problem, we propose a hard negative sample contrastive learning prediction model (HNCPM) with encoder module, GRU regression module and decoder module, used for feature embedding, regression RUL prediction and vibration data reconstruction, respectively. We introduce self-supervised contrast learning by constructing positive and negative samples of vibration data rather than constructing any health indicators. Furthermore, to avoid the subtle variability of vibration data in the health stage to aggravate the degradation features learning of the model, we propose the hard negative samples by cosine similarity, which are most similar to the positive sample. Meanwhile, a novel infoNCE and MSE-based loss function is derived and applied to the HNCPM to simultaneously optimize a lower bound on mutual information of the positive and negative sample over life cycle, as well as the discrepancy between true and predicted values of the vibration data, such that the model can learn the fine-grained degradation representations by predicting the future without any HIs as labels. The HNCPM is validated on the IEEE PHM Challenge 2012 dataset. The results demonstrate that the prediction performance of our model is superior to the state-of-the-art methods.

Keywords: positive and negative samples; contrastive learning; gated recurrent unit; remaining useful-life prediction

Citation: Xu, J.; Qian, L.; Chen, W.; Ding, X. Hard Negative Samples Contrastive Learning for Remaining Useful-Life Prediction of Bearings. *Lubricants* **2022**, *10*, 102. https://doi.org/10.3390/lubricants10050102

Received: 15 April 2022
Accepted: 18 May 2022
Published: 21 May 2022

1. Introduction

Bearings undergo an irreversible degradation process during use that eventually leads to bearing breakdown. Via predicting the Remaining Useful Life (RUL) of bearings, one can avoid missing the best time for maintenance [1] and remanufacturing [2]. However, in realistic industrial scenarios, due to noise, variation in life cycle and other prediction uncertainties [3], RUL prediction is a challenging issue [4].

Generally, RUL prediction of bearings is implemented in two different strategies [5]: physically based approaches [6–8] and data-driven [9–12] approaches. Compared with the former, data-driven RUL methods use historical data directly to build prediction models, which avoids the difficulties in modeling of the physical model and has become the prevalent approach in recent years [13–15]. For instance, Xia et al. used a denoising autoencoder to classify the original signal into different degradation stages and extracted representative features directly from the original signal using DNN. Then they obtained RUL values for each stage using deep regression models [16]. Guo et al. proposed a recurrent neural-network-based health indicator(RNN-HI) with fairly high monotonicity and correlation values, which is beneficial to bearing RUL prediction [17].

Among several data-driven methods, deep learning-based methods obtained considerable results in bearing RUL prediction, which do not require manual feature design and build an end-to-end deep neural network to map the relationship between the degradation process and the original sensory data [18].

The existing deep-learning-based approaches usually construct health indicators (HI) that can represent the degradation trend of bearing performance, then predict the degradation trend using deep-learning regression models [18]. The effectiveness of these approaches largely relies on these HIs accurately modeling the degradation trend of bearings. However, in real-world industrial applications, the degradation processes of different bearings are diverse. The existing HIs usually describe the overall trend of the whole life cycle, but fail to represent the changes of local details. Using these inappropriate HIs as model input will directly restrict the prediction performance of the model.

Noted that the original time-domain vibration data of bearings contains the sequential relationship with respect to the degradation trend of bearings, thus predicting that the original vibration data, instead of any designed HI, can more accurately evaluate the model performance. However, the time-domain vibration data changes slightly in the health stage; the existing well-developed methods are less capable of capturing the latent features of the bearing data at this stage.

To circumvent the aforementioned problems, we propose an end-to-end hard negative sample contrastive learning prediction model, termed HNCPM. We construct a three-layer, one-dimensional convolutional-based encoder module to map the high-dimensional original vibration data to a low-dimensional feature space to facilitate the model's computation. Next, we adopt a Gated Recurrent Unit with a decoder module in this feature space to learn the sequence relationship of vibration data for RUL prediction. Importantly, the HNCPM introduces self-supervised contrastive learning to construct positive and negative samples of vibration data, rather than supervised leaning via any HI labels. Considering there is no significant variability between the positive and negative samples in the health stage, thus we select the most similar negative sample to the positive sample as the hard negative sample via cosine similarity, to improve the fine-grained feature identification of the model. Finally, we design the novel loss function of the model combining the Mean Square Error(MSE) with infoNCE [19] for self-supervised training from the original vibration data.

The main contributions of this work are summarized as follows:

1. Unlike existing supervised RUL prediction methods, this study explores a more practical self-supervised-leaning RUL prediction method that directly learns the sequence relationship from the original vibration data instead of using any HIs as labels for the model supervised training. We propose the HNCPM with the encoder module, GRU regression module and decoder module, respectively used for feature embedding, regression RUL prediction and vibration data reconstruction.
2. To encounter this dilemma that the subtle variability between the positive and negative samples in the healthy stage makes the model fail to learn the latent sequence features, we select the negative sample that is most similar to the positive sample as the hard negative sample. Correspondingly, we design a novel loss function combining the MSE with infoNCE loss to improve the fine-grained feature representation of the model.
3. The performance of the proposed HNCPM is comprehensively evaluated on the IEEE PHM Challenge 2012 dataset. The comparative experimental results show that the HNCPM is superior for the excellent prediction accuracy than the state-of-the-art methods with respect to different bearings.

2. Related Works

2.1. Deep-Learning-Based Approaches for Rul Prediction

The existing deep-learning-based approaches usually include two steps: constructing the bearing HIs for model training that can represent the degradation trend of bearing performance, and designing the deep learning regression models to predict the degradation trend (usually the HIs).

In terms of model design, it generally includes: CNN [20], RNN [21] and AE [22]. Xu et al. used the SAE to extract features of bearing data to construct HI [23]. Wang et al. [24] applied deep separable convolutional networks to learn high-level representations from the original signal and then predict RUL. Wu et al. [25] proposed Deep Long Short-Term Memory (DLLSTM) networks, and Han et al. proposed a Transferable Convolutional Neural Network(TCNN) to accurately predict the bearing RUL under different failure behaviors [26].

The common model structure can no longer fulfill the demand of researchers for superior prediction, and many recent studies have combined the advantages of different models to propose model variants. For instance, Luo et al. proposed a novel convolution-based attention mechanism bidirectional long and short-term memory (CABLSTM) network to achieve the end-to-end lifetime prediction of rotating machinery [27]. Meng et al. proposed CLSTM by conducting convolutional operation on both the input-to-state and state-to-state transitions of the LSTM to learn high-level features in the time-frequency domain for RUL prediction [28].

With respect to the HI construction, i.e., the RUL labels, since the damage extent of bearings cannot be directly observed, the RUL labels are almost unavailable in real-world scenarios. Hence, there is no uniform criterion of HI construction for RUL prediction models. The existing studies usually extracted the different fault characteristics from the original sensory signal. She et al. [29] proposed a health indicator construction method based on a Sparse Auto-encoder with a Regularization (SAEwR) Model for rolling bearings. Zhang et al. used the summation of the mean maximum radius of the different datasets divided by the k-means clustering algorithm as the health indicator, and then used the local outlier coefficient algorithm to eliminate the outliers' influence [30]. Li et al. designed the generative adversarial network to learn the data distribution in the health states of machines, using the output of the discriminator as HI [31].

In summary, the performance of the existing HIs-based deep-learning approaches heavily relies on whether the HIs are constructed properly. Nevertheless, the models usually fail to represent the changes of local details, while capturing the overall gradation trend of the whole life-cycle of bearings, which restricts the prediction performance of the models.

Therefore, in this paper we introduce self-supervised contrastive learnin, directly using the original vibration data, rather than any HIs, to learn the sequential relationship with respect to the degradation trend of bearings.

2.2. Contrastive Learning

Self-supervised learning is committed to avoiding the manually annotating large-scale datasets by setting up pretext tasks to learn data representations [32–34]. The learning process is unsupervised and the trained model can be used for multiple downstream tasks. Among self-supervised learning, contrastive learning is the most widely concerned. It uses discriminative modeling to learn useful representations of unlabeled data. Several contrastive learning models, e.g., MoCo [35], SimCLR [36] and BERT [37] provide competitive results in comparison with supervised learning models within the pretext task of image classification or next-sentence prediction.

Specifically, contrastive learning is a discriminative method that aims to compact the similar samples and discriminate the dissimilar samples [38]. The models include: similarity and dissimilarity distributions for sampling positive and negative samples of the query, one or more encoders for each data pattern, and comparative loss functions for evaluating a batch of positive and negative pairs.

Given a set $X = \{x_i \mid i = 1, 2, \ldots, n\}$ of n samples, for any input sample x, q is the encoded representation of x, function $q = f(x)$, denote k^+, k^- as a positive and negative sample of the sample x.

$$s(q, k^+) \gg s(q, k^-) \quad (1)$$

s is a measure of the similarity between the embedded vectors, or it can be a calculation of the distance between the vectors.

To the best of our knowledge, there are limited studies that introduce contrastive learning to the RUL prediction. Mohamed Ragab et al. [39] propose a contrastive adversarial domain adaptation (CADA) method for cross-domain RUL prediction, which transfers knowledge for RUL prediction from one condition(distribution/domain) to another. That is no longer the problem dealt with in this paper. Our paper aims to discuss an unsupervised RUL prediction approach to learn degradation representations from original vibration data by using certain contrastive learning models.

Furthermore, with regard to existing contrastive learning approaches, the positive samples generated by data enhancement of the original data (e.g., rotating, segmenting), while samples are not generated from the same view, are negative samples from each other. Such approaches of constructing samples ignores the role of negative samples in model training. In fact, negative samples can teach learning models to correct their errors faster in representation learning. More importantly, the information-rich counterexamples are intuitively those that map far away from the positive sample but are closer to the positive sample than to other negative samples [40].

Inspired by this idea, in this paper we construct positive and negative samples based on the different temporal relationship of the vibration data, enabling the model to learn the sequence features by discriminating whether the temporal relationship of the vibration data is correct or not. Moreover, we construct hard negative samples to express the most similar to the positive sample to improve the fine-grained feature representation of the model.

3. Proposed Method

As shown in Figure 1, the complete framework of the proposed method includes construction of the positive sample and hard negative samples, the hard negative sample contrastive learning prediction model and corresponding optimization function.

3.1. Positive Sample and Hard Negative Sample Construction

Without loss of generality, we denote that the original vibration data of bearing is $D = \{b_i\}_{i=1}^{M}$, the positive sample dataset is $D_P = \{X_P, Y_P\}$, $X_P = \{x_1, x_2, \ldots, x_N\}$, and $Y_P = \{y_1, y_2, \ldots, y_N\}$, where $x_i = [b_i, \ldots, b_{i+timestep-1}]$ and $y_i = b_{i+timestep}$, respectively. The timestep is the timestep of the vibration sequence. In this paper, it is set to six.

Random $y_j (i \neq j)$ can form a negative sample with y_i. We use the cosine similarity to calculate the negative sample y_j with the highest similarity to y_i.

$$w(y_i, y_h) = max \frac{(y_i, y_j)}{||y_i|| \cdot ||y_j||} \quad y_j \in Y_P, i \neq j \quad (2)$$

when $y_j = y_h$, w is maximum, at which x_i and y_h form a hard negative sample, and x_i and y_i form a positive sample. We use (x_i, y_h) to construct the hard negative sample dataset $D_N = \{X_N, Y_N\}$, denote $X_N = \{x'_1, x'_2, \ldots, x'_N\}$, $Y_N = \{y'_1, y'_2, \ldots, y'_N\}$.

Given a dataset $C = \{c_1, c_2, \ldots, c_k\}$ of k random samples containing one sample from X_P and $k-1$ samples from X_N, corresponding prediction labels, the positive and negative sample labels are $P = \{p_1, p_2, \ldots, p_k\}$ and $T = \{t_1, t_2, \ldots, t_k\}$, respectively. Initialize the positive and hard negative sample label set $T = \{t_i\}_{i=1}^{N}$.

$$t_i = \begin{cases} 1, & c_i \in D_P \\ 0, & c_i \in D_N \end{cases} \quad (3)$$

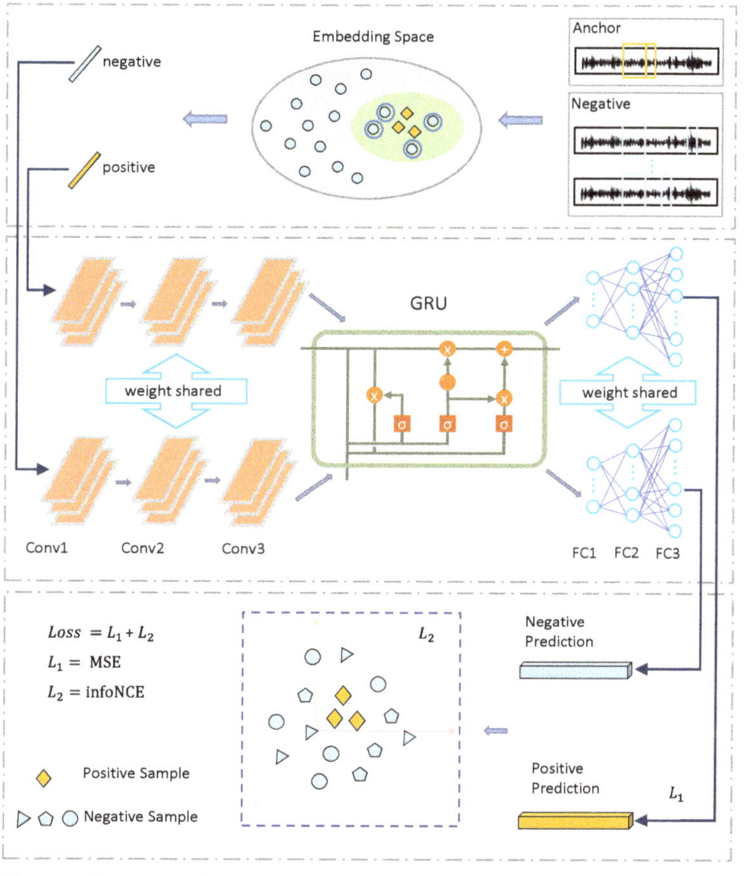

Figure 1. The proposed method.

3.2. Hard Negative Contrastive Prediction Model (Hncpm)

After the positive and negative sample pairs are generated based on the original vibration data of bearing $D = \{b_i\}_{i=1}^{M}$, the feature sequence are extracted through an encoder composed of three layers of a one-dimensional convolutional network, which compresses the original vibration data into a low-dimensional feature space. Then, we use a gated recurrent unit as a regression module to learn the sequence relationship of the feature sequence. Finally a decoder composed of three fully connected layers predicts the vibration data in future time-periods (i.e., RUL).

Unlike the existing supervised learning model to predict the HI labels for RUL, the HNCPM directly predicts the vibration data in future time-periods with the purpose of allowing the model to better fit the time domain graph curve of the original vibration data, so the potential representation of the context is captured in the gated recurrent units and the final prediction results are obtained through a decoder. The model structure is shown in Table 1.

The feature encoder module composed of three convolutional layers is formulated as follows:

$$\hat{x}_i = ReLU(W_c \otimes x_i) \tag{4}$$

where $\otimes$ is the valid cross-correlation operator and W_c is the weight.

Table 1. Model structural parameters.

No.	Symbol	Operator	Shape	Kernel Size	Stride
1	Input	Input signal	(62,560)	-	-
2	Conv1d1	Convolution	(6512)	4	4
3	Conv1d2	Convolution	(6256)	2	2
4	Conv1d3	Convolution	(650)	1	2
5	GRU	prediction	50	-	1
6	FC1	Fully-connected	256	-	-
7	FC2	Fully-connected	512	-	-
8	FC3	Fully-connected	2560	-	-

The regression module uses gated recurrent units, which can effectively suppress gradient disappearance or exploding gradient when capturing long sequence data association. It is better than traditional RNN and has less computational complexity than LSTM.

$$r_i = \sigma(W_r \cdot [h_{i-1}, \hat{x}_i]) \tag{5}$$

$$z_i = \sigma(W_z \cdot [h_{i-1}, \hat{x}_i]) \tag{6}$$

$$\widetilde{h}_i = tanh(W \cdot [r_i \odot h_{i-1}, \hat{x}_i]) \tag{7}$$

$$h_i = (1 - z_i) \odot h_{i-1} + z_i \odot \widetilde{h}_i \tag{8}$$

where h_i is the hidden state at time i. r_i, z_i, $\widetilde{h}_i$ are the reset, update, and new gates, respectively. W_r,W_z, and W are the r_i weight, z_i weight, $\widetilde{h}_i$ weight, respectively.

After the regression module, the decoder consisting of three fully connected layers is used for decoder enhancement of the dimensionality of h_i, with the following output:

$$\hat{y}_i = W_l \cdot h_i + b \tag{9}$$

where W_l is the weight of the fully connected layers.

3.3. Optimization Function

The loss function of the model in this paper consists of two parts: (1) the regression loss function L_1; (2) the contrastive prediction loss function L_2.

The regression loss function L_1 uses the mean square error to predict the bearing data of the next time period based on the vibration data of the previous time segment. The input dataset is X_P, Y_P for the model training. The loss function is as follows:

$$L_1 = \frac{1}{N} \sum_{i=1}^{N} (\hat{y}_i - y_i)^2 \tag{10}$$

where N is the total number of samples.

The contrastive prediction loss function L_2 is the loss function $infoNCE$ [19], which is commonly used in contrastive learning. Feed set C into HNCPM, the output is $\hat{P} = \{\hat{p}_1, \hat{p}_2, \ldots, \hat{p}_k\}$.

According to previous study [19], the higher the density ratio function $f(\hat{p}_i, p_i)$, the larger the mutual information of $\hat{p}_i$ and p_i, and the more capable of predicting p_i by $\hat{p}_i$. The formula is as follows:

$$f(\hat{p}_i, p_i) = exp(s(\hat{p}_i, p_i)) \tag{11}$$

$s(\hat{p}_i, p_i)$ measures the degree of similarity between the predicted result $F(c_i)$ and p_i.

$$s(\hat{p}_i, p_i) = \hat{c}_i \odot p_i \tag{12}$$

where $\odot$ is the Hadamard multiplier.

$$L_2 = -\mathbb{E}_P \left[log \frac{f(\hat{p}_i, p_i)}{\sum_{\hat{p}_i \in P} f(\hat{p}_i, p_i)} \right] \tag{13}$$

Finally, the total loss function of HNCPM is as shown:

$$L = L_1 + \alpha L_2 \qquad \alpha \in (0, 1) \tag{14}$$

where α is the hyperparameter.

The pseudo code of the algorithm for training is shown in Algorithm 1.

Algorithm 1 Hard Contrastive Prediction Model

Input:

original bearing samples: $D = \{b_1, b_2, \ldots, b_M\}$.
positive samples: $X_P = \{x_1, x_2, \ldots, x_N\}$, $y_i = [b_i, \ldots, b_{i+timestep-1}]$, $Y_P = \{y_1, y_2, \ldots, y_N\}$, $y_i = b_{i+timestep}$.
hard negative samples: $X_N = \{x'_1, x'_2, \ldots, x'_N\}$, $Y_N = \{y'_1, y'_2, \ldots, y'_N\}$, Y_N select by (2).
F consists of encoder, gated recurrent unit, decoder. θ is the proposed model parameters.
x_t is the test bearing data

while $i <$ number of iterations **do**
  for i to N **do**
    $\hat{x}_i = F(x_i)$
    Compute loss $L1$ by (8)
    Construct C and T by (3)
    $\hat{p}_i = F(c_i)$
    Calculate the density ratio: $f(\hat{p}_i, p_i)$ by (8),(9)
    Compute infoNCE loss function $L2$ by (11)
    Update θ by $L = L_1 + \alpha L_2 \quad \alpha \in (0, 1)$
  end for
end while
save F

Output: the model prediction $F(x_t)$

4. Experimental Section

In this section, we use the IEEE PHM Challenge 2012 bearing dataset to validate the effectiveness of the proposed method. The Mean Absolute Error (MAE) and Root Mean Square Error(RMSE) are selected as indicators to evaluate the model performance. The smaller the value of MAE and RMSE, the more superior the RUL model is. The MAE and RMSE, respectively, are expressed as follows.

$$MAE = \frac{1}{N} \sum_{i=1}^{N} \mid p_i - y_i \mid \tag{15}$$

$$RMSE = \sqrt{\frac{1}{N} \sum_{i=1}^{N} (p_i - y_i)^2} \tag{16}$$

where y_i denotes the true value of the sample, p_i denotes the predicted value, and N is the total number of samples.

4.1. Data Description

As shown in Figure 2, The PRONOSTIA test platform contains a rotating part, a load part, and a data collection part. The motor power is 250 W. The power is transferred to the bearing by the axis of rotation. The accelerated degradation experiments was conducted on this platform to generate the run-to-failure vibration signal. The acceleration sensors are placed in horizontal and vertical directions to collect vibration signals under three working conditions. The sampling frequency is 25.6 kHz. The vibration signal is recorded every 10 s, and each acquisition lasts 0.1 s. For example, under the first working condition, the vibration signals of six bearings were collected, namely, bearing 1_2, bearing 1_3, bearing 1_4, bearing 1_5, bearing 1_6 and bearing 1_7. The motor's rotation speed is 1800 rpm, and the load is 4000 N.

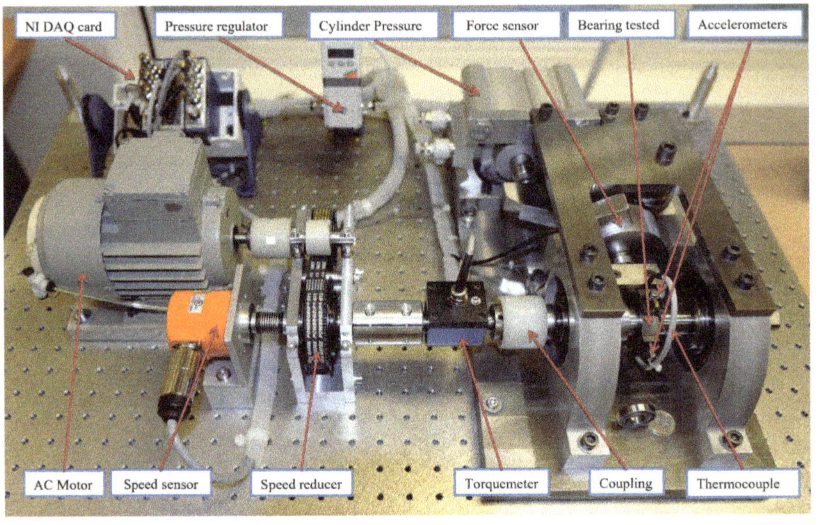

Figure 2. PRONOSTIA test platform.

In the following different experiments, the vibration signals of bearing 1_2 are the training set, while bearing 1_4, bearing 1_5, bearing 1_6, bearing 1_7, bearing 2_3, bearing 2_4, bearing 2_5, and bearing 2_6 are used as the testing set, respectively, as shown in Table 2.

4.2. *Weighting Factor Analysis*

In Equation (12), the weights α of the loss function are empirically determined. Hence, we perform several experiments to discuss their influence on the model performance.

Herein, the experiments α are taken as 0.1, 0.3, 0.5, 0.7, 0.9, 1, respectively. Bearing 1_4, bearing 1_5, bearing 2_4, and bearing 2_6 are used as the test set and RMSE is used as the model predictive ability index. For each series of experiments, we repeat the experiment ten times and report the average performance. Figure 3 depicts the box plot of RUL prediction results for four different bearings at different α.

When $\alpha = 0.1$, the weight of L_2 in the total loss L is too small and plays a low optimization role in the model, resulting in poor prediction results in bearing 1_4, bearing 2_4, bearing 2_6. In contrast, when $\alpha = 1$, the model is overly concerned with the fine-grained recognition of vibration data, resulting in overfitting; thus the model has an unsatisfactory generalization effect in testing bearing 1_5, bearing 2_4, and bearing 2_6. When $\alpha = 0.7$, our method performs best on bearing 2_6, and the maximum and minimum values of RMSE are lower than other weights. It also has the same superiority on bearing 1_5 and bearing 2_4 compared with other weights. Considered comprehensively, the prediction accuracy of the model is satisfactory when $\alpha = 0.7$.

Table 2. The details of PHM2012 dataset.

Data Set	Sample Number	Sample Dimension	Rotation Speed	Load	Division
Bearing 1_2	871	(8,712,560)	1800 rpm	4000 N	training
Bearing 1_3	1802	(18,022,560)			testing
Bearing 1_4	1139	(11,392,560)			testing
Bearing 1_5	2302	(23,022,560)			testing
Bearing 1_6	2302	(23,022,560)			testing
Bearing 1_7	1502	(25,022,560)			testing
Bearing 2_3	1202	(12,022,560)	1650 rpm	4200 N	testing
Bearing 2_4	612	(6,122,560)			testing
Bearing 2_5	2002	(20,022,560)			testing
Bearing 2_6	572	(5,722,560)			testing
Bearing 2_7	172	(1,722,560)			testing
Bearing 3_3	352	(3,522,560)	1500 rpm	5000 N	testing

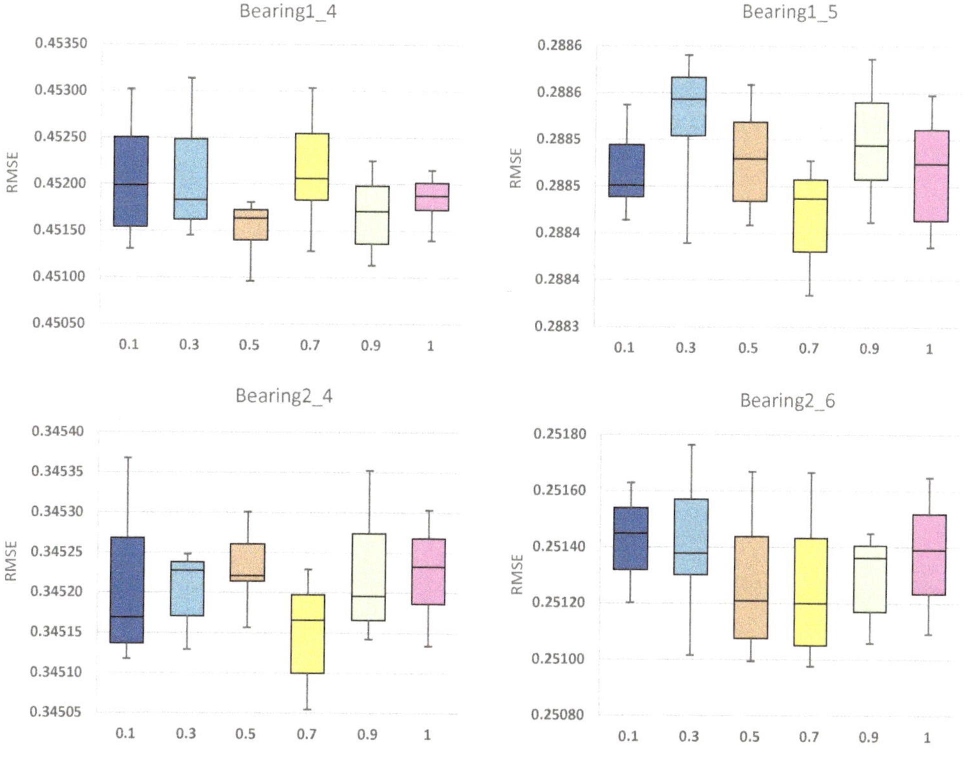

Figure 3. RUL prediction results with different weighting factor α.

4.3. Ablation Experiments

In this section, we perform ablation experiments to verify the contribution of each module in our model. We keep the model structure unchanged but without contrast learning (termed as no-contrast), and GRU model without encoder and decoder modules (termed as GRU) to compare with HNCPM. MAE is used as the predictive ability index of the models. The experimental results are demonstrated in Figure 4.

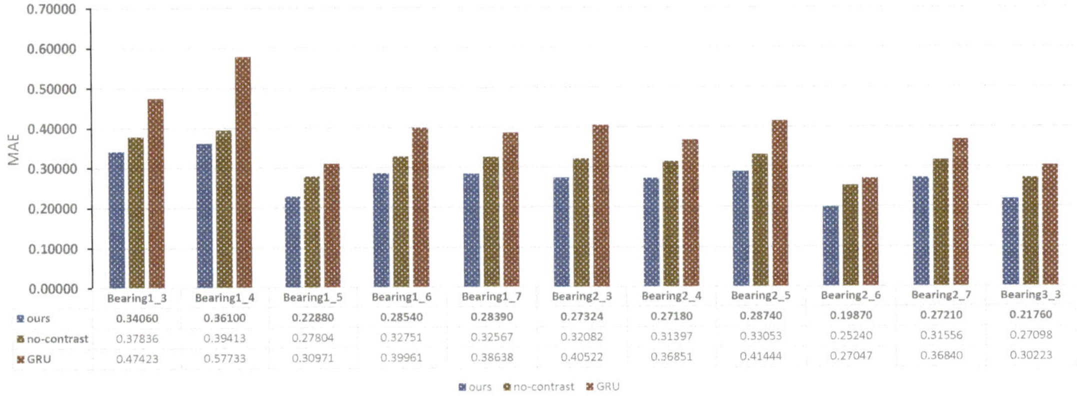

Figure 4. Ablation experiment results.

HNCPM has the best superiority on bearing 2_6, which is 0.05 lower than the no-contrast model and 0.07 lower than the GRU model. In addition, the model has the largest MAE value on bearing 1_4, which is 0.03 and 0.22 lower than the no-contrast and GRU model, respectively.

Moreover, compared to the GRU model, the no-contrast model has a better prediction performance because it contains the feature extraction layer to extract high-dimensional degradation features from the original vibration data, facilitating the RUL regression prediction. However, no-contrast ignores latent sequence features during the training, while HNCPM introduces contrast learning to improve the fine-grained model training and improve the prediction efficiency compared to no-contrast. Overall, the prediction results on all bearing data indicate that the proposed HNCPM method is significantly superior to the other two models.

4.4. Comparison with State-Of-The Art Methods

In this section, our proposed HNCPM is compared with the state-of-the-art novel rolling bearing health-prediction methods based on CNN and BiLSTM models [41], and BiLSTM with attention mechanism [42]. In addition, we compare general encode and regression model combinations (i.e., SAE+GRU, CNN+LSTM) to evaluate the model prediction performance. MAE is used as the predictive ability index of the models. The overall RUL prediction results using the aforementioned models are depicted in Figure 5.

It can be seen that SAE model has the worst performance. It is well known that SAE has strong signal denoising ability, but its feature extraction ability in time series data is not as good as CNN. Attention increases the weight of important features, but it also ignores the bearing sequence information that is masked by noise, which results in an unstable prediction performance on multiple bearings. Therefore, the performances of CNN-based models are superior to the other model.

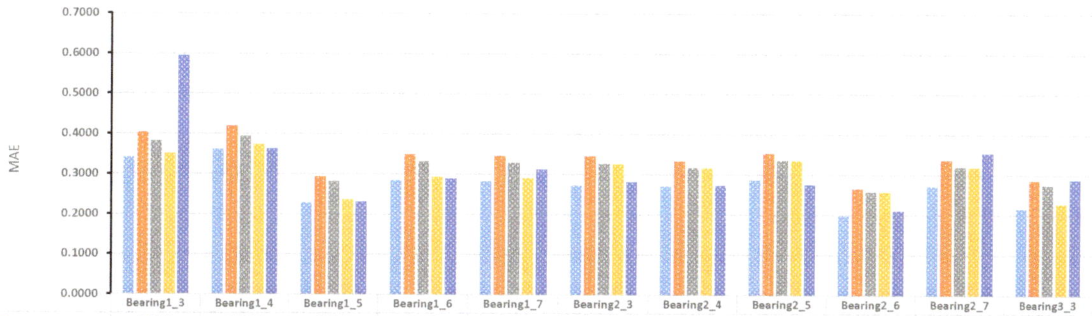

	Bearing1_3	Bearing1_4	Bearing1_5	Bearing1_6	Bearing1_7	Bearing2_3	Bearing2_4	Bearing2_5	Bearing2_6	Bearing2_7	Bearing3_3
ours	**0.3406**	**0.3610**	**0.2288**	**0.2854**	**0.2839**	**0.2732**	**0.2718**	**0.2874**	**0.1987**	**0.2721**	**0.2176**
SAE+GRU	0.4036	0.4192	0.2942	0.3489	0.3457	0.3459	0.3336	0.3528	0.2659	0.3368	0.2862
CNN+LSTM	0.3820	0.3937	0.2818	0.3316	0.3292	0.3264	0.3170	0.3356	0.2575	0.3194	0.2745
CNN+BiLSTM	0.3511	0.3741	0.2373	0.2940	0.2912	0.3264	0.3170	0.3356	0.2575	0.3194	0.2293
BiLSTM+Attention	0.5947	0.3623	0.2317	0.2903	0.3130	0.2815	0.2735	0.2768	0.2109	0.3547	0.2894

Figure 5. RUL prediction results of different comparison models.

In addition, The MAE value of CNN+BiLSTM is, on average, 0.2 lower than CNN+LSTM, as the BiLSTM improves the model prediction ability through backward prediction of past bearing data through future data. However, BiLSTM has one more backward hidden layer relative to LSTM, which boosts the number of model parameters and increases the model overhead, and the prediction performance is not as efficient as HNCPM.

Compared with the other models, HNCPM has the smallest MAE value at bearing 2_6 and is 0.06 lower than the second-best model (i.e., CNN+BiLSTM). Meanwhile, HNCPM has the largest MAE value at bearing 1_4, i.e., 0.361, but it is also at least 0.01 lower than the other models. The experiments demonstrate that HNCPM has a superior performance in RUL prediction.

4.5. Prediction Results Visualization

In this section, unlike the existing RUL prediction techniques which predict the degradation index (DI) [41] or the root mean square(RMS) [42] from the vibration data, we use manifold learning to visualize the fitting curve between the prediction and the original data to evaluate the prediction performance of the model more intuitively. We use the vibration data of the first 1800 time points to predict the remaining vibration data on bearing 1_3 and bearing 2_3, respectively. Figure 6a,b are the original time domain vibration data of bearing 1_3 and bearing 2_3, respectively. Each time point of vibration data is 2560 in dimension. In order to illustrate the prediction performance more intuitively, TSNE is used to transform the 2560 dimension data of each time point into 1 dimension on bearing 1_3 and bearing 2_3, as in Figure 6c,d. The blue curve is the real vibration data; the orange curve is the prediction data.

From Figure 6c, some of the prediction data are biased with the real data curve in the rapid degradation stage because the vibration data of bearing 1_3 changes drastically at the end of life, which is difficult for the prediction of the model. Meanwhile it can be found from Figure 6d that the vibration data of bearing 2_3 changes slowly in the whole life-cycle; thus, the prediction curve of model is more feasible to fit the real data curve.

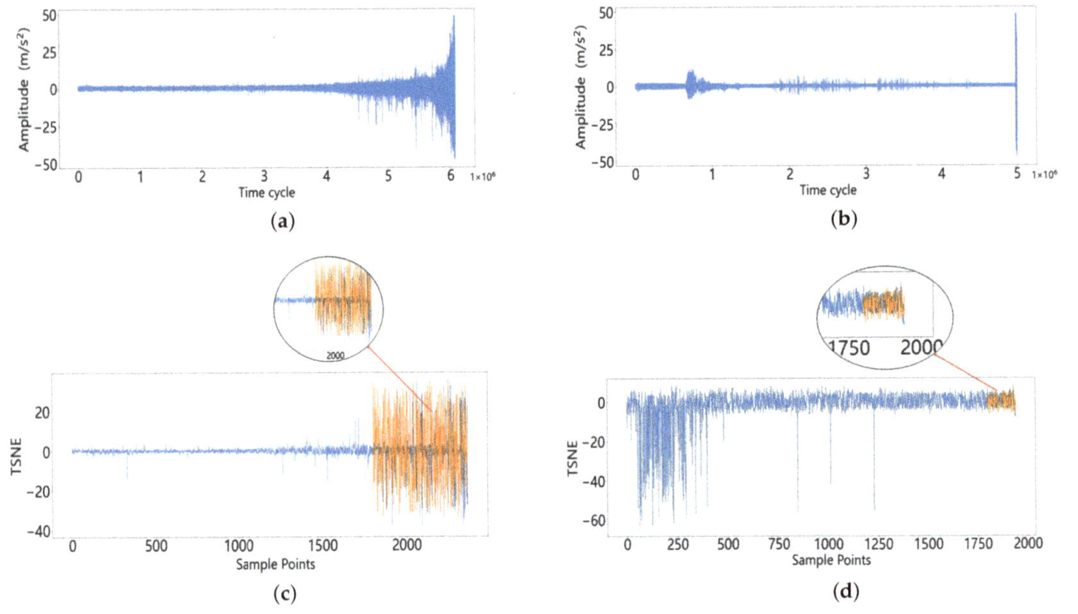

Figure 6. (**a**) Bearing1_3 data set; (**b**) bearing2_3 data set; (**c**,**d**) are the prediction fitting curves of TSNE corresponding to (**a**,**b**), respectively. The blue represents the real data; the orange represents the prediction.

5. Conclusions

In this article, we proposed a novel RUL prediction approach of bearings, termed HNCPM, which introduces self-supervised contrast learning to directly use the original vibration data for model training rather than any HIs as the RUL label, improving the model's ability to extract sequential relationships from the original data. Meanwhile, we construct a hard negative sample and further design a novel loss function, combining the MSE with infoNCE loss to address the dilemma for the model to learn fine-grained degradation features due to the insignificant variation of positive and negative samples in the bearing health stage. Our experimental results demonstrate the superiority of the proposed HNCPM method for RUL prediction and the fine fit of the model predictions to true RUL observed by manifold learning.

Author Contributions: Conceptualization, J.X.; writing—original draft preparation, L.Q.; formal analysis, W.C.; validation, X.D. All authors have read and agreed to the published version of the manuscript.

Funding: This work was supported in part by the National Key Research And Development Plan under Grant 2018YFB2000505, in part by the Key Research and Development Plan of Anhui Province under Grant 202104a04020003, and in part by the Fundamental Research Funds for the Central Universities under Grant PA2021KCPY0045.

Institutional Review Board Statement: Not applicable.

Informed Consent Statement: Not applicable.

Data Availability Statement: Public datasets used in our paper: https://github.com/wkzs111/phm-ieee-2012-data-challenge-dataset (accessed on 10 April 2022).

Conflicts of Interest: The authors declare no conflicts of interest.

References

1. Uckun, S.; Kai, G.; Lucas, P. Standardizing research methods for prognostics. In Proceedings of the International Conference on Prognostics & Health Management, Denver, CO, USA, 25 March 2008.
2. Zhang, M.; Amaitik, N.; Wang, Z.; Xu, Y.; Maisuradze, A.; Peschl, M.; Tzovaras, D. Predictive Maintenance for Remanufacturing Based on Hybrid-Driven Remaining Useful Life Prediction. *Appl. Sci.* **2022**, *12*, 3218. [CrossRef]
3. Jardine, A.; Lin, D.; Banjevic, D. A review on machinery diagnostics and prognostics implementing condition-based maintenance-ScienceDirect. *Mech. Syst. Signal Process.* **2006**, *20*, 1483–1510. [CrossRef]
4. Heng, A.; Zhang, S.; Tan, A.C.; Mathew, J. Rotating machinery prognostics: State of the art, challenges and opportunities. *Mech. Syst. Signal Process.* **2009**, *23*, 724–739. [CrossRef]
5. Lei, Y.; Li, N.; Guo, L.; Li, N.; Yan, T.; Lin, J. Machinery health prognostics: A systematic review from data acquisition to RUL prediction. *Mech. Syst. Signal Process.* **2018**, *104*, 799–834. [CrossRef]
6. Li, N.; Gebraeel, N.; Lei, Y.; Bian, L.; Si, X. Remaining useful life prediction of machinery under time-varying operating conditions based on a two-factor state-space model. *Reliab. Eng. Syst. Saf.* **2019**, *186*, 88–100. [CrossRef]
7. Prakash, G.; Narasimhan, S.; Pandey, M. A probabilistic approach to remaining useful life prediction of rolling element bearings. *Struct. Health Monit.* **2019**, *18*, 466–485. [CrossRef]
8. Rai, A.; Kim, J.M. A novel health indicator based on the Lyapunov exponent, a probabilistic self-organizing map, and the Gini-Simpson index for calculating the RUL of bearings. *Measurement* **2020**, *164*, 108002. [CrossRef]
9. Jiang, J.R.; Lee, J.E.; Zeng, Y.M. Time Series Multiple Channel Convolutional Neural Network with Attention-Based Long Short-Term Memory for Predicting Bearing Remaining Useful Life. *Sensors* **2020**, *20*, 166. [CrossRef]
10. Cui, L.; Wang, X.; Wang, H.; Ma, J. Research on Remaining Useful Life Prediction of Rolling Element Bearings Based on Time-Varying Kalman Filter. *IEEE Trans. Instrum. Meas.* **2019**, *69*, 2858–2867. [CrossRef]
11. Cheng, C.; Ma, G.; Zhang, Y.; Sun, M.; Teng, F.; Ding, H.; Yuan, Y. A Deep Learning-Based Remaining Useful Life Prediction Approach for Bearings. *IEEE/Asme Trans. Mechatron.* **2020**, *25*, 1243–1254. [CrossRef]
12. Chen, Y.; Peng, G.; Zhu, Z.; Li, S. A novel deep learning method based on attention mechanism for bearing remaining useful life prediction. *Appl. Soft Comput.* **2019**, *86*, 105919. [CrossRef]
13. Shao, S.Y.; Sun, W.J.; Yan, R.Q.; Peng, W.; Gao, R.X. A Deep Learning Approach for Fault Diagnosis of Induction Motors in Manufacturing. *Chin. J. Mech. Eng.* **2017**, *30*, 1347–1356. [CrossRef]
14. Kim, J.M.; Sohaib, M. Reliable Fault Diagnosis of Rotary Machine Bearings Using a Stacked Sparse Autoencoder-Based Deep Neural Network. *Shock Vib.* **2018**, *2018*, 2919637.1–2919637.11.
15. Bienefeld, C.; Kirchner, E.; Vogt, A.; Kacmar, M. On the Importance of Temporal Information for Remaining Useful Life Prediction of Rolling Bearings Using a Random Forest Regressor. *Lubricants* **2022**, *10*, 67. [CrossRef]
16. Xia, M.; Li, T.; Shu, T.; Wan, J.; de Silva, C.W.; Wang, Z. A Two-Stage Approach for the Remaining Useful Life Prediction of Bearings Using Deep Neural Networks. *IEEE Trans. Ind. Inform.* **2019**, *15*, 3703–3711. [CrossRef]
17. Guo, L.; Li, N.; Jia, F.; Lei, Y.; Lin, J. A recurrent neural network based health indicator for remaining useful life prediction of bearings. *Neurocomputing* **2017**, *240*, 98–109. [CrossRef]
18. Rui, Z.; Yan, R.; Chen, Z.; Mao, K.; Gao, R.X. Deep learning and its applications to machine health monitoring. *Mech. Syst. Signal Process.* **2019**, *115*, 213–237.
19. Van den Oord, A.; Li, Y.; Vinyals, O. Representation learning with contrastive predictive coding. *arXiv* **2018**, arXiv:1807.03748.
20. Krizhevsky, A.; Sutskever, I.; Hinton, G.E. Imagenet classification with deep convolutional neural networks. *Commun. ACM* **2017**, *60* , 84–90. [CrossRef]
21. Graves, A.; Mohamed, A.R.; Hinton, G. Speech recognition with deep recurrent neural networks. In Proceedings of the 2013 IEEE International Conference on Acoustics, Speech and Signal Processing, Vancouver, BC, Canada, 26–31 May 2013; IEEE: Piscataway, NJ, USA, 2013; pp. 6645–6649.
22. Vincent, P.; Larochelle, H.; Lajoie, I.; Bengio, Y.; Manzagol, P.A.; Bottou, L. Stacked denoising autoencoders: Learning useful representations in a deep network with a local denoising criterion. *J. Mach. Learn. Res.* **2010**, *11*, 3371–3408.
23. Xu, F.; Huang, Z.; Yang, F.; Wang, D.; Tsui, K.L. Constructing a health indicator for roller bearings by using a stacked auto-encoder with an exponential function to eliminate concussion. *Appl. Soft Comput.* **2020**, *89*, 106119. [CrossRef]
24. Wang, B.; Lei, Y.; Li, N.; Yan, T. Deep separable convolutional network for remaining useful life prediction of machinery. *Mech. Syst. Signal Process.* **2019**, *134*, 106330. [CrossRef]
25. Wu, J.; Hu, K.; Cheng, Y.; Zhu, H.; Shao, X.; Wang, Y. Data-driven remaining useful life prediction via multiple sensor signals and deep long short-term memory neural network. *Isa Trans.* **2020**, *97*, 241–250. [CrossRef]
26. Cheng, H.; Kong, X.; Chen, G.; Wang, Q.; Wang, R. Transferable convolutional neural network based remaining useful life prediction of bearing under multiple failure behaviors. *Measurement* **2021**, *168*, 108286. [CrossRef]
27. Luo, J.; Zhang, X. Convolutional neural network based on attention mechanism and Bi-LSTM for bearing remaining life prediction. *Appl. Intell.* **2022**, *52*, 1076–1091. [CrossRef]
28. Ma, M.; Mao, Z. Deep-convolution-based LSTM network for remaining useful life prediction. *IEEE Trans. Ind. Inform.* **2020**, *17*, 1658–1667. [CrossRef]
29. She, D.; Jia, M.; Pecht, M.G. Sparse auto-encoder with regularization method for health indicator construction and remaining useful life prediction of rolling bearing. *Meas. Sci. Technol.* **2020**, *31*, 105005. [CrossRef]

30. Li, X.; Zhang, W.; Ma, H.; Luo, Z.; Li, X. Data alignments in machinery remaining useful life prediction using deep adversarial neural networks. *Knowl.-Based Syst.* **2020**, *197*, 105843. [CrossRef]
31. Zhang, S.; Zhang, Y.; Li, L.; Wang, S.; Xiao, Y. An effective health indicator for rolling elements bearing based on data space occupancy. *Struct. Health Monit.* **2018**, *17*, 3–14. [CrossRef]
32. Jing, L.; Tian, Y. Self-supervised visual feature learning with deep neural networks: A survey. *IEEE Trans. Pattern Anal. Mach. Intell.* **2020**, *43*, 4037–4058. [CrossRef]
33. Goodfellow, I.; Pouget-Abadie, J.; Mirza, M.; Xu, B.; Warde-Farley, D.; Ozair, S.; Courville, A.; Bengio, Y. Generative Adversarial Nets. *Adv. Neural Inf. Process. Syst.* **2014**, *27*, 2672–2680.
34. Kingma, D.P.; Welling, M. Auto-Encoding Variational Bayes. *arXiv* **2014**, arXiv:1312.6114.
35. He, K.; Fan, H.; Wu, Y.; Xie, S.; Girshick, R. Momentum contrast for unsupervised visual representation learning. In Proceedings of the IEEE/CVF Conference on Computer Vision and Pattern Recognition, Seattle, WA, USA, 13–19 June 2020; pp. 9729–9738.
36. Chen, T.; Kornblith, S.; Norouzi, M.; Hinton, G. A simple framework for contrastive learning of visual representations. In Proceedings of the International Conference on Machine Learning, PMLR, Virtual, 13–18 June 2020; pp. 1597–1607.
37. Devlin, J.; Chang, M.W.; Lee, K.; Toutanova, K. Bert: Pre-training of deep bidirectional transformers for language understanding. *arXiv* **2018**, arXiv:1810.04805.
38. Jaiswal, A.; Babu, A.R.; Zadeh, M.Z.; Banerjee, D.; Makedon, F. A survey on contrastive self-supervised learning. *Technologies* **2020**, *9*, 2. [CrossRef]
39. Ragab, M.; Chen, Z.; Wu, M.; Foo, C.S.; Kwoh, C.K.; Yan, R.; Li, X. Contrastive adversarial domain adaptation for machine remaining useful life prediction. *IEEE Trans. Ind. Inform.* **2020**, *17*, 5239–5249. [CrossRef]
40. Robinson, J.; Chuang, C.Y.; Sra, S.; Jegelka, S. Contrastive learning with hard negative samples. *arXiv* **2020**, arXiv:2010.04592.
41. Cheng, Y.; Hu, K.; Wu, J.; Zhu, H.; Shao, X. A convolutional neural network based degradation indicator construction and health prognosis using bidirectional long short-term memory network for rolling bearings. *Adv. Eng. Inform.* **2021**, *48*, 101247. [CrossRef]
42. Chen, Y.; Liu, Z.; Zhang, Y.; Zheng, X.; Xie, J. Degradation-trend-dependent Remaining Useful Life Prediction for Bearing with BiLSTM and Attention Mechanism. In Proceedings of the 2021 IEEE 10th Data Driven Control and Learning Systems Conference (DDCLS), Suzhou, China, 14–16 May 2021; pp. 1177–1182.

MDPI
St. Alban-Anlage 66
4052 Basel
Switzerland
Tel. +41 61 683 77 34
Fax +41 61 302 89 18
www.mdpi.com

Lubricants Editorial Office
E-mail: lubricants@mdpi.com
www.mdpi.com/journal/lubricants

www.ingramcontent.com/pod-product-compliance
Ingram Content Group UK Ltd.
Pitfield, Milton Keynes, MK11 3LW, UK
UKHW060123010826
14089UKWH00006B/117